Switching Phenomena in High-Voltage Circuit Breakers

ELECTRICAL ENGINEERING AND ELECTRONICS

A Series of Reference Books and Textbooks

1. Rational Fault Analysis, *edited by Richard Saeks and S. R. Liberty*
2. Nonparametric Methods in Communications, *edited by P. Papantoni-Kazakos and Dimitri Kazakos*
3. Interactive Pattern Recognition, *Yi-tzuu Chien*
4. Solid-State Electronics, *Lawrence E. Murr*
5. Electronic, Magnetic, and Thermal Properties of Solid Materials, *Klaus Schröder*
6. Magnetic-Bubble Memory Technology, *Hsu Chang*
7. Transformer and Inductor Design Handbook, *Colonel Wm. T. McLyman*
8. Electromagnetics: Classical and Modern Theory and Applications, *Samuel Seely and Alexander D. Poularikas*
9. One-Dimensional Digital Signal Processing, *Chi-Tsong Chen*
10. Interconnected Dynamical Systems, *Raymond A. DeCarlo and Richard Saeks*
11. Modern Digital Control Systems, *Raymond G. Jacquot*
12. Hybrid Circuit Design and Manufacture, *Roydn D. Jones*
13. Magnetic Core Selection for Transformers and Inductors: A User's Guide to Practice and Specification, *Colonel Wm. T. McLyman*
14. Static and Rotating Electromagnetic Devices, *Richard H. Engelmann*
15. Energy-Efficient Electric Motors: Selection and Application, *John C. Andreas*
16. Electromagnetic Compossibility, *Heinz M. Schlicke*
17. Electronics: Models, Analysis, and Systems, *James G. Gottling*
18. Digital Filter Design Handbook, *Fred J. Taylor*
19. Multivariable Control: An Introduction, *P. K. Sinha*
20. Flexible Circuits: Design and Applications, *Steve Gurley, with contributions by Carl A. Edstrom, Jr., Ray D. Greenway, and William P. Kelly*
21. Circuit Interruption: Theory and Techniques, *Thomas E. Browne, Jr.*
22. Switch Mode Power Conversion: Basic Theory and Design, *K. Kit Sum*
23. Pattern Recognition: Applications to Large Data-Set Problems, *Sing-Tze Bow*
24. Custom-Specific Integrated Circuits: Design and Fabrication, *Stanley L. Hurst*
25. Digital Circuits: Logic and Design, *Ronald C. Emery*
26. Large-Scale Control Systems: Theories and Techniques, *Magdi S. Mahmoud, Mohamed F. Hassan, and Mohamed G. Darwish*
27. Microprocessor Software Project Management, *Eli T. Fathi and Cedric V. W. Armstrong (Sponsored by Ontario Centre for Microelectronics)*
28. Low Frequency Electromagnetic Design, *Michael P. Perry*
29. Multidimensional Systems: Techniques and Applications, *edited by Spyros G. Tzafestas*
30. AC Motors for High-Performance Applications: Analysis and Control, *Sakae Yamamura*

31. Ceramic Materials for Electronics: Processing, Properties, and Applications, *edited by Relva C. Buchanan*
32. Microcomputer Bus Structures and Bus Interface Design, *Arthur L. Dexter*
33. End User's Guide to Innovative Flexible Circuit Packaging, *Jay J. Miniet*
34. Reliability Engineering for Electronic Design, *Norman B. Fuqua*
35. Design Fundamentals for Low-Voltage Distribution and Control, *Frank W. Kussy and Jack L. Warren*
36. Encapsulation of Electronic Devices and Components, *Edward R. Salmon*
37. Protective Relaying: Principles and Applications, *J. Lewis Blackburn*
38. Testing Active and Passive Electronic Components, *Richard F. Powell*
39. Adaptive Control Systems: Techniques and Applications, *V. V. Chalam*
40. Computer-Aided Analysis of Power Electronic Systems, *Venkatachari Rajagopalan*
41. Integrated Circuit Quality and Reliability, *Eugene R. Hnatek*
42. Systolic Signal Processing Systems, *edited by Earl E. Swartzlander, Jr.*
43. Adaptive Digital Filters and Signal Analysis, *Maurice G. Bellanger*
44. Electronic Ceramics: Properties, Configuration, and Applications, *edited by Lionel M. Levinson*
45. Computer Systems Engineering Management, *Robert S. Alford*
46. Systems Modeling and Computer Simulation, *edited by Naim A. Kheir*
47. Rigid-Flex Printed Wiring Design for Production Readiness, *Walter S. Rigling*
48. Analog Methods for Computer-Aided Circuit Analysis and Diagnosis, *edited by Takao Ozawa*
49. Transformer and Inductor Design Handbook, Second Edition, Revised and Expanded, *Colonel Wm. T. McLyman*
50. Power System Grounding and Transients: An Introduction, *A. P. Sakis Meliopoulos*
51. Signal Processing Handbook, *edited by C. H. Chen*
52. Electronic Product Design for Automated Manufacturing, *H. Richard Stillwell*
53. Dynamic Models and Discrete Event Simulation, *William Delaney and Erminia Vaccari*
54. FET Technology and Application: An Introduction, *Edwin S. Oxner*
55. Digital Speech Processing, Synthesis, and Recognition, *Sadaoki Furui*
56. VLSI RISC Architecture and Organization, *Stephen B. Furber*
57. Surface Mount and Related Technologies, *Gerald Ginsberg*
58. Uninterruptible Power Supplies: Power Conditioners for Critical Equipment, *David C. Griffith*
59. Polyphase Induction Motors: Analysis, Design, and Application, *Paul L. Cochran*

60. Battery Technology Handbook, *edited by H. A. Kiehne*
61. Network Modeling, Simulation, and Analysis, *edited by Ricardo F. Garzia and Mario R. Garzia*
62. Linear Circuits, Systems and Signal Processing: Advanced Theory and Applications, *edited by Nobuo Nagai*
63. High-Voltage Engineering: Theory and Practice, *edited by M. Khalifa*
64. Large-Scale Systems Control and Decision Making, *edited by Hiroyuki Tamura and Tsuneo Yoshikawa*
65. Industrial Power Distribution and Illuminating Systems, *Kao Chen*
66. Distributed Computer Control for Industrial Automation, *Dobrivoje Popovic and Vijay P. Bhatkar*
67. Computer-Aided Analysis of Active Circuits, *Adrian Ioinovici*
68. Designing with Analog Switches, *Steve Moore*
69. Contamination Effects on Electronic Products, *Carl J. Tautscher*
70. Computer-Operated Systems Control, *Magdi S. Mahmoud*
71. Integrated Microwave Circuits, *edited by Yoshihiro Konishi*
72. Ceramic Materials for Electronics: Processing, Properties, and Applications, Second Edition, Revised and Expanded, *edited by Relva C. Buchanan*
73. Electromagnetic Compatibility: Principles and Applications, *David A. Weston*
74. Intelligent Robotic Systems, *edited by Spyros G. Tzafestas*
75. Switching Phenomena in High-Voltage Circuit Breakers, *edited by Kunio Nakanishi*

Additional Volumes in Preparation

Electrical Engineering-Electronics Software

1. Transformer and Inductor Design Software for the IBM PC, *Colonel Wm. T. McLyman*
2. Transformer and Inductor Design Software for the Macintosh, *Colonel Wm. T. McLyman*
3. Digital Filter Design Software for the IBM PC, *Fred J. Taylor and Thanos Stouraitis*

Switching Phenomena in High-Voltage Circuit Breakers

edited by
Kunio Nakanishi
Tokyo Denki University
Tokyo, Japan

Marcel Dekker, Inc. New York • Basel • Hong Kong

Library of Congress Cataloging-in-Publication Data

Switching phenomena in high-voltage circuit breakers / edited by Kunio Nakanishi.
p. cm. -- (Electrical engineering and electronics; 75)
Includes bibliographical references and index.
ISBN 0-8247-8543-6
1. Electric-circuit breakers. 2. High voltages. 3. Switching theory. I. Nakanishi, Kunio. II. Series.
TK2842.S95 1991
621.31'7–dc20 91-22145
CIP

This book is printed on acid-free paper.

MARCEL DEKKER, INC.
270 Madison Avenue, New York, New York 10016

Current printing (last digit):
10 9 8 7 6 5 4 3 2 1

PRINTED IN THE UNITED STATES OF AMERICA

PREFACE

The switching behavior of mechanical circuit breakers is always accompanied by the occurrence of arc discharge between contacts during their operation. In former times, when the properties of the electric arc were not clearly understood, the development and design of new products were mainly derived from long-term accumulated experience. Moreover, the capacity of testing facilities was insufficient to confirm the performance of switching devices. In the 1940s, dynamic arc equations that express arc characteristic with arc parameters were presented by Mayr, Cassie, and Browne and introduced the procedure to treat switching phenomena analytically in conjunction with the circuit equations of power systems. On the other hand, the physical properties of interrupting arcs were revealed through the development of measurement techniques. Also, practical computer analysis has made it possible to clarify the behavior of interrupting arcs in detail by utilizing precise arc models constructed by the combination of thermodynamics and fluid dynamics. For design purposes, computer-aided engineering (CAE) such as electric field analysis, pressure distribution analysis, and mechanical (including operating force and seismic response) analysis are widely applied to achieve optimally designed circuit breakers.

During recent years there has been a remarkable growth of demand for higher-rated voltage and larger-rated breaking capacity of circuit breakers to cope with the increasing demands in power systems. To meet these trends, high-capacity breaking units having a very reliable performance have been and continue to be developed, aided by the utilization of computer and opto-electronic techniques covering the vast areas of analysis, design, measurement, and testing.

To conform with the technical trends, this book is intended to give a description of switching phenomena, test procedures, and applications concerning modern high-voltage circuit breakers, especially SF_6 gas-blast type and vacuum type, used predominantly in high-voltage and medium-voltage ranges respectively.

Chapter 1 is a comprehensive treatment of general problems concerning current interruption. Presented is the mathematical arc model that represents the dynamic arc characteristics as a one-terminal-pair resistive circuit element in the form of an ordinary differential equation.

Chapter 2 deals with the physical arc model intended for the analysis of the internal construction and mechanism of a spatial dynamic arc, by using partial differential equations originated from thermo-fluid dynamics. Electric field analysis and mechanical analysis are also involved in the design of circuit breakers.

Chapter 3 reviews miscellaneous types of modern switching apparatus, such as gas circuit breakers, gas-insulated switch-gear, vacuum circuit breakers and high-voltage direct-current circuit breakers.

Chapter 4 discusses relevant standards, direct and synthetic testing, and diagnosis, which are indispensible in the production of circuit breakers of high performance and high reliability.

This book is written to be easily understood by the reader who might be unfamiliar with recent circuit breaker technology; the outstanding technical trends of the 1980s are emphasized. Great care has been taken to avoid duplication of material in *Circuit Interruption* by Thomas E. Browne, Jr. (Marcel Dekker, 1984), which covers the comprehensive field of switching apparatus in detail and which can supplement this book.

The editor greatfully acknowledges the contributors, who, despite their very busy situations as managers in promoting research, design, manufacturing, or testing at Hitachi Ltd., Mitsubishi Electric Corporation, or Toshiba Corporation, gave their assistance in the publication of this book.

Acknowledgment is also given to the editing staff members of the publisher.

Kunio Nakanishi

CONTENTS

CONTRIBUTORS

Eiichi Haginomori Chief Specialist, Power Transmission and Substation Engineering Department, Toshiba Corporation, Tokyo, Japan

Kunio Hirasawa Chief Research Engineer, Hitachi Research Laboratory, Hitachi Ltd., Hitachi, Japan

Koji Ibuki Deputy Manager, Switchgear Department, Itami Works, Mitsubishi Electric Corporation, Amagasaki, Japan

Hisatoshi Ikeda Manager, High Power Laboratory, Toshiba Corporation, Kawasaki, Japan

Eiji Kaneko Manager of Switchgear Technology, Heavy Apparatus Engineering Laboratory, Toshiba Corporation, Tokyo, Japan

Yukio Kurosawa Senior Research Engineer, The 2nd Department, Hitachi Research Laboratory, Hitachi Ltd., Hitachi, Japan

Hiroshi Kuwahara* Deputy General Manager, Itami Works, Mitsubishi Electric Corporation, Amagasaki, Japan

Susumu Nishiwaki Senior Specialist, High Power Laboratory, Toshiba Corporation, Kawasaki, Japan

Youichi Ohshita Senior Research Engineer, The 2nd Department, Hitachi Research Laboratory, Hitachi Ltd., Hitachi, Japan

Shuichi Sakuma Manager, Energy Engineering Section, Development Department, Itami Works, Mitsubishi Electric Corporation, Amagasaki, Japan

Masanori Tsukushi Senior Research Engineer, The 2nd Department, Hitachi Research Laboratory, Hitachi Ltd., Hitachi, Japan

Takao Yamauchi Manager, Switchgear Department, Itami Works, Mitsubishi Electric Corporation, Amagasaki, Japan

Satoru Yanabu Senior Manager, Transmission and Substation Engineering Department, Toshiba Corporation, Tokyo, Japan

Yoshio Yoshioka Senior Chief Engineer, Kokubu Works, Hitachi Ltd., Hitachi, Japan

Toshiaki Yoshizumi Manager of Switchgear Engineering, Switchgear Department, Itami Works, Mitsubishi Electric Corporation, Amagasaki, Japan

Current affiliation: Managing Director, SPC Electronics Corporation, Tokyo, Japan

Switching Phenomena in High-Voltage Circuit Breakers

1
OVERVIEW

Yoshio Yoshioka, Kunio Hirasawa, Yukio Kurosawa, Masanori Tsukushi, and Youichi Ohshita Hitachi Ltd., Hitachi, Japan

1.1 FUNCTIONS AND REQUIREMENTS FOR CIRCUIT BREAKERS (Y. Yoshioka, K. Hirasawa, Y. Ohshita)

Circuit breakers, which have the function of opening and closing a circuit, are indispensable for protecting other transmission devices in a power system. When a short-circuit fault occurs, they quickly eliminate the fault to secure system stability. They are also used for modifying the system configuration under normal conditions. The International Electrotechnical Commission (IEC) defines the circuit breaker as a *mechanical switching device, capable of making, carrying and breaking currents under normal circuit conditions and also making, carrying for a specified time and breaking currents under specified abnormal circuit conditions such as those of short-circuit.* It adds, as a note, that a *circuit-breaker is usually intended to operate infrequently, although some types are suitable for frequent operation.*

Circuit breakers must have the following basic functions:

1. When closed, they are good conductors, and they withstand normal currents as well as short-circuit currents, thermally and mechanically.
2. When open, they are excellent insulators, and they withstand the voltage to ground or to the other phase and the voltage between contacts.
3. When closed, they can interrupt a rated, or lower, interrupting current quickly without generating an abnormal voltage.
4. When opened, they can close a shorted circuit quickly and safely without incidental contact erosion.

Some of these functions have mutually contradictory requirements, and therefore the development of a circuit breaker entails overcoming several difficulties, but they are interesting and challenging subjects, technically and theoretically.

Power-system load capacities are increasing annually, in line with economic growth, and, as shown in Fig. 1.1, this tendency is expected to continue. To transmit such huge amounts of energy, the transmission capacity per line must also continue to increase. This has been promoted by various means, such as the use of multiple larger diameter wires and the introduction of higher voltage class power systems. Through similar processes, power systems should continue to grow and, with increases in the short-circuit capacity, the breaking capacity of circuit breakers must also continue to increase.

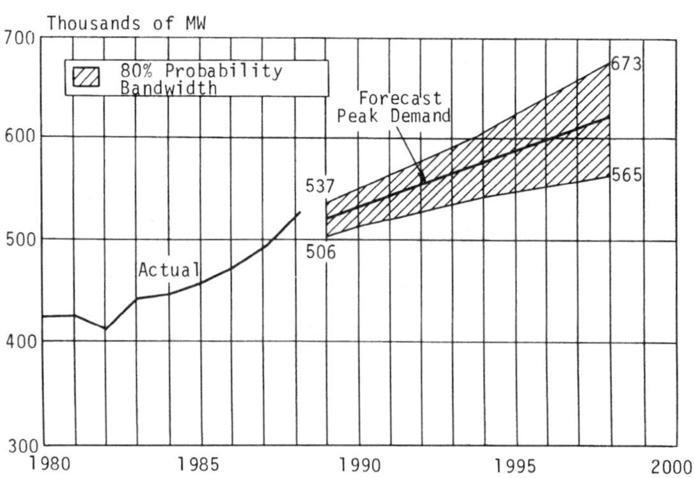

Fig. 1.1 Trend forecast of summer peak load (United States). (IEEE, *Power Engineering Review,* 1989. © IEEE.)

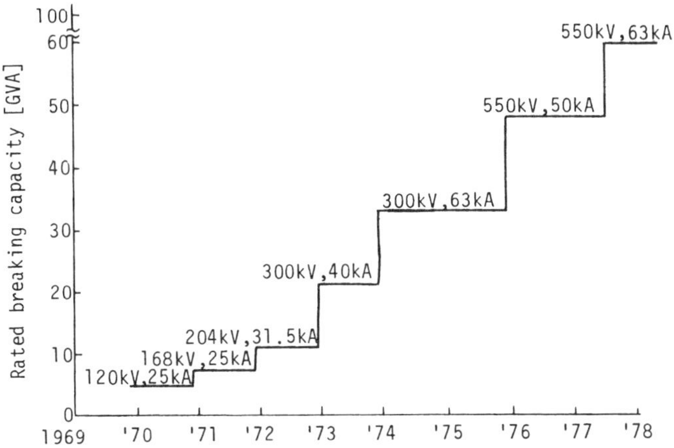

Fig. 1.2 Trend of voltage and capacity rise of puffer-type gas circuit breaker.

Figure 1.2 shows the trend in the breaking capacity of power system gas circuit breakers. There was a tenfold increase in the decade of the 1970s. A general high-voltage circuit breaker consists of several interrupter units connected in series; for economy, the breaking capacity per interrupter unit is increased to reduce the number of units required. Performance requirements for circuit breakers also include high-speed interruption, ensuring power system stability, high-speed closing, and suppression of switching surge, particularly for ultrahigh-voltage (UHV) systems. Resistance to mechanical vibration, low-noise operation, and easy maintenance and checkup are also desirable features.

This chapter stresses the basic principles and theory of circuit-breaker techniques. The reader is encouraged to refer to the quoted references for details of other topics.

1.2 BASIC PHYSICS OF INTERRUPTING ARCS (Y. Yoshioka, Y. Ohshita, Y. Kurosawa)

1.2.1 General Description

The basic function of the power circuit breaker is a switching action consisting of opening and closing an electric circuit, that is, making and breaking the current under any conceivable circuit conditions. In that sense, it can be described as a circuit element whose function is intrinsically similar to that of the switches routinely met in daily life. However, major differences are that the current or circuit voltage used in a power transmission system are considerably large and the switching time must be very short, which raises technical problems. Thus circuit-breaker development constitutes one of the fields where the most advanced techniques are required.

"Switching action" refers to the mutual conversion of conductor and insulating material in a part of the circuit at a given potential. In addition to arc plasma, several physical phenomena are capable of being utilized as circuit elements; examples are semiconduction (i.e., change of conductivity by control of electron energy level in a solid material), superconductivity (i.e., temperature control in a superconductive area), and vacuum electron current (i.e., potential control). However, the switching capacity of these phenomena is somewhat lower than that of arc plasma. In power circuit breakers, the intensity of the current that flows through the conductors and the voltage level that must be sustained by the insulating material are very high, and at present the only practical approach may be the use of arc plasma.

Present commercial circuit breakers resort to plasma temperature control as the basis for their function. To be more precise, arc discharge is used as a circuit element for switching.

Historically, the first industrial uses of circuit breakers did not actively take advantage of arc plasma. When parts of the conductor are separated in order to interrupt a circuit current, an arc is generated before the current is extinguished; it is considered that *current breaking* refers to extinguishing this arc. Arc discharge was thought to be an obstacle to current interruption, and the technical concern in circuit interruption was how quickly and effectively to extinguish the arc. In fact, "current breaking" and "arc extinction" were used as synonyms. However, from the basic standpoint that switching is the interconversion of conductor and insulating materials, arc plasma has very desirable characteristics, and high-voltage and large-current circuit switching is possible only with an arc.

The desirable characteristics of the arc plasma of circuit breakers are (1) as large a range as possible in which the conductivity changes; and (2) as quick a conductivity change as possible. As for the range of conductivity change, the characteristics depend essentially on the plasma material, and the introduction of new extinguishing media constitutes a major step in technical progress. On the other hand, the speed with which conductivity changes depends largely on the plasma material and is also affected by the plasma control method. This is the reason the extinction chamber is a vital part of a circuit breaker.

In many circuit breakers, the arc plasma conductivity is changed mainly through temperature control. Therefore, current breaking or the quick reduction of arc conductivity often relies on some technical means for cooling the arc. The technical basis for circuit breakers lies in creating arc plasma of as high a conductivity as possible, carrying a large current, cooling it effectively and quickly, and converting it into an insulating gas space in a short time.

On ac circuit breakers, the current crosses zero every half-cycle of the power frequency. Therefore, it is technically easy to quickly change the plasma conductivity near the current zero, and, as described in Section 1.3, such an approach is effective for avoiding surge trouble such as flashovers in the power system equipment. Therefore, current breaking by ac circuit breakers is usually *current-zero breaking,* and as for the arc behavior and its role, the characteristics near the current zero are particularly important.

The remainder of this chapter provides basic information on characteristics of power circuit breaker arc space, arc extinction—the conversion from plasma to insulating space—and information on insulating space characteristics.

1.2.2 General Characteristics of Arc Plasma

(A) Gas Thermal Dissociation and Thermal Ionization

In addition to the three states gas, liquid, and solid, a substance may have a fourth state that is referred to as *plasma.* In space, plasma is most general and overwhelming in volume, but on earth it is experienced as a rather special state. The four states are distinguished by the conbinational status of their constituent particles. Maintaining a state or changing it generally depends on temperature. When thermal energy is given to solid molecules, the molecular kinetic energy exceeds the combinational energy required for maintaining the solid state, resulting in a change to a liquid. The energy required at this time is on the order of 10^{-2} eV per molecule. Likewise, by providing enough energy to the liquid molecules to overcome the van der Waals forces, the liquid becomes a gas. Again, energy on the order of 10^{-2} eV per molecule is required. When a gaseous molecule receives more than the bond energy between its component atoms, it dissociates into single atoms, and, furthermore, if the atom receives more than its ionization potential energy, orbital electrons ionize, leading to the plasma state.

The energy necessary for ionization is 1–30 eV per atom, which is high. The liquid, gaseous, or plasma states are maintained by the temperature. This means that the resulting state when the combination is released can be maintained as long as the average thermal kinetic energy of the particles at that temperature is beyond the combinational energy. When an equilibrium is maintained for the state, it is referred to as a *thermodynamic equilibrium state,* and it is governed by the statistical mechanical rules for that state. A gas, for example, is described by the classic statistical mechanics of Maxwell and Boltzmann.

Next, the transition from gas to arc plasma is considered. At high temperature, the kinetic energy of gaseous particles increases and energy transfer through collisions causes *thermal dissociation* and/or *thermal ionization*. The rate of thermal dissociation or ionization of a particle (molecule or atom) into several particles (atoms, ions, or electrons) is theoretically given as a function of temperature; that is, the thermal dissociation or ionization equilibrium is determined by an equilibrium coefficient that increases considerably with temperature. Nitrogen, which is the main component of air, undergoes dissociation from the molecule to single atoms ($N_2 \rightarrow N + N$) at 5000 K or higher; at about 8000 K, it ionizes ($N \rightarrow N^+ + e$). As shown in Section 2.1, dissociation of SF_6 molecules into F atoms and F_2 molecules becomes prominent at about 1800 K. At 5000–6000 K, fluorine ionization becomes prominent and the conductivity abruptly starts to increase.

The existence of electrons and ions produced by ionization gives conductivity to a gaseous space, and this space conductivity can be described as a function of temperature. Thus, space conductivity is also a kind of thermodynamic state variable.

Space conductivity can be described theoretically by using classical statistical mechanics. The degree of space ionization, assuming thermal equilibrium, can be derived thermodynamically. It is well known as Saha's thermal ionization equation.

Recently, various studies were undertaken on the composition and conductivity of a thermal equilibrium space containing many kinds of particles. Generally speaking, space conductivity shows a saturation characteristic dependent on temperature. Figure 1.3 shows the relation of nitrogen conductivity versus temperature. The conductivity starts rising progressively at about 5000 K and tends to saturate beyond 10,000 K.

The arc plasma in a high-temperature area can considerably change the conductivity, provided the temperature can be controlled, and it is clear that the extinction of arc plasma in a circuit breaker can be performed by cooling that arc space.

Even if the space conductivity is given as shown in Fig. 1.3, as the result of an ionization equilibrium uniquely determined by temperature or kinetic energy, the time delay in reaching space conductivity equilibrium must be considered. If the time delay is too long, the change in conductivity cannot follow the plasma cooling quickly enough, and in the case of a quick transient change, the conductivity cannot be considered a thermodynamic state variable, which would otherwise be uniquely determined by temperature.

The conductivity time constant can be considered separately as the time required to reach a particle velocity distribution equilibrium and an

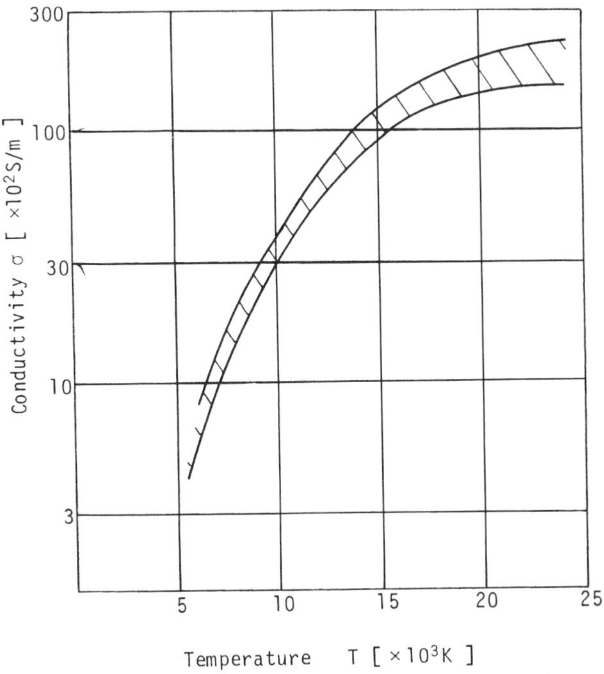

Fig. 1.3 Nitrogen plasma conductivity versus temperature. (SF_6 is described in Fig. 2.3 of Chapter 2.) (Nakanishi et al., 1975). Copyright JIEE.)

ion—electron recombination speed. As for the former, if each particle experiences several elastic collisions with other particles, the particle velocity distribution reaches a local equilibrium, and therefore several times the reciprocal of the number of particle collisions may be regarded as the approximate time to reach equilibrium. This is about 10^{-8} sec for molecules or atoms and 10^{-10} sec for electrons under the temperature and pressure conditions of an ordinary arc plasma.

From the possible ion electron recombination mechanisms, three types of recombinations—dissociative recombination, three-body recombination, and two-step recombination—are considered dominant. From experimental and theoretical considerations, 10^{-7}–10^{-8} sec is given as the time required for an ion–electron recombination.

The speed to reach ionization equilibrium is higher than the transient phenomenon speed encountered in circuit breakers, so it is reasonable to assume that the arc plasma phenomenon contributing to current interruption is almost always at an ionization equilibrium state.

(B) Heat Capacity and Cooling of Plasma

Current interruption is dominated by the cooling of the plasma in an ionization equilibrium. There are several cooling processes; the four most important mechanisms are

1. Isentropic cooling
2. Cooling by thermal conduction
3. Cooling by mixing with cool background gas
4. Cooling by radiation loss

In mechanism 4 the energy is radiated in the form of light (either ultraviolet, visible, or infrared). This light is absorbed by the surrounding gas, which is usually optically dense, and this helps in transferring the heat energy. Which process is dominant among these depends on the characteristics of the medium. Isentropic cooling is important for a medium with a high *specific heat ratio* γ such as air. γ is defined as C_p/C_v, where C_p is *constant-pressure specific heat* and C_v is *constant-volume specific heat;* for example; $\gamma = 1.4$ for air and $\gamma = 1.07$ for SF_6 (see Appendix 1). Processes 2 and 3 become dominant in an SF_6 arc because of the high thermal conductivity and high dissociation energy of the gas.

Isentropic cooling occurs during the flow of plasma along the pressure gradient. Adiabatic expansion during the flow causes loss of internal energy, and cooling is given by $-\text{div}\,(p\vec{u})$ (where $\vec{u}$ is flow velicity and p is pressure).

Cooling by thermal conduction involves the transfer of thermal energy due to the temperature gradient. Thermal conduction in plasma is not as straightforward as that in the solid body, but it is dependent on various states in the gas. Therefore, an equivalent thermal conductivity κ is convenient in expressing effective cooling as $-\text{div}(\kappa \cdot \text{grad}\,(T))$ per unit volume (where T denoted temperature).

The processes of isentropic cooling and cooling by thermal conduction are observed without having any mass flow through the boundary of the plasma during its motion because the thermal flow in the latter process does not require it. The process of cooling by mixing with background gas is a result of the difference in temperature of the masses flowing into and out of the volume with the velocity u. This is given by a term expressed in the form $\vec{u}\text{grad}(\rho h)$ (where ρ is gas density and h is total thermal energy or enthalpy in plasma).

The processes of exhausting the energy into the surroundings are as mentioned, and the rate of cooling is also dominated by the specific heat of the plasma, C_p. A low specific heat *capacity* is good for rapid cooling as well as high energy dissipation. The specific heat of the gas is a complex function of temperature; it is given by $C_p = dh/dT$.

The energy dissipated by the dissociation of molecules into atoms and the energy of ionization appear as a heat sink, and when these processes are active the specific heat appears to be high. Specific heat as a function of temperature is shown in Fig. 1.4 for air. The curve has two prominent peaks corresponding to dissociation and ionization. Sulfur hexafluoride (SF_6) gas is an important interruption medium in modern technology; its specific heat is shown in Fig. 1.5.

The specific heat C_p is calculated on the assumption that the gas is in a thermal equilibrium state. However, in actual arc plasma, the temperature change cannot always be assumed to be slow enough to ensure thermal equilibrium. Especially in the current-decreasing phase of the arc, the recombination of the atoms into molecules is rather slow (typically 10^{-3} sec), whereas deionization is fast (typically 10^{-7}–10^{-8} sec). Therefore, the dynamic C_p in the process of cooling of the arc is represented by C_{pf} rather than the quasi-static C_p curve in Fig. 1.4 (air).

The nonequilibrium state in the process of cooling caused by the difference between the time constants of deionization and association is important in understanding the interruption of an air-blasted arc. Deionization ($N^+ + e \rightarrow N$) occurs before association ($N + N \rightarrow N_2$), and this process makes the conductivity of the plasma low without interference from the slow association. The specific heat of SF_6 (Fig. 1.5), on the other hand,

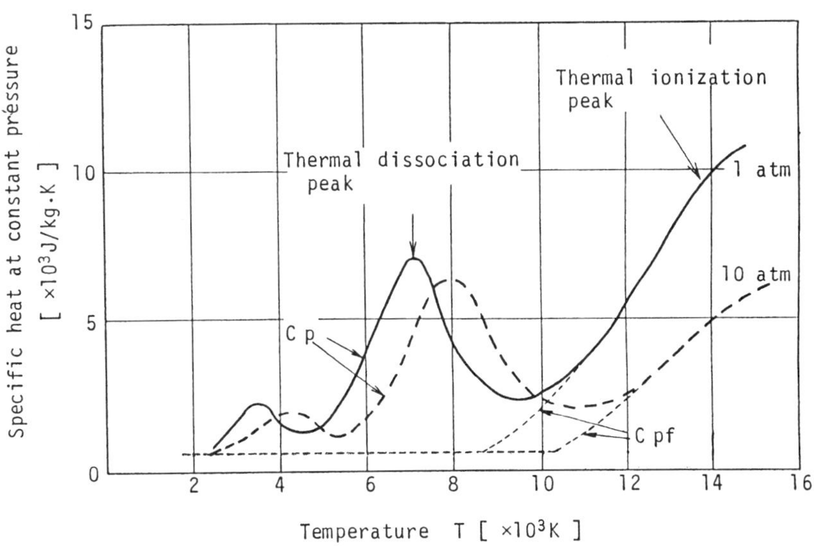

Fig. 1.4 Specific heat at constant air pressure versus temperature. (Nakanishi et al., 1975. Copyright JIEE.)

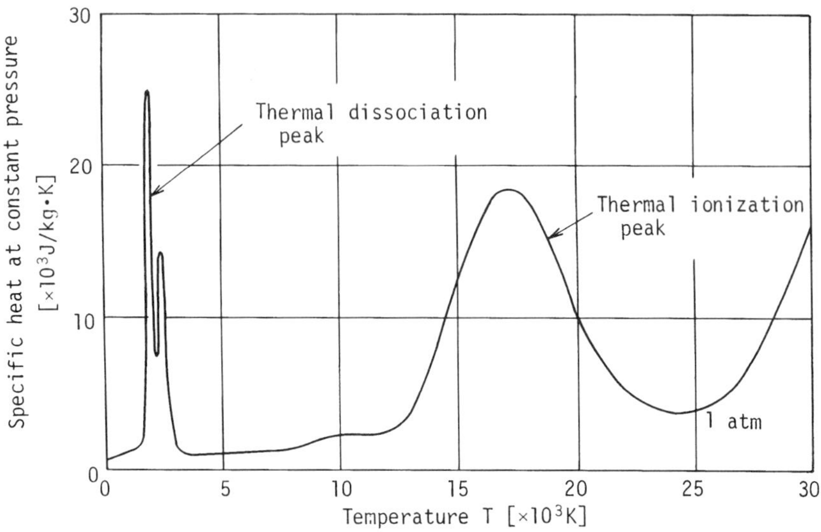

Fig. 1.5 Specific heat at constant SF_6 pressure versus temperature.

shows different behavior, as the peak corresponding to dissociation appears below 3000 K, and at 5000 K ionization becomes significant. This means that the nonequilibrium state during the cooling process caused by the difference between the time constants of deionization and association does not particularly help in lowering the electrical conductivity of the plasma. However, the high specific heat corresponding to dissociation in comparison with the peak corresponding to ionization suggests a different process of cooling. A peculiar cooling is caused when low-temperature gas below 1000 K, which is not dissociated, is injected into the hot plasma. The low-temperature gas absorbs significant dissociation energy from the high-temperature gas. Selection of the interruption medium requires a deeper understanding of the detailed process of plasma behavior.

1.2.3 Mathematical Arc Models

(A) Energy Equilibrium in Plasma

The phenomenon by which the arc is actually extinguished in circuit breakers is very complicated and cannot be explained accurately and quantitatively. It is useful, however, to employ a model that is quantitatively based on the general characteristics of arc plasma stated above, in order to understand the basics of circuit breakers. The dynamic characteristics of arc plasma lie in an energy equilibrium relationship that is given by the general expression

$$EI = \frac{dQ}{dt} - \mathrm{div}(p\vec{u}) - \mathrm{div}(\kappa \cdot \mathrm{grad}(T)) + \vec{u}\,\mathrm{grad}(\rho h) + S(T) \tag{1.1}$$

where

EI	=	input power at unit length [W]
Q	=	energy stored in arc [J]
t	=	time [sec]
$-\mathrm{div}(p\vec{u})$	=	energy loss by isentropic cooling [W]
$-\mathrm{div}(\kappa \cdot \mathrm{grad}(T)$	=	thermal condition loss [W]
$\vec{u}\,\mathrm{grad}(\rho h)$	=	loss by mass flow (convection loss) [W]
$S(T)$	=	radiation loss [W]

This relation indicates that the heat generated is equal to the sum of the stored energy increase and the heat loss.

The analytical models of Mayr and Cassie that deal with the arc phenomenon macroscopically under such a relationship are well known. In these models, the arc is regarded as a single circuit element that can be expressed by arc resistance R or arc conductance G. Such arc models are known as *mathematical arc models,* in contrast to *physical arc models,* in which the internal structure of an arc is analyzed as described in Chapter 2. The Mayr and Cassie mathematical models are outlined below. Table 1.1 compares the two.

(B) Mayr Arc Model

In the Mayr arc model, the arc is regarded as a homogeneous cylindrical column of constant diameter, and the arc conductance transition is attributed to a transition in the degree of ionization. Further, a constant heat loss N_0, supposing thermal conduction, is assumed in solving the energy equilibrium equation. As arc conductance G, the ionization degree obtained from Saha's expression is considered, and to facilitate the development of subsequent expressions the function is approximated somewhat as

$$G = K \exp\left(\frac{Q}{Q_0}\right) \tag{1.2}$$

where K and Q_0 are constants. The energy equilibrium equation is

$$EI = \frac{dQ}{dt} + N_0 \tag{1.3}$$

Table 1.1 Comparison of the Mayr Arc Model and the Cassie Arc Model

Model name	Mayr	Cassie
Arc conductance	By change of ionization degree	By change of arc diameter
	$G \propto \exp\left(\frac{Q}{Q_0}\right)$	$G \propto Q$
Heat loss	By thermal conduction	By thermal convection
	$N = N_0$ (Constant)	$N \propto Q$
Dynamic characteristic expression	$\frac{1}{G} \cdot \frac{dG}{dt} = \frac{1}{\theta}\left(\frac{EI}{N_0} - 1\right)$	$\frac{1}{G} \cdot \frac{dG}{dt} = \frac{1}{\theta}\left(\frac{E^2}{E_0{}^2} - 1\right)$
	Small current region	Large current region
Conceptual diagram	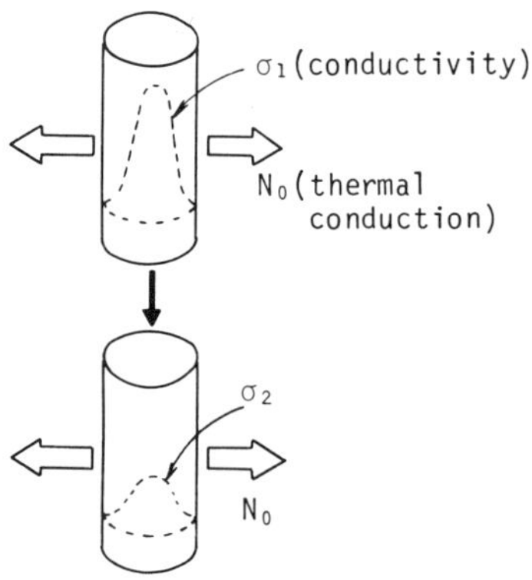	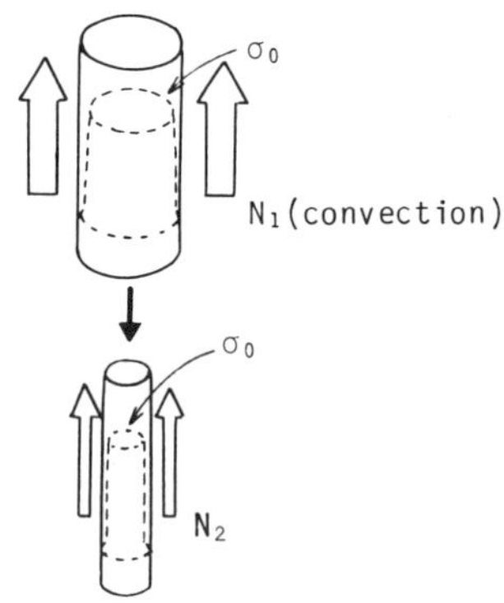

Source: JIEE (1986), Copyright JIEE.

Hence, Mayr's expression is deduced as

$$\frac{1}{G}\frac{dG}{dt} = \frac{1}{\theta}\left(\frac{EI}{N_0} - 1\right) \tag{1.4}$$

where $\theta = Q_0/N_0$ has a time dimension and is called the *arc time constant.* It coincides with the time constant of arc cooling when the arc thermal input is removed. Mayr's arc model is suitable for the analysis of a relatively small current area.

(C) Cassie Arc Model

In Cassie's arc model, the arc ionization degree is sufficiently high, and the arc conductance transition is attributed to a change in arc cross-sectional area A. The energy equilibrium equation is solved on the assumption that the heat loss N depends mainly on convection loss and is proportional to the arc cross-sectional area A. Likewise, arc conductance G and stored heat Q are deemed to be proportional to the arc cross-sectional area A. Then, considering the unit length gives

$$Q = hA \tag{1.5}$$

$$N = N'A \tag{1.6}$$

where h is enthalpy and N' is the heat loss at unit cross section. Hence, the energy equilibrium equation (1.1) becomes

$$EI = h\frac{dS}{dt} + N'A \tag{1.7}$$

Substituting for A according to the relation

$$G = \sigma A \tag{1.8}$$

(where σ is conductivity) gives

$$\frac{1}{G}\frac{dG}{dt} = \frac{1}{\theta}\left(\frac{E^2}{E_0^{\,2}} - 1\right) \tag{1.9}$$

Here again, $\theta = Q/N$ is the same arc time constant as defined for (1.4). $E_0^{\,2} = N/G$ holds and corresponds to arc voltage at the steady state where heat generated by the arc and arc loss are equal. Cassie's arc model is suitable for the analysis of a relatively large current area.

(D) Alternative Mathematical Model

The classic arc models described so far have often been used for studies of interruption behavior. They explain interruption quite well qualitatively but not quantitatively. For a gas-blast circuit breaker with SF_6 as an extinction medium, calculated results obtained with these classic models have only poor agreement with experimental results. Therefore, several new models have been proposed. One of these arc models is shown below.

Swanson et al. (1972) introduced the *heat flux potential* S for an arc column and deduced an arc dynamic characteristic expression of normalized form known as the *thermal arc model,* which reduces to the Cassie and Mayr equations when the electrical conductivity varies linearly and exponentially with the heat flux potential ($S = \int_0^T \kappa dT$, where T is temperature and κ is thermal conductivity). First, the thermal equation is given as follows:

$$\underset{\text{storage}}{\text{Thermal}} = \underset{\text{losses}}{\text{Diffusion}} + \underset{\text{heating}}{\text{Joule}}$$

$$\frac{1}{\alpha}\frac{\partial S}{\partial t} = (1+\beta)\left(\frac{\partial^2 S}{\partial r^2} + \frac{1}{r}\frac{\partial S}{\partial r}\right) + \frac{\sigma I^2}{(2\pi \int_0^{r_a} \sigma r\, dr)^2} \tag{1.10}$$

where

I = arc current

r = radial coordinate

r_a = arc radius

t = time

α = thermal diffusivity ($\alpha = \kappa/\rho C_p$, where ρ is gas density and C_p is specific heat)

β = a turbulent diffusion factor

σ = electrical conductivity

To develop a thermal arc model, Swanson et al. assume that the heat flux potential and electrical conductivity both have Bessel function radial distributions. Then expression (1.10) is normalized as

$$\frac{d\tilde{S}}{dt} = -\frac{\tilde{S}}{\Theta} + \alpha\tilde{\sigma}E^2 \tag{1.11}$$

where

$\tilde{S}$ = centerline heat flux potential

E = arc voltage

$\tilde{\sigma}$ = centerline electrical conductivity

Θ = is the normalized arc time constant defined to obtain a more useful equation form.

(E) Measured Value Feedback Model

As measurement techniques for phenomena near the current zero have been advanced, it has become clear that the arc time constant is not necessarily constant with time. Other models have therefore been proposed by Regros, Hochrainer, and others in which the arc time constant, loss, or other arc constants are determined as functions of parameters (current, arc resistance, etc.) based on measured values.

Hochrainer and others introduced the relationship between the arc conductance g and the steady-state arc conductance G,

$$\frac{dg}{dt} = \frac{1}{\theta}(G - g) \tag{1.12}$$

where θ is an arc time constant determined from the measured values of g and dg/dt.

In another of these models the relation between the arc voltage V_a and current I as a static characteristic of the arc is assumed to be

$$V_a = NI^{-\alpha} \tag{1.13}$$

and the time constant θ is given as a dynamic characteristic of arc,

$$\theta = \theta_0 R^{-n} \tag{1.14}$$

where θ is supposed to be a function of arc resistance R on the basis of many measurements. The arc characteristic is expressed by the four parameters N, α, θ_0, and n. The energy loss is assumed to be proportional to the mth power of stored energy Q. Using the relation

$$\frac{dQ}{dt} = -kQ^m + VI \tag{1.15}$$

and expressions (1.13)–(1.15), the following dynamic characteristic expression is obtained:

$$\frac{dR}{dt} = \frac{R}{\theta}\left[1 - \left(\frac{VI^{\alpha}}{V_a}\right)^{2/(1+\alpha)}\right] \tag{1.16}$$

This model indicates that the arc time constant is a function of resistance as shown in Eq. (1.14). According to these methods, further simulation can be done with an electric circuit that is an expression of the model. The arc resistance can be obtained by calculating the current flowing through a series circuit of resistance and inductance with a universal circuit-analyzing program. Since arc characteristics are introduced in the circuit-analyzing program, the interrupting characteristics under given circuit conditions can easily be analyzed.

1.2.4 Dielectric Strength Recovery After Current Interruption

The circuit interruption is complete when the circuit breaker poles have recovered sufficient dielectric strength to a circuit recovery voltage after arc extinction near current zero as described above. Here, the dielectric strength recovery characteristics of post-plasma space play an important role.

(A) Cold Recovery and Hot Recovery

The dielectric strength recovery characteristics between circuit-breaker contacts when the circuit breaker has been opened under no load or has turned off a small current are inherent to that circuit breaker, and the process of recovering normal dielectric strength is referred to as a *cold recovery*. The circuit-breaker electrode shape, contact distance, and dynamic gas density distribution in the insulated space all affect the recovery characteristics. They generally increase with distance between contacts, and saturate to a certain value at the final contact position.

The dielectric strength recovery after a current is interrupted by a circuit breaker is a *hot recovery* action and is slightly different from a cold recovery. Even after the current is interrupted at current zero and the plasma has disappeared, the insulating gas in the post-plasma space is rather hot, and its insulation characteristics differ considerably from those of a cold gas. The dielectric strength of the post-plasma space is restored by means of the hot gas cooling. It is related to the residual hot gas quantity in the post-plasma space. Therefore, the recovery is quick when the space-cooling force is high, and the initial rate of rise of the dielectric strength recovery characteristics becomes slow as the breaking current or arc duration time increases.

(B) Analysis of Dielectric Strength Recovery Characteristics

As already stated, the dielectric recovery characteristics are affected by the local electric field concentration on the electrode, local gas density distribution in high-speed gas flow between contacts, the history of the hot gas heated by a large current arc, and other factors. These characteristics are being studied by means of computer simulation. As an example, analyzed results of cold-recovery characteristics are shown in Fig. 1.6. These results were obtained by axisymmetric two-dimensional analysis, but three-dimensional analyses can also be done. Figure 1.6a represents a field analysis, and Fig. 1.6b indicates the results of gas-flow analysis. The results allow prediction of dielectric strength recovery characteristics with a reasonable degree of accuracy. The results are compared with measured data in Fig. 1.6c, and they conclude well with each other. As for the hot recovery, the large current arc energy is difficult to estimate, and a practical tolerable accuracy has not yet been obtained. Presently an energy-injection method simulating an arc column and several other devices is under study, and further developments are expected in this field (see Section 2.3).

1.2.5 Typical Arc Characteristics

Current interruption by arc plasma and dielectric recovery characteristics of post-plasma space have been outlined above. Features of typical arc plasmas that appear in present-day commercial circuit breakers are described in this section.

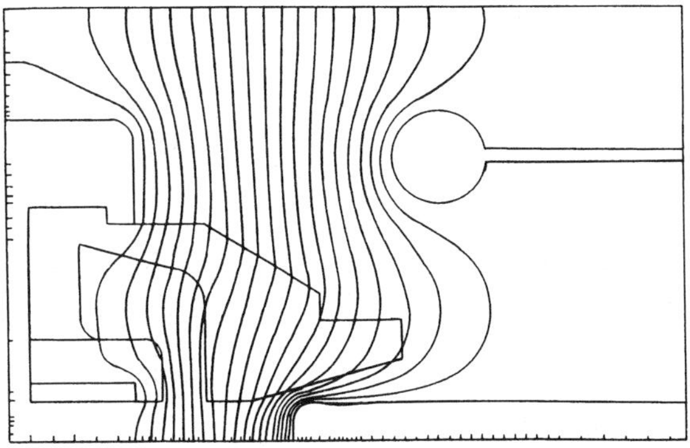

(a) Field analysis results

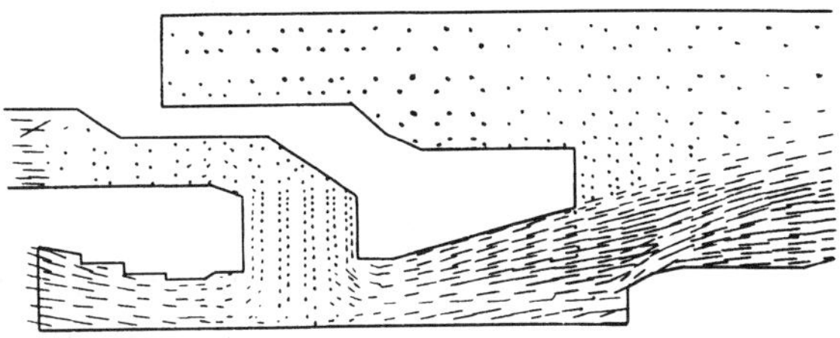

(b) Gas flow analysis results

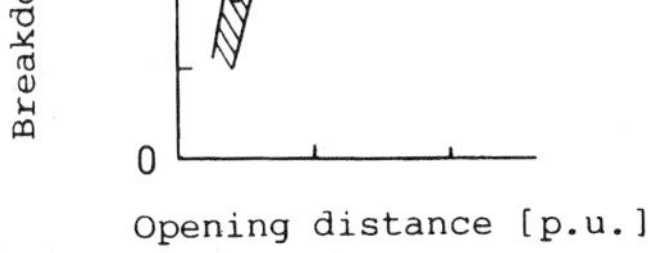

(c) Comparison between analyzed and experimental results
(no-load dielectric strength)

Fig. 1.6 Analysis of dielectric strength recovery characteristic. (Nakanishi et al., 1988. Copyright CIGRE.)

(A) Features of SF_6 Arcs

Sulfur hexafluoride gas (SF_6) is commonly used as an extinction medium for large-capacity circuit breakers. It has excellent insulation and extinction characteristics. Its arc characteristics are somewhat unusual compared with the arc plasma characteristics of other gases, particularly air and nitrogen.

Sulfur hexafluoride was initially introduced as an insulating gas because its dielectric strength is several times that of air. Its physical characteristics are described in Chapter 2. Its extinction ability is considerable, and the speed at which it changes from a conductive arc to an insulating material is very high. Its measured arc time constant is 1/100 that of an arc in static air, which agrees with the fact that its breaking capacity is 100 times that of air for current as well as for voltage.

The extremely large extinction ability of the SF_6 arc cannot be explained by a single mechanism. Two remarkable characteristics can be cited as particular evidence of its peculiar thermal nature and its strong electronegativity.

(1) Thermal Characteristics of SF_6

Sulfur hexafluoride, which is a polyatomic molecular gas, has a very complicated dissociative reaction at high temperatures that has not been completely clarified. However, reasonable thermal characteristics at high temperature can be assigned on the basis of experimental and statistical mechanical techniques.

Suppose there is an arc in the gas and that the heat flow emitted radially from its center is Γ. In static gas, Γ depends mainly on thermal conduction and can be described as

$$\Gamma = \frac{-\kappa \, dT}{dr} \tag{1.17}$$

where dT/dr is the temperature gradient of the arc in the radial direction. Γ is considered nearly constant along the radial direction within the temperature range of concern.

As is evident later from Figs. 2.2 and 2.3, the temperature (4000–5000 K) at which the conductivity increases abruptly in SF_6 is rather higher than the temperature (2000 K) at which the thermal conductivity peak due to dissociation appears. Therefore, for the arc column at 5000 K or higher where an arc current flows, κ is small, and therefore dT/dr is large. The flame on the column exterior is at 2000 K, κ is large, and therefore dT/dr decreases considerably in the temperature distribution. As shown in Fig. 1.7, the central column develops a conspicuous peak, the temperature distribution in the outer flame is very low and flat, and the current flow concentrates at the

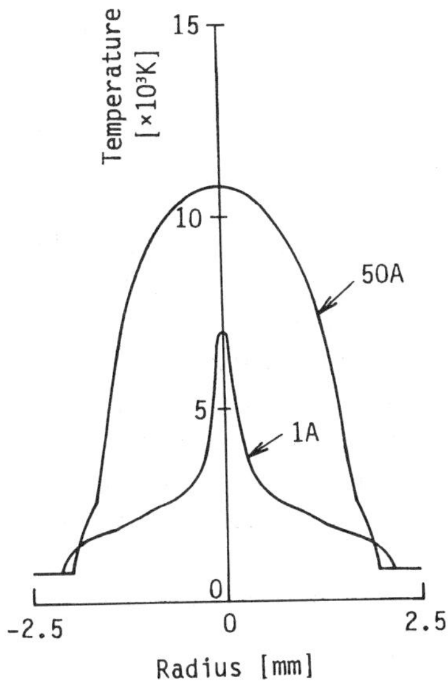

Fig. 1.7 Temperature distribution of SF_6 gas plasma.

column. This is why, with SF_6, a very thin and bright arc is seen and hardly any outer flame is visible.

With an arc in air or nitrogen, on the other hand, the temperature at which the conductivity starts increasing and the temperature at which κ rises almost coincide, and therefore dT/dr is almost the same at the column and in the outer flame, and the outer flame section has a large cross-sectional area and a rather high temperature.

With the SF_6 arc, an extremely small arc time constant is obtained because of cooling by a thermal conduction mechanism due to the dissociation peculiar to this gas mentioned earlier. Responding quickly to increases and decreases in current, the arc cross-sectional area changes, and near the current zero it drops abruptly, leaving a thin arc column, which exists down to a very low current level.

The fact that a narrow arc column exists down to a low current in SF_6 is a very advantageous characteristic when the gas is considered as a current-breaking element of a circuit breaker. It provides an ideal element for switching action or a quick change at current zero from a good conductor to an insulating material. Therefore, even when breaking a small current, the column contracts continuously down to the current zero and the current is not forced to turn off nor does current chopping occur, and switching surges are limited to a low level.

(2) *Electrical Negativity of* SF_6

The halogen element and halogen compound gases have high electronegativities, so that molecules and dissociated atoms tend to form negative ions easily. This tendency is very strong in the case of SF_6. Electrons dissociated by the arc and contributing to conduction are strongly attracted by SF_6 and the halogen molecules and atoms produced by its decomposition. The mobility of the charged particles drops considerably, and, through a two-step recombination process, negative ions combine with positive ions to return to neutral molecules and atoms, whereupon the space conductivity disappears very quickly. This apparently has the same effect as when the cooling force in the ionization space is very strong. The transition rate of the space conductivity becomes very high near the current zero of the arc, and, along with the thermal characteristics of a very fine arc stated above, the arc constant becomes small.

This process can be simplified to typical reactions as follows.

$$\begin{aligned} &SF_6 \rightarrow SF_6^+ + e \\ &SF_6 + e \rightarrow SF_6^- \\ &SF_6^- + SF_6^+ \rightarrow 2\,SF_6 \end{aligned} \tag{1.18}$$

Here the relationship between particle densities is

$$\frac{(SF_6^-)}{(SF_6)(e)} = \beta T^{1/2} \exp\left(-\frac{\beta}{kT}\right) \tag{1.19}$$

For arc extinction, the arc should preferably be in touch with as many neutral molecules as SF_6 as possible. In a high-velocity air flow, the arc is extinguished because of the thermodynamic isentropic cooling of air, but in the SF_6 atmosphere the arc is extinguished by the SF_6 flow.

Attachment of electrons in the case of SF_6 is marked by a phenomenon called *resonance attachment* by which a very strong resonating electron attachment is carried out at a rather low energy. The resonance attachment phenomenon occurs not only with SF_6 molecules but also with F, F_2, and other species transiently produced by SF_6 decomposition. Measured results are shown in Fig. 1.8. Near the current zero, in particular , many F_2^- ions are produced with this resonance. Figure 1.9 shows the negative-ion intensity change near current zero. On account of this phenomenon, when the arc is maintained stably down to a rather low current level, the energy level of the electrons is reduced to less than a certain level (for example, 0.1 A down to a low resonance level), they attach quickly, and the space conductivity changes abruptly. This fact provides a fairly good explanation of the arc-extinction capacity of F_6.

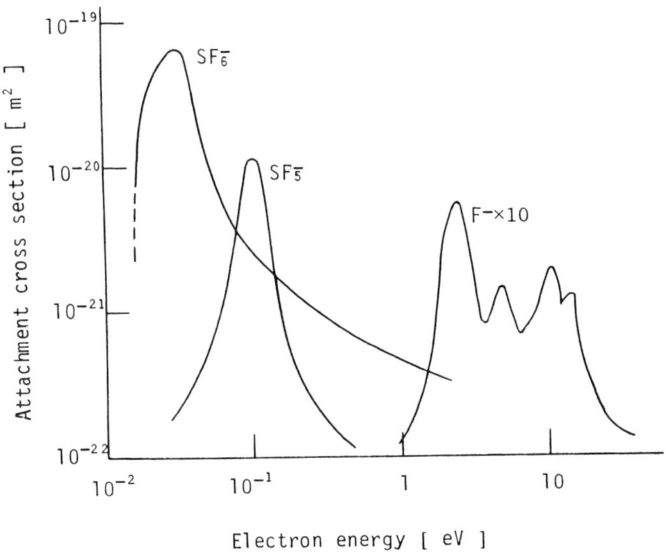

Fig. 1.8 Attachment cross section versus electron energy. (Nakanishi et al., 1975. Copyright JIEE.)

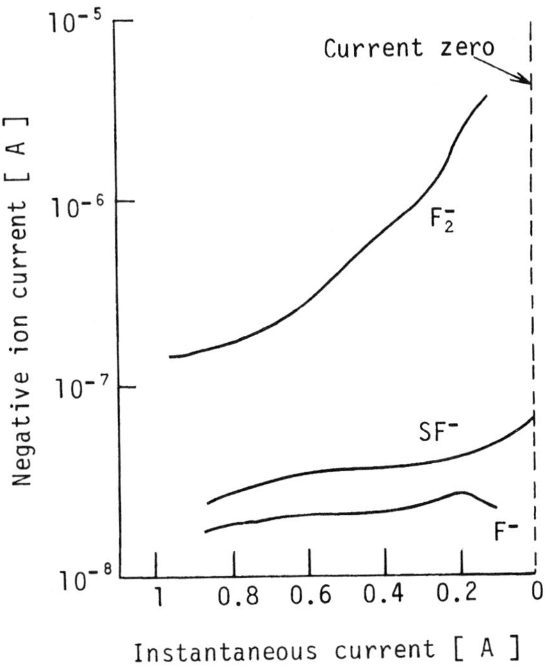

Fig. 1.9 Negative-ion current near current zero. (Nakanishi et al., 1975. Copyright JIEE.)

(B) Features of Vacuum Arcs

(1) Arc Behavior

The substance that forms an arc in vacuum is a metal vapor emitted from the cathode, whereas it is the gas itself in an arc formed in gas or air-blast circuit breakers. Therefore, the vacuum arc is called a *metal vapor arc*. The metal vapor pressure of a vacuum circuit breaker is quite low compared with the pressure of a gas circuit breaker. It is estimated to be on the order of several hundred pascals. This low-pressure arc shows various peculiar features. There is no voltage drop in the arc column or the anode. There exists only a cathode voltage drop. Therefore the arc voltage of a vacuum arc is very low and is kept constant independently of arc current. It depends on the cathode material and not on the electrode gap. The typical arc voltage of a copper electrode is about 20 V.

In vacuum, a lot of parallel arcs appear in proportion of a cathode to a current of about 100 A, moving rapidly and repelling each other. This arc mode is called the *diffuse arc mode*. However, the movement of the arcs slows, and many cathode spots form groups when the arc current is increased beyond about 10 kA. Finally, they converge to a thick arc in which the vapor pressure becomes high, and the arc shows the characteristics of high-pressure arcs like gaseous arcs, where voltage drops exist in the arc column and the anode. This arc mode is called the *constricted arc mode*.

Furthermore, vacuum arc behavior can be observed in a magnetic field, especially a perpendicular or axial magnetic field. When a perpendicular magnetic field is applied to a small-current vacuum arc, the arc moves in the direction against the $\mathbf{j} \times \mathbf{B}$ force. This phenomenon is known as *retrograde motion* of a vacuum arc. When the arc current becomes large, the retrograde motion disappears. The movement of a large-current vacuum arc under a perpendicular magnetic field obeys the $\mathbf{j} \times \mathbf{B}$ force. An axial magnetic field has a very useful effect that prevents formation of the constricted arc mode. It makes the vacuum arc maintain the diffuse arc mode at currents greater than 100 kA.

(2) Arc Extinction in Vacuum

A circuit is broken by extinction of a vacuum arc at a current-zero period in the same way as in other circuit breakers. However, extinction in vacuum does not need any additional measure like the puffer action in gas circuit breakers. Arc extinction can be achieved by diffuse action in a vacuum. There are residues of metal vapor and plasma in the interelectrode space at current-zero time, which diffuse from the interelectrode space to outer space and condense or recombine on wall surfaces. The residual plasma also enables a significant post-arc current to flow under the transient recovery voltage buildup. An ion sheath forms in front of the new cathode due to the post-arc current. This ion sheath can bring about large dielectric strength. The residual metal vapor also affects the dielectric recovery.

The dielectric recovery scheme of a vacuum arc can be classified into three categories according to the residual metal vapor pressure and plasma:

1. Dielectric recovery that is governed by the interelectrode gap and the electrode material because of very thin residual metal vapor and plasma.
2. Dielectric recovery that is governed by the thickness of the ion sheath because of thick residual plasma
3. Dielectric recovery that is governed by Pashen's law because of thick residual metal vapor

The actual recovery characteristics are combined in a complicated fashion with these mechanisms. The vacuum arc has a prominent feature relative to electrical breakdown during the recovery period—there is no thermal breakdown. Reignition in a vacuum arc occurs only by dielectric breakdown.

Since the vacuum arc is quite different from the gaseous arcs described earlier, the mathematical arc models presented for gaseous arcs cannot be used for it. Extensive fundamental studies are needed to produce a mathematical model. However, a vacuum arc can be extinguished very quickly in a very large current, and the vacuum circuit breaker is the most successful application of the vacuum arc.

1.3 SWITCHING CONDITIONS IN THE ELECTRIC POWER SYSTEM (M. Tsukushi, K. Hirasawa, Y. Kurosawa)

1.3.1 Interrupting Current

(A) Types of Short-Circuit Faults and Short-Circuit Currents

A circuit breaker must be capable of making, carrying, and interrupting the current under both normal and abnormal conditions, especially in the case of a short-circuit fault. Before installing a circuit breaker, it is necessary to estimate the maximum short-circuit current that can occur in the electric power system and then select a circuit breaker that can interrupt and make the estimated current.

Many types of short-circuit faults occur in electric power systems. Table 1.2 shows the relation between the short-circuit current and the type of short-circuit fault. In this table, E is the phase voltage, and X_1, X_2,

Table 1.2 Types of Short-Circuit Faults and Short-Circuit Currents

Type of short circuit	Symbol of current	Short circuit current
Three phase un-grounded	I_{3S}	E/X_1
Two phase un-grounded	I_2	$\sqrt{3}\,E/(X_1 + X_2)$
Two phase grounded	I_{2e}	$\sqrt{3}\,E\,\frac{\sqrt{X_2^2 + X_2X_0 + X_0^2}}{X_1X_2 + X_1X_0 + X_2X_0}$
One phase grounded	I_{1e}	$3E/(X_1 + X_2 + X_0)$

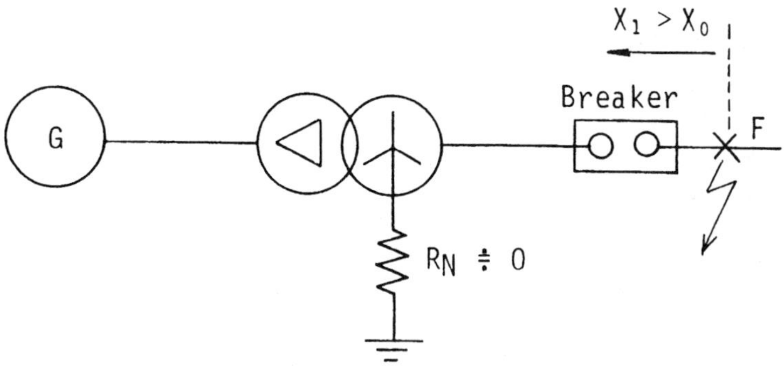

Fig. 1.10 Model of electric power system in the case of $I_{1e} > I_{3s}$.

and X_0 are positive-sequence, negative-sequence, and zero-sequence reactances as viewed from the fault point.

In many electric power systems, the short-circuit current has its maximum value during a three-phase short-circuit fault, but in a solidly grounded system where X_0 is less than X_1, the current of a one-phase grounded fault is larger than that of a three-phase short-circuit fault. This case is shown in the electric power system of Fig. 1.10, where the one-phase grounded fault current is

$$I_{1e} = \frac{3E}{X_1 + X_2 + X_0}$$

and the three-phase short-circuit fault current is

$$I_{3S} = \frac{E}{X_1}$$

(1) Voltage Phase Angle at Time of Short Circuit and Current Waveform

Figure 1.11 shows the *short-circuit current waveforms* in the first few cycles just after the occurrence of a short-circuit fault. A decaying *dc component* is superimposed on a decaying sinusoidal *ac component* in accordance with the voltage phase at the instant the short-circuit fault occurs and an *asymmetrical short-circuit current* flows. The current waveform when the short circuit occurred in the neighborhood of the voltage phase zero degree is also shown in the figure. The entire waveform is referred to as an *asymmetrical short-circuit current,* and the component remaining after subtracting the ac component from the asymmetrical current is the dc component.

Both ac and dc components decay with time, and the decay of the former is greatly influenced by the time for generator reactance on the dynamic reactance as viewed from the short-circuit point. Then the ac component decay offurs quickly in the short-circuit current near the power station.

However, dc component decay is influenced by the time constant of the electric power system as viewed from the short-circuit point, so the dc component decays rapidly for a circuit with a short time constant *L*/*R*, for example, a line between the short-circuit point and the power station.

The circuit breaker must withstand the electrical and mechanical stresses produced by the short-circuit current and interrupt it.

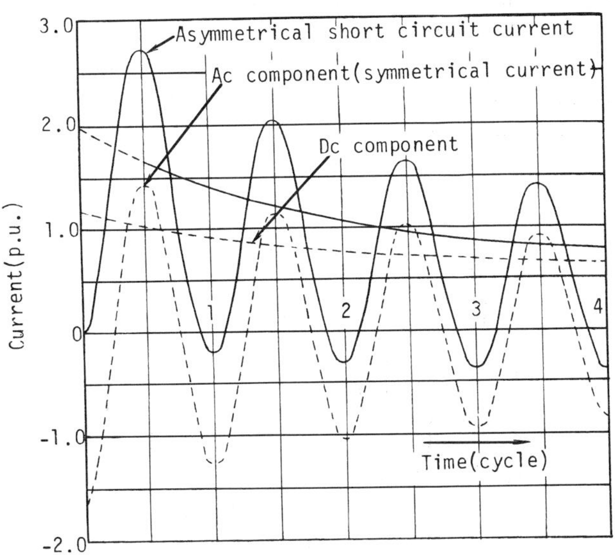

Fig. 1.11 Short-circuit current waveform.

(2) Calculation of Short-Circuit Current

Formerly, many simple manual calculation methods were used, but now analytical methods implemented on a computer are in general use because of the great complexity of today's power system generators and series electrical equipment. The *Electro Magnetic Transients Program* (*EMTP*), developed by the Bonneville Power Administration, is a typical program that is used worldwide.

(3) Rated Interrupting Current

The current-interrupting ability of the circuit breaker is expressed as the rated interrupting current. This current involves the dc component that decays exponentially with a time constant of 45 msec.

(B) Short-Circuit Current Under Exceptional Conditions

(1) Current-Zero-Missing Passing Current

When a fault occurs in the neighborhood of a large-capacity power station, an extremely asymmetrical current flows. In the case of a low-resistance line, the ac component decay is faster than the dc component decay, and the current does not cut across the zero line until several cycles after the fault occurrence, as shown in Fig. 1.12. The initial value $I(0)$ depends on the instantaneous value of load current at the instant the short-circuit fault occurs and takes a positive value for the capacitive load and a negative value for the inductive load. An ac circuit breaker can interrupt the current only at the instant when it passes the zero point; therefore, it cannot interrupt until several cycles have passed. Then a special technique is

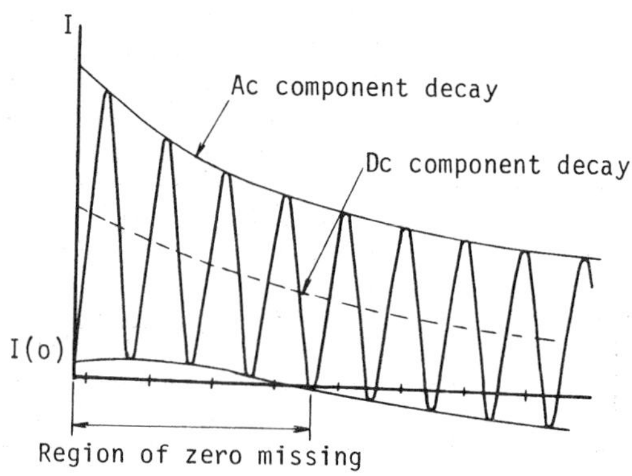

Fig. 1.12 Zero-missing passing short-circuit current waveform.

needed such as delaying the opening of the circuit breaker until the current zero occurs. Tests to be made under this condition must also use special techniques because the time constant of the dc component deviates from that of general standards.

(2) Evolving Fault

When the fault-point condition varies during the current-interrupting process, the current through the circuit breaker sometimes increases rapidly from a small current to a large one. Such a fault is known as an *evolving fault;* for example, when a transformer-exciting current interruption occurs, an overvoltage is generated because of current chopping (see Section 1.4), and this overvoltage induces a dielectric breakdown of the transformer terminal, causing a short-circuit fault. Another example of an evolving fault is when a one-phase fault is varied to a three-phase short-circuit fault under ungrounded system conditions.

In actual situations, evolving faults rarely occur, so they seldom influence the interrupting ability of the circuit breaker. At present, this fault condition is standardized only in France, but the circuit breaker must be able to interrupt an evolving fault.

(C) Small-Current Interruption

All circuit breakers should be able to interrupt a small current, such as a load current, line-charging current, or transformer-exciting current. Except for certain types of older circuit breakers, such small currents can be easily interrupted, but the overvoltage just after interruption becomes a problem in some cases. This problem will be mentioned later in the section on switching surge.

1.3.2 Transient Recovery Voltage

(A) Power Frequency Recovery Voltage

Whenever a circuit breaker interrupts a current, a voltage is generated between the contacts of the circuit breaker. In the case of a short-circuit interruption, this voltage has two components as shown in Fig. 1.13. One is a transient recovery voltage (TRV) just after current interruption, and the other is a power frequency recovery voltage (PFRV) that oscillates with the power frequency after the TRV oscillation has decreased.

The TRV has a direct effect on the interrupting ability of the circuit breaker, but the PFRV is also important because it determines the center of the TRV oscillation. The PFRV magnitude differs for different system conditions and fault conditions. For example, in the first interrupting

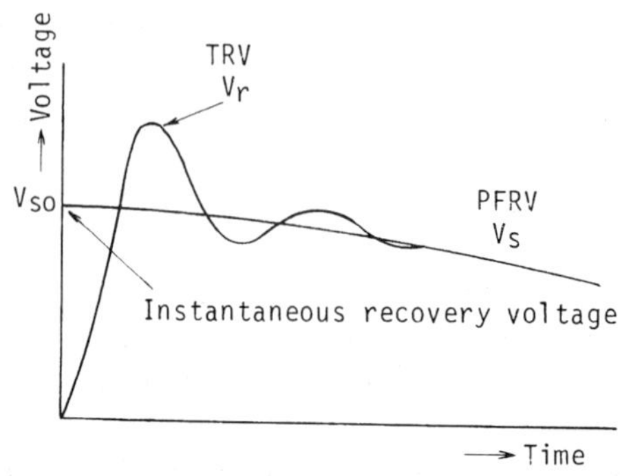

Fig. 1.13 Voltage waveform just after current interruption.

phase, the peak value is $1.5\sqrt{2}\ E$ in high-resistance grounded systems, whereas it is $1.3\sqrt{2}\ E$ in solidly grounded systems.

(B) Transient Recovery Voltage

Transient recovery voltage (TRV) is the *transient natural oscillation* that is impressed between the contacts just after current interruption. It is as important a factor as interrupting current in evaluating the interrupting ability of a circuit breaker.

(1) Transient Recovery Voltage of Power Systems

The waveform and magnitude of a TRV vary according to many factors such as system voltage, system constitution, constants of the equipment, installed position of the circuit breaker, and fault types. The waveform determined by the system circuit parameter alone is termed the *inherent transient recovery voltage.*

Figure 1.14 shows the representative waveform for each kind of system configuration and explains why there is such a shape for that circuit condition. Configurations 1–3 are for a power station circuit breaker. A high natural frequency TRV appears for them. Number 4 is the case of a transformer bank circuit breaker, and again a high natural frequency TRV appears. Numbers 5–9 are cases of a line-side circuit breaker for a power station or substation. Number 8 shows an inverse *short-line fault* (SLF) interruption in which a short overhead line is present between the circuit breaker and the station, Number 9 shows the SLF interruption in which a short overhead line is present between the circuit breaker and the fault point. Number 8 and 9 both have a high-frequency voltage generated according to a reciprocating voltage reflection on the line, which is the most severe fault condition for the circuit breaker. The voltage waveform of SLF interruption is explained in the next subsection.

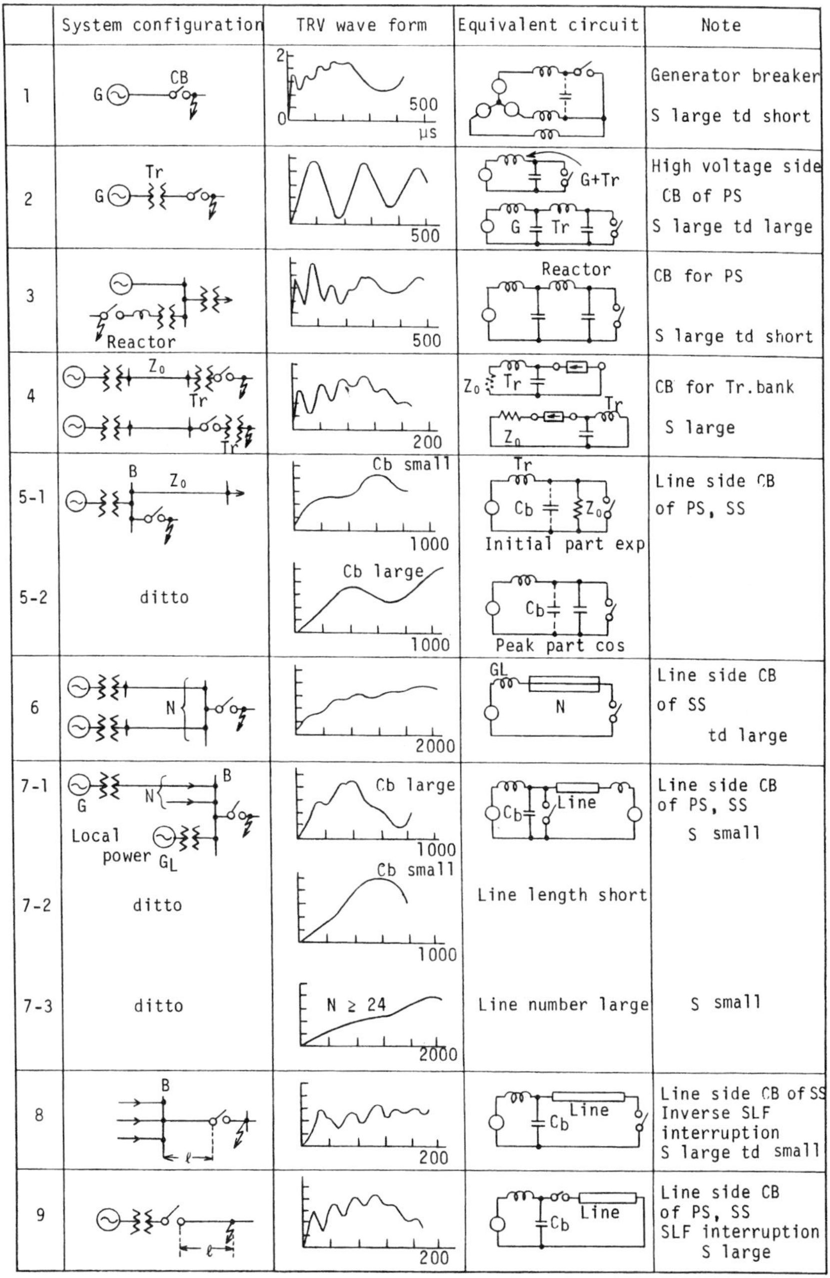

S: Initial rate of rise of TRV td: Time delay

Fig. 1.14 System configuration and representative TRV waveforms.

(2) Transient Recovery Voltage Waveform of Short-Line Fault Interruption

When the circuit breaker interrupts a line fault that occurs several kilometers distant from the circuit breaker, an SLF interruption occurs. The SLF interruption is more difficult to execute than the *breaker terminal fault* (BTF) interruption for the line-side circuit breaker, in spite of the de-

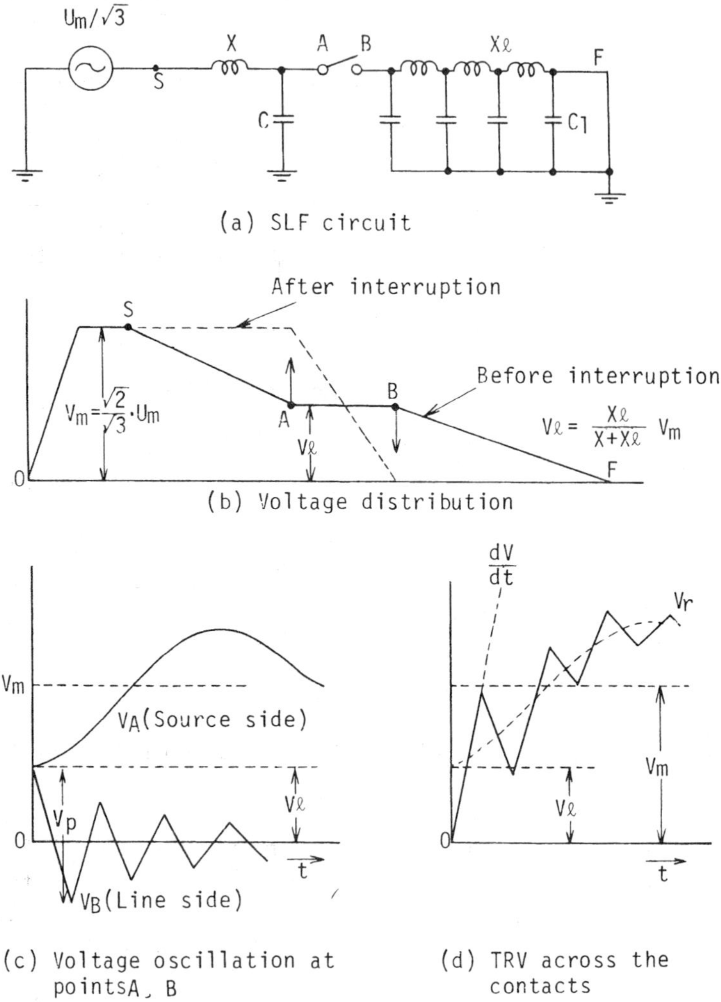

Fig. 1.15 Short-line-fault transient recovery voltage. (Nakanishi et al., 1975. Copyright JIEE.)

creased current, because voltage with a steep slope is generated owing to the presence of the short line.

Figure 1.15a shows an SLF circuit. The power-source-side circuit is approximated by a lumped-constant circuit with single natural frequency, and the line-side circuit is approximately by the lossless distributed constant line. The frequency of the short-circuit current is the same as the power frequency, so the influence of the electrostatic capacity can be neglected and the voltage distribution of the line side is linear as shown in Fig. 1.15b. When this voltage distribution is free from the restriction of the source-side voltage after current interruption, transient voltage oscillation is generated according to a shift of the distribution as shown by the dashed line. When the voltage becomes free of the restriction, the *coefficient of the voltage reflection k* is 1 at the opening end of the circuit breaker and −1 at the fault point, so the waveform of the voltage *traveling wave* goes to the right side and left side with each half value of the initial voltage as shown in Fig. 1.16a. The synthetic waveform on the line side is trapezoidal, and its peak value changes with time as shown in Fig. 1.16b–d.

Therefore, at the entrance of the line side, the voltage waveform is a triangular oscillation as shown in Fig. 1.16e, and at the central part of the line the voltage waveform is trapezoidal as shown in Fig. 1.16f.

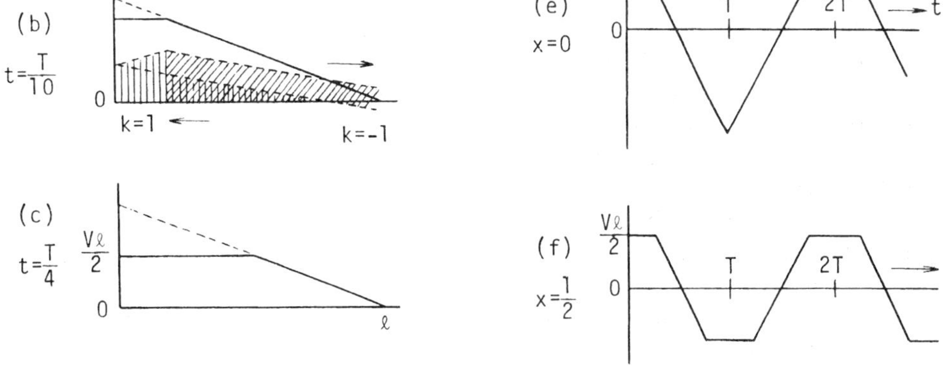

Fig. 1.16 Time variation of line voltage distribution after SLF interruption. (Nakanishi et al., 1975. Copyright JIEE.)

The slope of the triangular oscillation is the initial rate of rise of voltage and can be expressed by the equation

$$\frac{dV}{dt} = Z\frac{dI}{dt} = Z\,\omega\,\sqrt{2}\,I_m$$

where Z is *line surge impedance,* ω is source power frequency, and I_m is the short-circuit current (effective value). The voltage drop on the line side can be expressed as

$$V_l = X_l\sqrt{2}\,I_m$$

The voltage oscillations at A and B are as shown in Fig. 1.15c, and the voltage impressed between the contacts is as shown in Fig. 1.15d. It is characteristic of SLF interruption that a high dV/dt is imposed on the circuit breaker just after current interruption. Since this dV/dt is proportional to the interrupting current, SLF interruption becomes a serious problem when the fault point and a power station with a large short-circuit capacity are connected by a short overhead line. For a gas circuit breaker, the most severe condition exists when a current is suppressed to 90% of the BTF current by the line impedance.

(3) Initial Transient Recovery Voltage

A more precise evaluation of TRV has become necessary as advances have been made in increasing the performance and capacity of circuit breakers.

All regions of the TRV have an effect on the breaker's interrupting ability, but the initial part (ITRV) has a particular effect on the interrupting ability in some kinds of circuit breakers. The difficulty of SLF interruption is also due to the steep slope of the voltage impressed in the ITRV region.

Therefore an IEC standard ITRV has been set by the International Conference on Large High Voltage Electric Systems (CIGRE) based on the results of many TRV investigations. This phenomenon is similar to the inverse SLF interruption mentioned earlier.

A small-amplitude oscillation generated by reflection of the traveling voltage wave between the circuit breaker and the first main discontinuous point determines the bus line structures of the station on the main source side, and this oscillation voltage determines the ITRV. The small oscillation of the ITRV is similar to the SLF phenomenon, but its peak value is less than that of the SLF, and the time from the current zero to the peak value is less than 1 μsec.

Figure 1.17 shows the relation between the TRV represented by a four-parameter method (see Section 4.1 of Chapter 4) and the ITRV. Previously, in the rising part of the TRV, a *time delay of rising* was permitted,

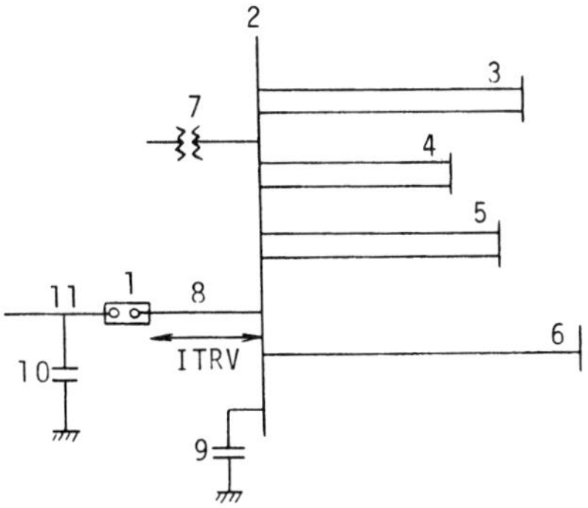

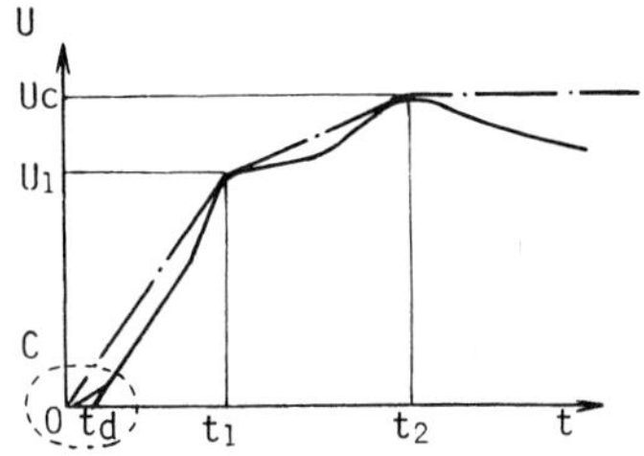

1:Breaker 8:Junction bus
2:Main bus 9:Equivalent bus capacitance
3-6:Line 10:Equivalent line side capacitance td:Time delay C:Initial part
7:Transformer 11:Line

(a) Typical high ITRV circuit configuration. (b) TRV of typical BTF interruption.

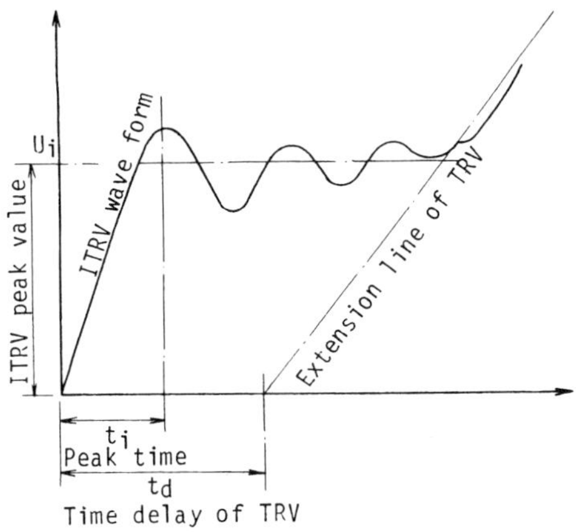

(c) Enlargement of initial part (ITRV).

Fig. 1.17 Voltage oscillation in ITRV region.

considering the electrostatic capacity, by the IEC Pub.56-2(1972) standard, but for the new standard a small-amplitude oscillation is superimposed on this initial region of TRV. This ITRV has a serious influence on the thermal breakdown characteristic depending on types of the circuit breaker.

1.3.3 Switching Surge

This section describes the occurrence of overvoltages that might have an effect on the power system insulation when the circuit breaker operates.

There are two kinds of overvoltages that threaten the insulation of electric power system. One is *lightning surge,* which determines the basic insulation level (BIL) of the insulation design, and the other is the switching surge caused by the circuit-breaker operation. There are many kinds of switching surges. They are a serious problem, because the circuit breaker should protect the power system equipment from faults but its operation generates an overvoltage that threatens the insulation. Many studies have been made on this phenomenon and means of counteracting it.

(A) Causes and Classifications

Circuit-breaker operations that can cause switching surges include

1. Switching operation of an un-load long transmission line and an un-load transformer at a distant terminal
2. Switching operation of an un-load or reactor load transformer
3. Switching operation of a shunt reactor

The TRV mentioned earlier is also a kind of a switching surge, but its overvoltage is usually less than twice the system voltage and generally does not become a serious problem. In ultrahigh-voltage class systems, considering the extreme supression of switching surge from an economic viewpoint, surge control is accomplished even for the TRV at fault-circuit interruption by applying a resistor breaking.

Figure 1.18 shows an electric power system model in which electric power is carried from A to B. Figure 1.19 illustrates the classification of circuit-breaker operations that generate switching surges.

(B) Small-Current Interruption

(1) Line-Charging-Current Interruption

The interruption of line-charging current was a major cause of switching overvoltages for the older oil circuit breakers. The interrupting circuit of

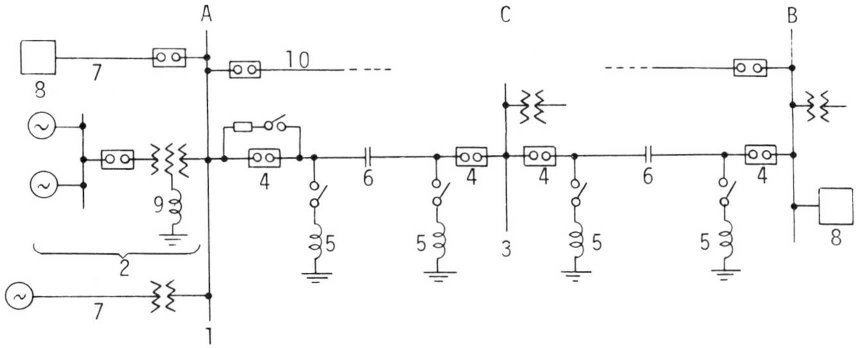

A,B,C:PS and SS
1:Bus
2:Local source
3:Switching station
4:Breaker
5:Shunt reactor
6:Series capacitor
7:Line
8:Load
9:Reactor
10:Parallel line

Fig. 1.18 Model of an electric power system.

an un-load transmission line is shown in Fig.1.20. After current interruption, the voltage across the circuit breaker is twice the phase voltage crest value in a half-cycle of the power frequency. Here, the interrupting current level is less than a few hundred amperes. An interelectrode breakdown (*restriking*) sometimes happens when the contact is opening. The waveforms of voltage and current during restriking are also shown in Fig. 1.20. If the circuit breaker interrupts a high-frequency current caused by restriking, an overvoltage of twice the phase voltage crest is charged at the line. With repeated restriking, the overvoltages become much larger and threaten the system insulation. As the newer types of circuit breakers do not experience restriking, the problem was been eliminated. In the case of a disconnecting switch, however, attention should be paid to such overvoltages.

(2) Small-Inductive-Current Interruption

In small-inductive-current interruption, circuit breakers with a strong arc-extinguishing ability cause a phenomenon in which the current is interrupted suddenly at some level before the ac natural current zero. This phenomenon is known as *current chopping* (Fig.1.21). The magnetic energy stored in the inductance of $(\frac{1}{2})LI^2$ is converted to a capacitive energy

Switching operation	Basic circuit
1. Line closing	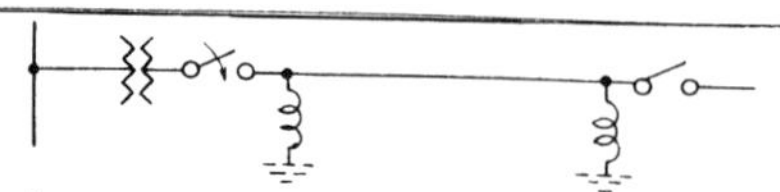
2. Line reclosing (trapped charge on the line)	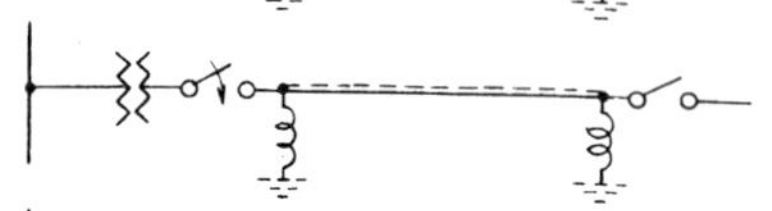
3. Line closing at low voltage side	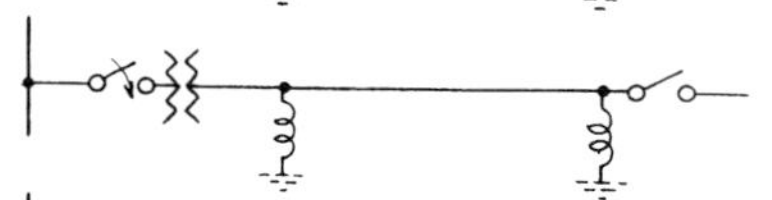
4. Line closing with unload transformer connected to the end	
5. Load interruption at the end of line	
6. Line interruption after load interruption at the end	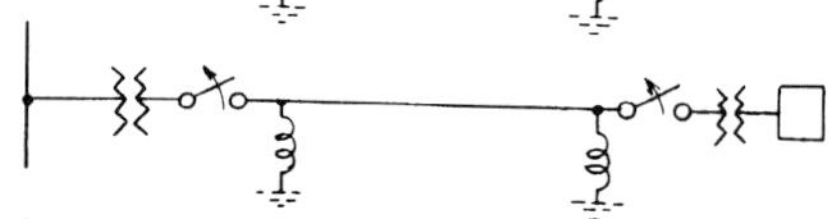
7. Un-load line interruption	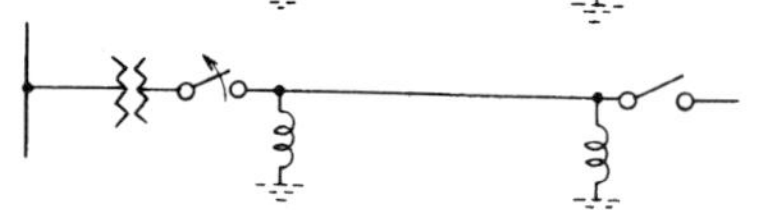
8. Un-load transformer switching	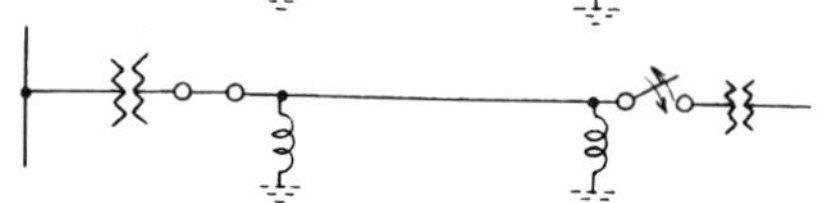
9. Reactor load transformer switching	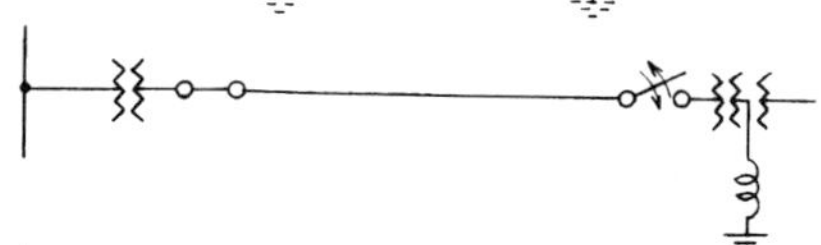
10. High voltage side reactor switching	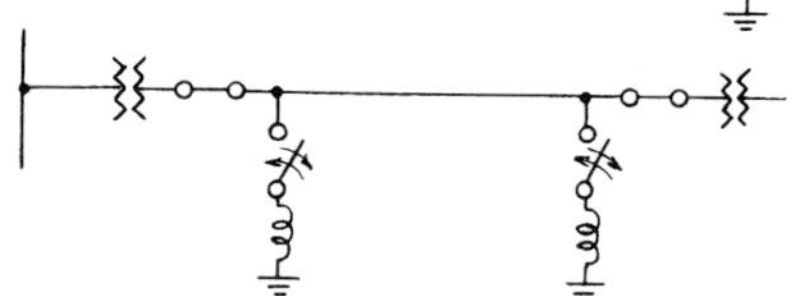
11. Switching at the midway switching station	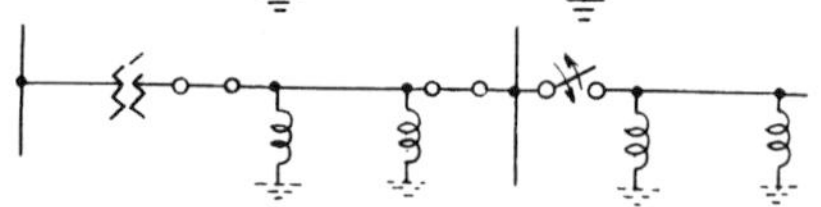
12. One phase grounded fault	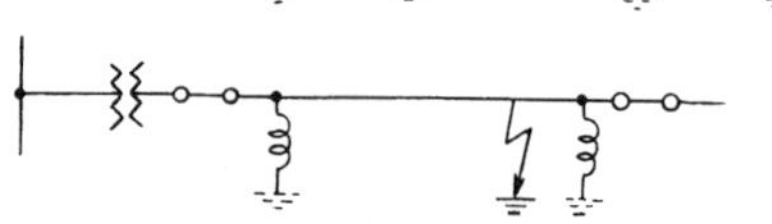

Fig. 1.19 Switching operation inducing switching surge. (Nakanishi et al., 1975. Copyright JIEE.)

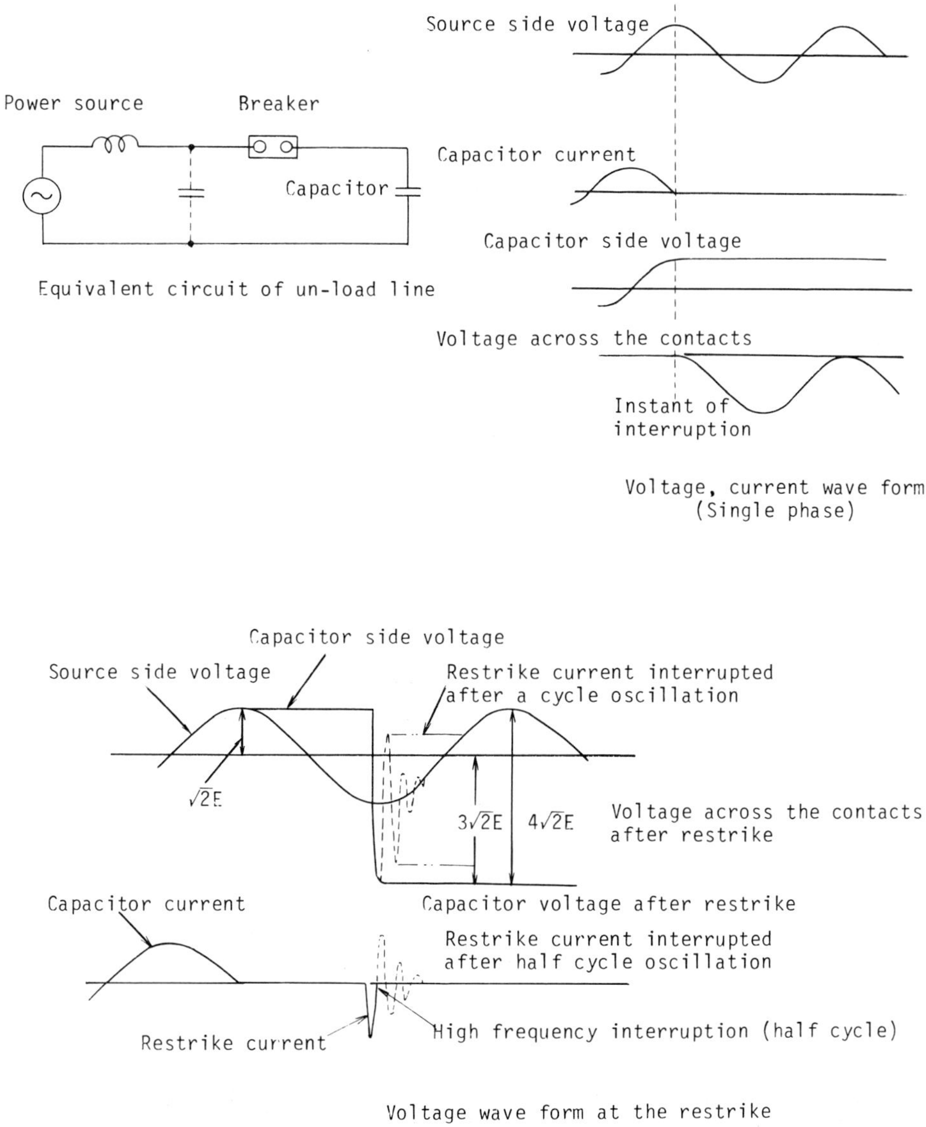

Fig. 1.20 Interruption of un-load line.

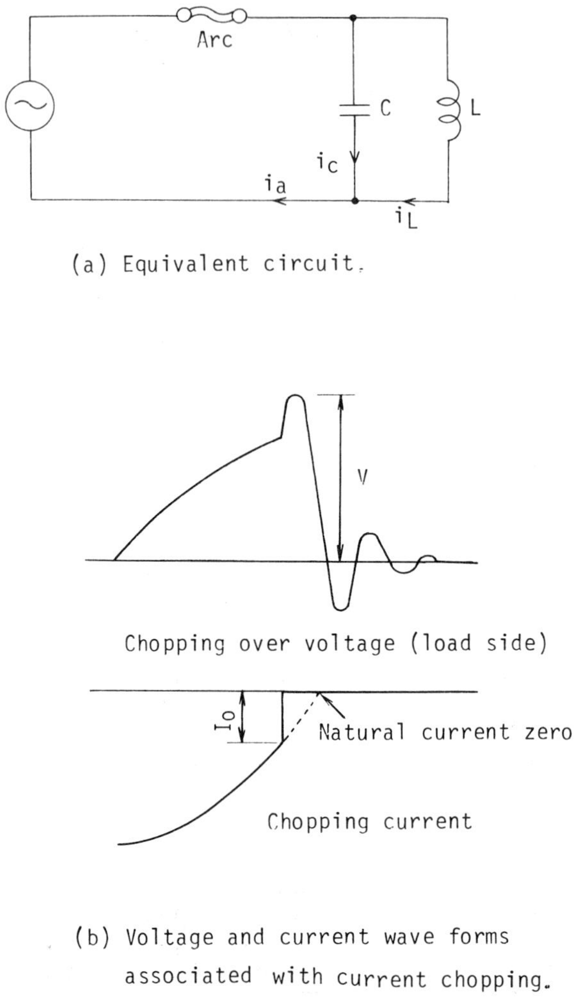

Fig. 1.21 Current chopping upon interruption of the exciting current.

of $(\frac{1}{2})CV^2$ in the transient oscillation, and overvoltage V occurs. Assuming that current chopping occurs with a chopping current I prior to current zero, voltage V generated on the load side after current chopping is given by

$$C\frac{dV}{dt} + \frac{1}{L}\int V\,dt = 0 \tag{1.20}$$

With the chopping phase angle written as θ, the initial conditions are given by

$$V(0) = E_m \cos\theta \tag{1.21}$$

$$I(0) = -\frac{E_m}{\omega L}\sin\theta \tag{1.22}$$

Solving Eq. (1.20) by using Eqs. (1.21) and (1.22), we have

$$V = E_m \cos\theta \cos\beta t + \sqrt{\frac{L}{C}}\, I(0) \sin\beta t \tag{1.23}$$

The first term of this equation represents the voltage generated even without chopping, with a frequency in agreement with the natural frequency of the circuit on the load side. The second term is chopping surge voltage generated in chopping by the energy stored in the inductance. Its peak value is represented by $ZI(0)$ where $Z\ (= \sqrt{L/C})$ is the surge impedance on the load side. Since the first and second terms are the same frequency, the frequency of the surge generated by chopping equals the natural frequency of the circuit on the load side. Generally, unless $I(0)$ is very large, this voltage presents hardly any problem.

As vacuum circuit breakers, which are widely utilized at medium voltages, have relatively high chopping current levels, careful attention should be paid to application of transformers with varnished paper insulation, for which the insulation level is lower than oil-immersed-type transformers. In such cases, some surge absorbers are installed with the circuit breaker. Recently, a low-surge type of vacuum circuit breaker has been developed: its current-chopping level is very small compared with conventional breakers, and it can be used without a surge absorber.

In *magnetizing current interruption,* the current can be interrupted with a small arcing time in the same manner as capacitive current interruption. The transient recovery voltage after interruption of magnetizing current is decided by the inherent frequency of the transformer, which is 500–1000 Hz. If the circuit breaker does not withstand the TRV, a reignition occurs. When reignition occurs, the circuit breaker usually interrupts the current at the next power frequency current-zero point. Some kinds of circuit breakers, such as vacuum circuit breakers, have the ability to interrupt *high-frequency current.* They can interrupt the current at the current-zero point of a higher frequency oscillation due to the resonance of stray capacitance and inductance near the circuit breaker. In such a case, the TRV builds again and another reignition occurs; thus reignitions and high-frequency-current interruptions are repeated many times. This is known as the *multiple reignition phenomenon,* and it happens in vacuum circuit breakers and air-blast breakers. Recently it was reported that puffer-type gas circuit breakers also experience this phenomenon during *shunt-reactor switching.* Figure 1.22 is a schematic of a shunt-reactor switching circuit. L_L denotes the shunt-reactor inductance. C_s, C_l, L_s, and L_l denote power-source-side capacitance, line-side capacitance, power-source-side stray inductance, and line-side stray inductance, respectively. The phenomenon during shunt-reactor current interruption by a gas circuit breaker is explained as follows, with reference to Fig. 1.23.

Region I: The current in the shunt reactor is usually several hundred amperes. Current chopping is assumed to happen at time t_1. The instantaneous value of the power-source voltage is denoted V_0.

Region II: The magnetic and capacitive energies stored in L_L and C_l at the instant of current chopping are $\frac{1}{2}L_L[I_L(0)]^2$ and $\frac{1}{2}C_l[V_L(0)]^2$. Here, $I_L(0)$ and $V_L(0)$ are reactor current and capacitor voltage, respectively, at that time. After current chopping, these energies oscillate between L_L and C_l.

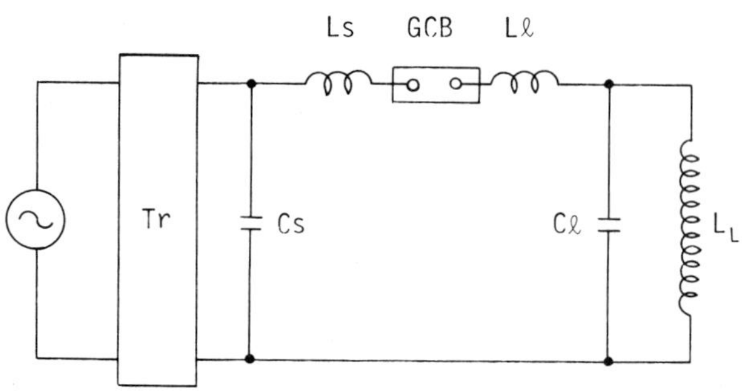

Tr : Transformer
GCB : GCB for shunt reactor switching
Cs : Source side total capacitance
Cℓ : Load side total capacitance
Ls : Source side total inductance
Lℓ : Load side total inductance
L_L : Shunt reactor inductance

Fig. 1.22 Equivalent circuit of shunt-reactor circuit.

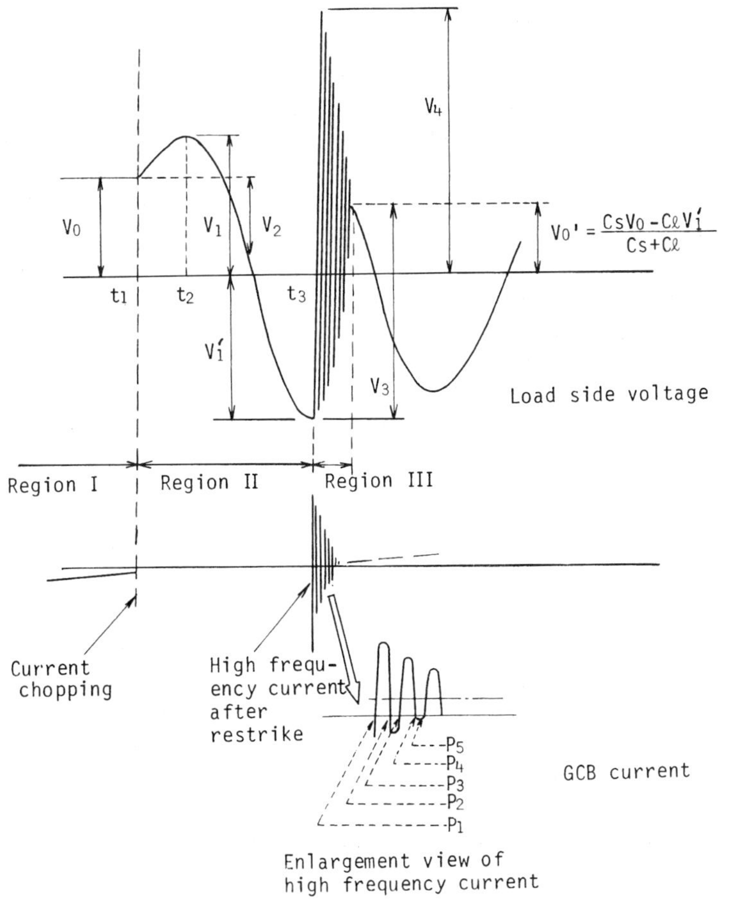

Fig. 1.23 Voltage and current waveforms when the shunt-reactor current is interrupted.

The load-side breaker terminal voltage has its maximum value when the total energy concentrates in C_l, that is, when

$$\frac{1}{2}C_lV_1^2 = \frac{1}{2}L_L[I_L(0)]^2 + \frac{1}{2}C_l[V_L(0)]^2 \tag{1.24}$$

From this equation, it can be seen that there is a possibility that the voltage V_l becomes higher than the phase voltage crest V_0. The voltage V_l is called the *chopping surge*. The transient recovery voltage that accompanies the chopping surge oscillates with the oscillation center of the load-side neutral voltage at a frequency of several hundred hertz to several kilohertz. The amplitude of the chopping surge is almost proportional to the chopping current. On the other hand, voltage V_2 is applied to the circuit-breaker terminals. Reignition occurs when the interelectrode gap length does not become large enough to withstand the voltage V_2 after current chopping occurs. In Fig. 1.23, reignition occurs at time t_3 and interelectrode voltage V_3. The reignition surge voltage is denoted V_4.

Region III: Reignition causes a short circuit between the electrodes of the circuit breaker. The load-side oscillating voltage converges to the voltage V_0', which is decided by C_s and C_l. Here, the maximum amplitude V_3 is $(V_0' + V_1')$, and it increases with the chopping current surge and C_s. The oscillation frequency at reignition is on the order of several hundred kilohertz. During the reignition, the current through the inductance changes. Therefore, the current that flows in the circuit breaker consists of a high-frequency current component and a power-frequency current component. Figure 1.23 shows the circuit-breaker current on an enlarged scale. Due to the imposition of the high-frequency current, new current-zero points appear and the current can be interrupted. If it is interrupted, it is a *high-frequency-current interruption*. The magnetic and capacitive energies at the instant of a high-frequency-current interruption are decided in the same manner as the current chopping. Looking at the magnetic energy, it decreases with repeated reignition when current chopping occurs before the current-zero point. Therefore, the inherent amplitude of the magnetic and capacitive energies tends to decrease with reignitions at pre-current-zero chopping and tends to increase at post-current-zero chopping. Nevertheless, the reignition surges increase to some extent with reignitions because the interelectrode gap becomes larger with time due to the electrode separating movement. Repetition of reignition stops when the interelectrode gap becomes large enough to withstand the escalated transient recovery voltage or the circuit breaker cannot interrupt the reignition high-frequency current.

(C) Making Surge of Un-load Transmission Lines

Switching surges are affected by the characteristics of the circuit breakers. Due to various improvements in the suppression of switching-surge voltages, most problems concerned with switching surges have been solved. Those problems that remain are concerned with switching surges in extra-high-voltage (EHV) and ultrahigh-voltage (UHV) transmission systems known as *making surges*. Making surge is the main factor that decides the system insulation level in UHV systems. It corresponds to item 1 in Fig.1.19, where system structure has a large and complicated influence on

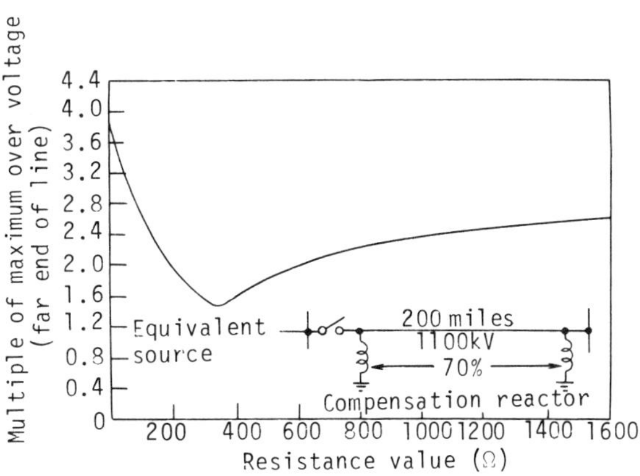

Fig. 1.24 Effect of preclosing resistance on switching surge.

making surge. In principle, its occurrence is due to the energizing of an *L-C* series circuit. Therefore, suppression of the surge voltage is completed by inserting a resistor circuit before the circuit-breaker contact is closed. With the resistor inserted, the transient oscillation is effectively suppressed. As a result, the surge voltage becomes low. Then the circuit-breaker contact is closed. Figure 1.24 shows the effectiveness of resistor insertion. With a resistor of about 350 ohms, the maximum overvoltage is reduced to 1.4 times the phase voltage crest.

1.4 ARC-INTERRUPTION PHENOMENA (Y. Yoshioka, M. Tsukushi, Y. Kurosawa)

1.4.1 Interaction Between Arc and Circuit

In Section 1.3, many kinds of interrupting conditions in the systems were described. Now interactions between arc and circuit must be considered. If the circuit breaker is an ideal switch in which arc resistance is zero initially and infinite after interruption, the phenomena that were explained in the previous section are valid. But in reality arc resistance is not zero; rather, it has some low value. After interruption, it does not change rapidly from the low value to the infinite value. For several microseconds after current interruption, a post-arc current flows in the circuit breaker, and it affects the waveform of the transient recovery voltage.

(A) During Arcing Period

Arc resistance acts like a resistor in a circuit. The time constant of the dc component in the short-circuit current becomes smaller because of the arc resistance. Effectively utilizing this effect might give a countermeasure for zero-missing current, which results because the decrease of the ac component is larger than that of the dc component. A fast decrease of the dc component can solve the zero-missing-current problem. This is actually done in gas circuit breakers (see Chapter 4, section 4.2).

(B) Around Current Zero

As the arc current approaches a current-zero point, the ratio of arc heat energy loss to electrical input energy increases and the arc voltage rises abruptly. Then the charging currents to the parallel capacitances of the breaker increase. But the total current—the arc current plus the charging current—flows through the power-source inductance L and cannot change rapidly. As a result, the arcing current decreases relative to the fast increment of the charging current. This causes distortion, or *squeeze,* of the arc-current waveform. That is, a fast increase in arc resistance squeezes the current before current zero, the arc voltage rises abruptly, and an extinction peak of the arc voltage, which is the maximum of the arc voltage, appears (Fig.1.25). After the arc is extinguished, the transient recovery voltage, which has a polarity opposite to that of the arc voltage, appears. Since the interelectrode space does not change to a completely insulating state, a slight current flows through the breaker as the TRV builds up, and soon disappears. This current is called *post-arc current.* It is usually very small. It sometimes distorts the inherent waveform of the TRV. When the extinguishing ability of the circuit breaker is small, the post-arc current does not decrease, and it increases with time. Finally, interruption failure—that is, thermal failure—occurs. Arc models proposed by Mayr and others were derived by simulating the phenomena during the post-arc-current period. They were expressed by a simple differential equation as described in Section 1.2, so an analysis can be done by solving simultaneous equations of combined circuit equations using the Runge–Kutta–Gill calculation method and so on. Figure 1.26 shows examples of current and voltage waveforms of short-line fault interruption with an air-blast breaker and a gas circuit breaker. Figure 1.27 shows examples of the short-line-fault interruption characteristics calculated with Mayr's arc model (Nakanishi and Tsukushi, 1974; Yamagata and Nakanishi, 1980). From Fig.1.27a, it can be seen that a dangerous region having a possibility of interruption failure exists at the line length where the fault current is about 75% of the rated interruption current because of the relationship between interruption ability and circuit characteristics determined by the line length. In other words, for the case of no parallel resistance and a bus short-circuit current of 20 kA, an interrupting current of 15 kA (75%) corresponding to a line length of about 2 km is the

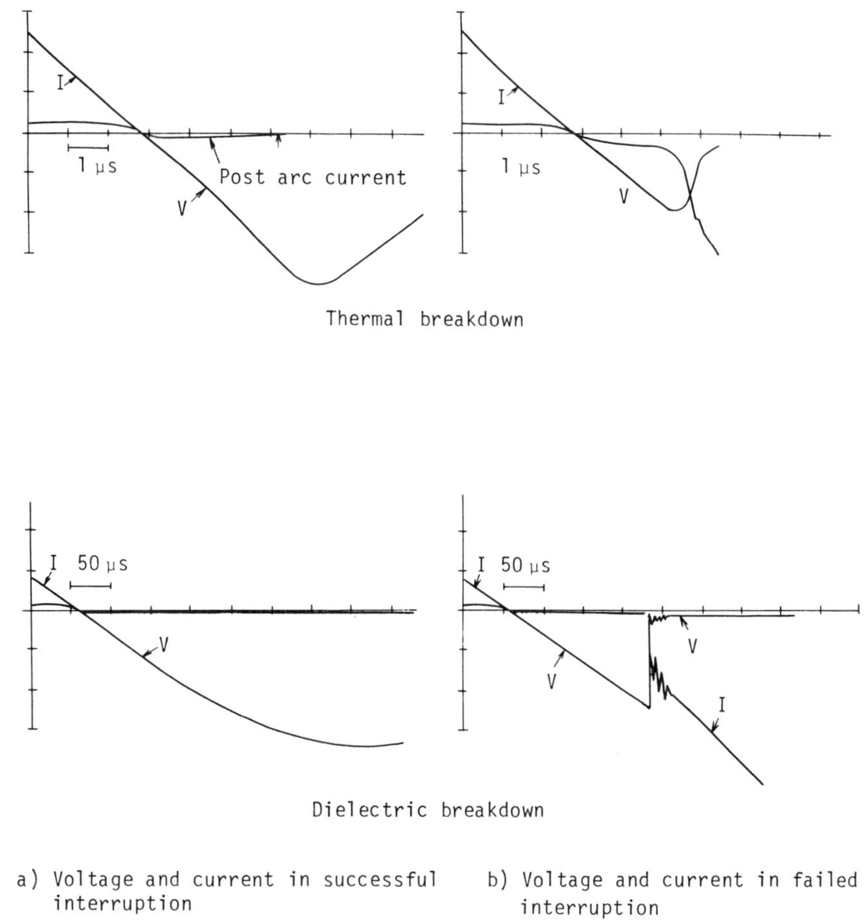

Fig. 1.25 Voltage and current waveforms in the neighborhood of current zero.

most severe condition. In the example in the figure, an air-blast breaker is used. The dangerous region depends on what kind of circuit breaker is used, because the rate of rise of the line-side TRV becomes steeper as the line is shortened. Conversely, the crest value of the triangular waveform becomes smaller. For a gas circuit breaker with a small arc time constant (less than 1 μsec), the dangerous region is at about 90% of the rated interrupting current as shown in Fig. 1.27b. Regarding arcs, various models have been presented, and the measured-value feedback model is a representative one. The arc time constant is not necessarily constant, so the following equation has been proposed, with the arc time constant and power loss defined as functions of

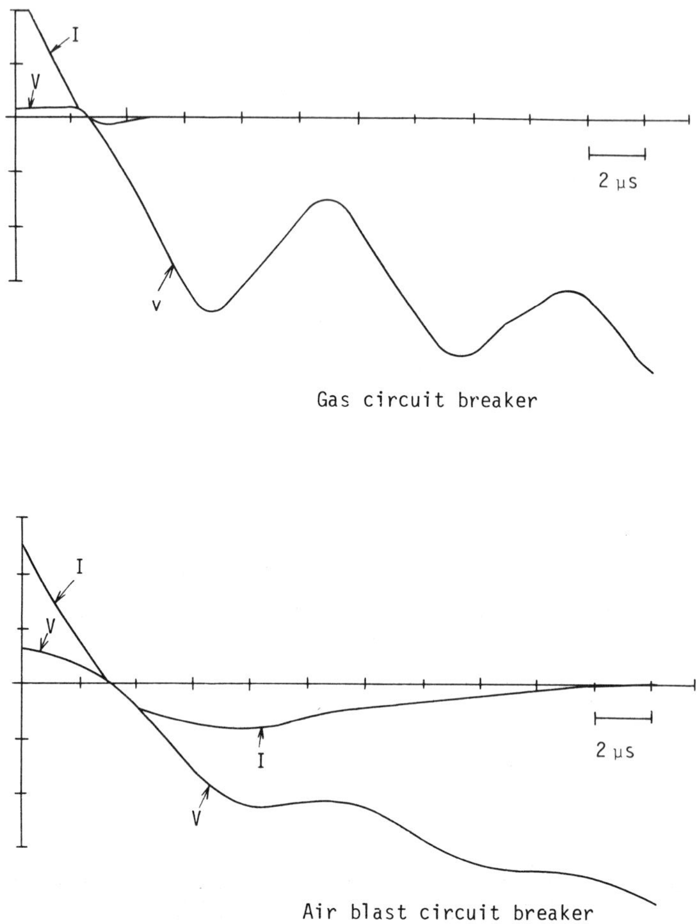

Fig. 1.26 Interruption of SLF condition by differential arcs.

some parameters (e.g., arc conductance) in order to keep it as simple as possible.

$$\frac{1}{G}\frac{dG}{dt} = \frac{1}{\theta(G)}\left(\frac{EI}{N(G)} - 1\right) \tag{1.25}$$

1.4.2 Current Chopping and Dynamic Characteristic of Arc

The current-chopping phenomenon mentioned previously also occurs because of the instability of the interrupting arc influenced by circuit condi-

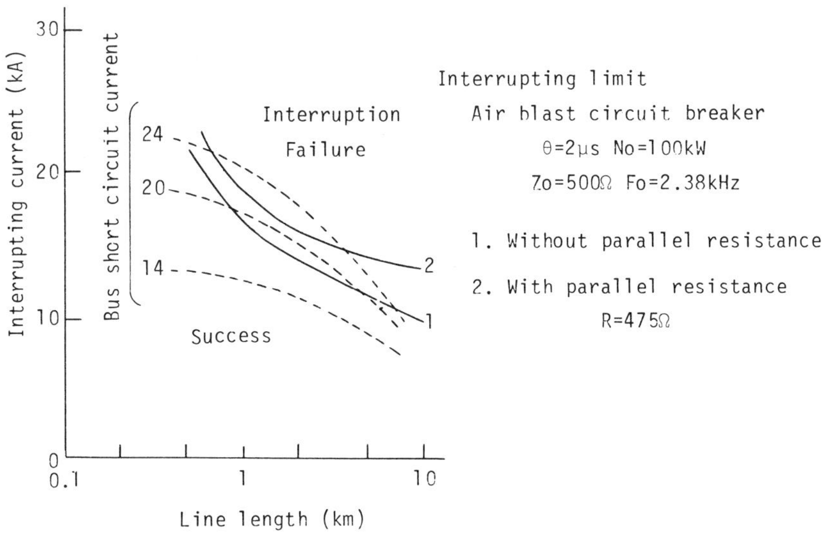

(a) Limit curves of air blast circuit breaker.

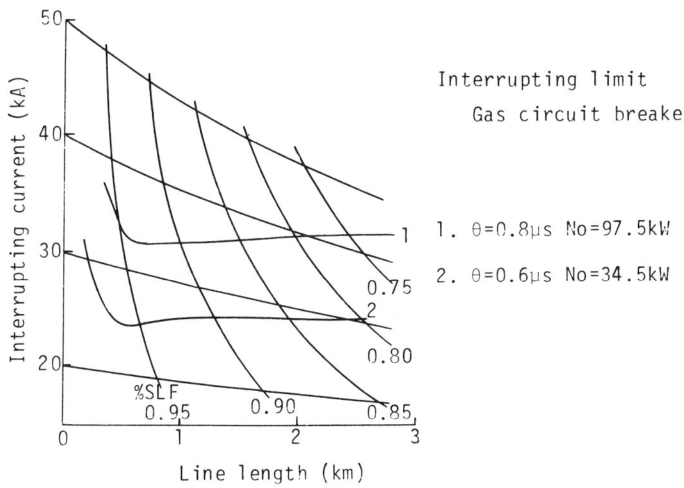

(b) Limit curves of gas circuit breaker.

Fig. 1.27 Limit curves for SLF interruption. Sources: (a) Nakanishi and Tsukushi, (1974). Copyright JIEE. (b) Yamagata and Nakanishi, (1980). Copyright JIEE.

tions. What kind of mechanism causes the arc instability depends on the type of interrupting arc. Here, taking high-frequency current interruption by the circuit breaker by means of a gaseous fluid such as air or SF_6 gas as an example, the modeling of current chopping is described.

High-frequency-current interruption is brought about by escalating the oscillation of the arc current, and it occurs in any interrupting arc where thermal equilibrium can be assumed. The current-chopping phenomenon of the air-blast breaker and gas circuit breaker are mainly caused by high-frequency-current litigation. In the steady-state arc, the following equation is satisfied.

$$E_a I_a^{\alpha} = \text{const} \tag{1.26}$$

where E_a is arc voltage, I_a is arc current, and α is a constant.

The equation of transient response is given by Rizk (1964), adding a small perturbation of E and I to E_a, I_a, respectively, as $E_a + E$, $I_a + I$ in the modified Mayr equation

$$\frac{1}{R}\frac{dR}{dt} = \frac{1}{\theta}\left[- \left(\frac{(E_a + E)(I_a + I)^{\alpha}}{N_0} \right)^{2/(1+\alpha)} \right] \tag{1.27}$$

Neglecting the higher order derivatives, the above equation can be written as

$$\frac{1}{I_a}\frac{dI}{dt} - \frac{1}{E_a}\frac{dE}{dt} = \frac{1}{\theta}\frac{2}{1+\alpha}\left(\frac{E}{E_a} + \alpha \frac{I}{I_a} \right) \tag{1.28}$$

The equivalent circuit for the perturbed arc is given in Fig.1.28. When an external capacitance exists in parallel with the arc, the instability limit is

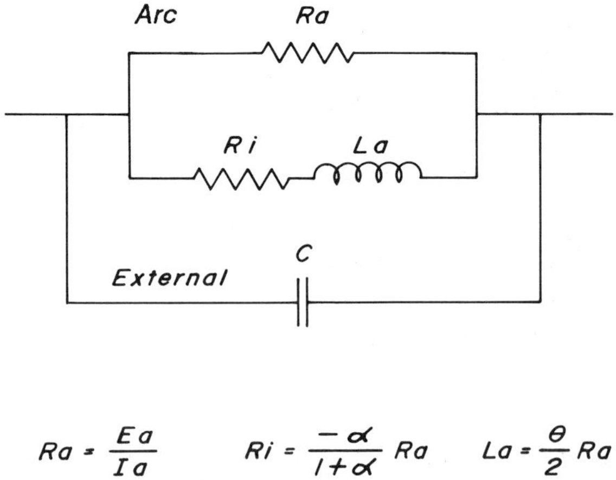

Fig. 1.28 Circuit for calculation of arc instability (IEC 17A(S)308, 1990).

calculated under the assumption that the damping coefficient for the characteristic loop impedance function will be positive.

The chopping current that corresponds to the instability limit current is derived as

$$I_{\mathrm{ch}}^{\alpha+1} = \frac{2\alpha}{I + \alpha} \frac{N_0}{\theta} C \tag{1.29}$$

The experimental results for various circuit breakers other than the vacuum type have shown that α is approximately unity, so the above equation simplifies to

$$I_{\mathrm{ch}} = \sqrt{\frac{N_0}{\theta}} \sqrt{C} = K \sqrt{C} \tag{1.30}$$

where the constant K is the *chopping number,* of which typical values are $(7\text{–}10) \times 10^4$ for an oil circuit breaker, $(15\text{–}20) \times 10^4$ for an air circuit breaker, and $(4\text{–}17) \times 10^4$ for a puffer-type gas circuit breaker [from IEC 17A(S)308(1990)].

As described above, it is necessary to consider the influence of arcing features when system phenomena are explained in terms of circuit-breaker characteristics.

REFERENCES

Hochrainer, A., Grutz, A., Schwartz, J., and Tiel, H.G. (1972). Study of Arcs in Breakers with the Help of Cybanetic Model and under the Influence of Turbulence, CIGRE Int. Conf. on Large High Tension Electric Systems, Session 13-10.

Hosokawa, M., and Hirasawa, K. (1987). Breaking Ability and Interrupting Phenomena of a Circuit Breaker Equipped with a Parallel Resistor or Capacitor, *IEEE Trans. Power Delivery, PWRD-2*(2): 382–392.

IEEE (1989). *Power Eng. Rev., 9*(12): 7–10.

JIEE (1986). Application of the Arc Model to the Breaker Vicinity Fault, Tech. Report in Japanese, Vol. 2, No.232.

Nakanishi, K., and Tsukushi, M. (1974). Calculation Method of Interrupting Limit of Circuit Breaker, *JIEE SPD-74-7.* In Japanese.

Nakanishi, K., Ushio, T., Ohtani, K., and Yamazaki, S. (1975). Power Circuit Breaker, *JIEE.* In Japanese.

Nakanishi, K., Ishikawa, M., Abe, Y., Andoh, K., Yanabu, S., Ohnuma, A., and Kurosawa, Y. (1988). Small Current Interruption Phenomena and Recent Progress of Gas Circuit-Breaker, CIGRE Int. Conf. on Large High Voltage Electric Systems, Session 13-06.

Regros, W. P., and Genon, A. M. (1979). Circuit Breaker Stochastic Behavior Simulation, Symposium on High Voltage Switching Equipment, pp. 38–43.

Rizk, F. (1964). Arc Instability and Time Constant in Air Blast Circuit Breaker, CIGRE paper No. 107.

Saha, M. N. (1920). *Phil. Mag.,* 40.

Swanson, B. W., Roidt, R. M., and Browne, T. E., Jr. (1972). A Thermal Arc Model for Short-Line Fault Interruption, *ETZ-A, 93:* 375–380.

Yamagata, Y., and Nakanishi, K. (1980). Analysis of Fault Interrupting Characteristics Using Thermal Arc Model, *JIEE, 100B:* 577–584. In Japanese.

2
ANALYSIS AND DESIGN

Koji Ibuki, Toshiaki Yoshizumi, Shuichi Sakuma, and Takao Yamauchi Mitsubishi Electric Corporation, Amagasaki, Japan

2.1 DETAILED ARC STRUCTURE (K. Ibuki, T. Yoshizumi, S. Sakuma)

In normal operating conditions, a circuit breaker is in the closed position, and some current is usually flowing through the closed contacts. The circuit breaker opens its contacts when a tripping signal is sent to it. The separation of the contacts causes the generation of an *electric arc,* which consists of metal vapor surrounded by ionized environmental gas. In most cases, this arc is established within a narrow and limited space so that the arc can be quickly cooled by the flow of fluid or gas.

In most SF_6 gas circuit breakers, the arc is generated in a cylindrical space surrounded by insulating material. The boundary of the arc in cylindrical space is formed by an insulator wall, and the diameter and amount of gas flow are limited by the size of the space. The interaction of the wall material and the arc through the exchange of heat should not be overlooked. The dissipation of thermal energy by vaporization of the wall material (Kuwahara et al., 1983) and changes in the thermodynamic characteristics of the gas in the arc by due to the mixing in of wall material (Kovitya, 1984) are typical examples of this interaction. A basic description of the arc in the SF_6 gas flow will be given in the following sections.

An arc located in a gas background structures itself by balancing the *electrical energy* put into the arc space and the *thermal energy* dispersed into the surrounding space. The *temperature distribution* makes the thermal energy flow, and the flow in turn determines the temperature distribution. The energy transfer into the surroundings is strongly dominated by turbulence and radiation. The most important factors in deciding its structure are the *thermodynamic characteristics* of the gas. The gas density, for example, is a factor that dominates the energy that can be sustained in the limited space. Figure 2.1 shows the *gas density* of SF_6 as a function of temperature at various pressures. It is seen that the density can be described by the Boyle–Charles law at temperatures below 1600 K, but at

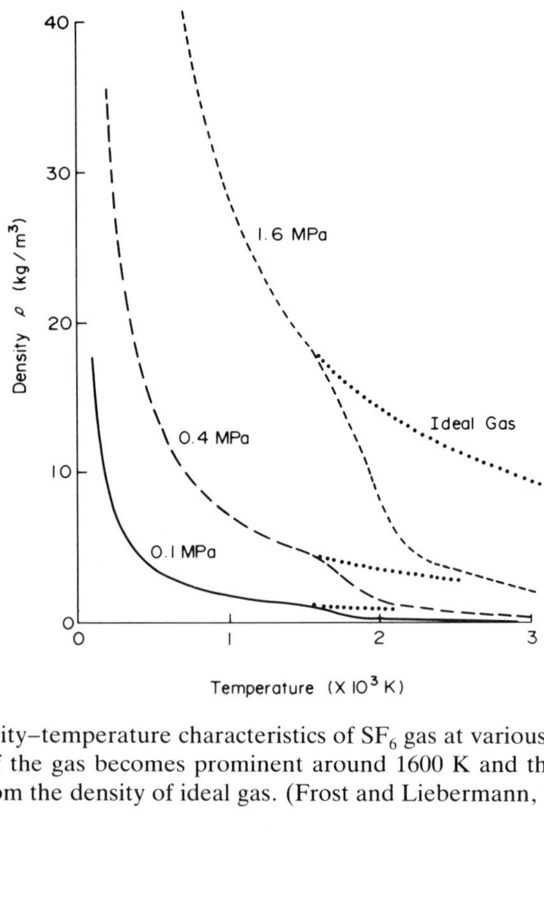

Fig. 2.1 Density–temperature characteristics of SF_6 gas at various pressures. Dissociation of the gas becomes prominent around 1600 K and the density drops significantly from the density of ideal gas. (Frost and Liebermann, 1971). © IEEE.)

temperatures above this the density is less than the value expected from this law. This is because the *dissociation of the molecule* becomes active at temperatures above 1600 K and the energy absorption of a molecule like SF_6 is a complex function of temperature. The macroscopic behavior of the gas can be conveniently described by thermodynamic properties such as enthalpy and transport properties. These properties have been determined through theoretical calculations from thermochemical properties and heats of formation of species (Frost and Liebermann, 1971).

Figure 2.2 shows that the first prominent peak of *thermal conductivity* appears at 1800 K, which corresponds to the temperature at which the portion of F + F_2 exceeds 50% of total volume. The *electrical conductivity* of SF_6 begins to build up (as shown in Fig. 2.3) when the electrons become more than 1% of the total volume at around 8000–9000 K. It is important for the interruption characteristics that the temperature at which the electrical conductivity becomes prominent is higher than the temperature at which the thermal conductivity peak appears. In the temperature range above 8000 K, both thermal and electrical conductivity are high. This means that active dispersion of thermal energy is high at a temperature where electrical conductivity is not yet significant, and this is an important characteristic of SF_6 as an arc-quenching medium. The energy that can be

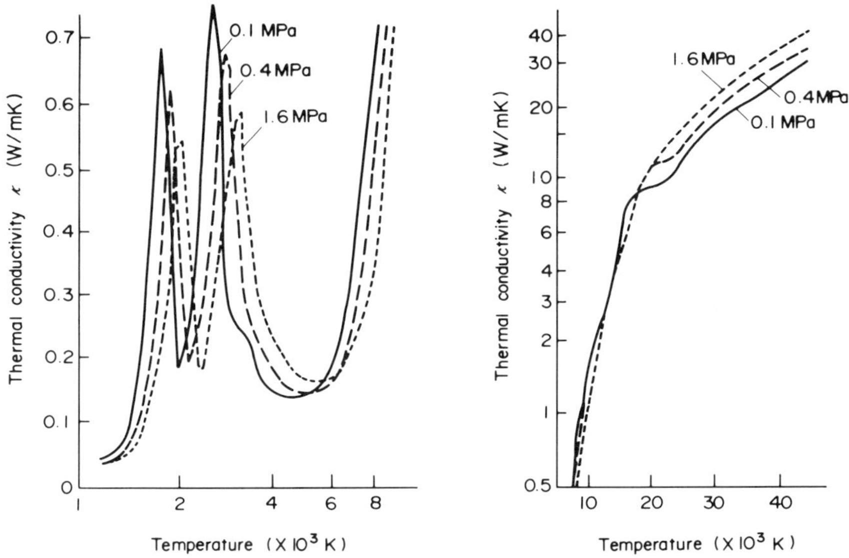

Fig. 2.2 Thermal conductivity at SF_6 gas at various temperatures. (Frost and Liebermann, 1971. © IEEE.)

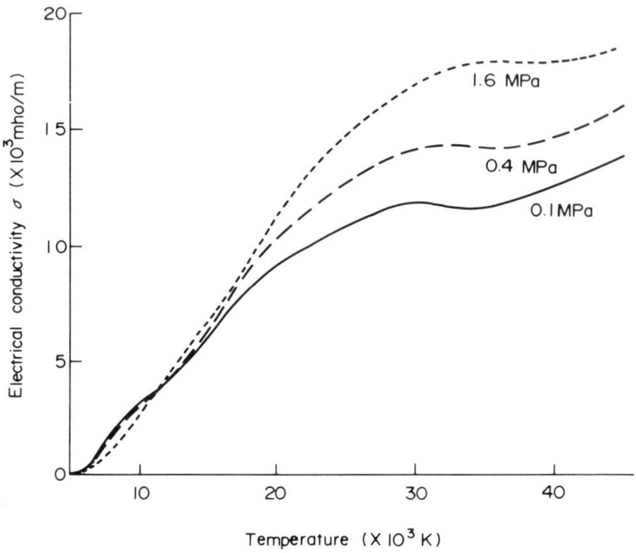

Fig. 2.3 Electrical conductivity of SF_6 gas at various temperatures. (Frost and Liebermann, 1971. © IEEE.)

sustained in a given mass increases as the temperature rises, and it increases rapidly at a temperature where the dissociation of SF_6 is significant, as shown in Fig. 2.4.

An important parameter in understanding the arc characteristics is ρh, which is interpreted as the energy sustained in a given volume; $\rho\ \partial h/\partial T$ is also an important parameter that shows the changeability of the energy in a given volume. These parameters are graphed in Figs. 2.5 and 2.6, respectively. The curves in Fig. 2.5 show that SF_6 gas at a temperature around 2000 K keeps a high energy content and that changing the temperature of this gas requires a high energy input or dissipation. This explains the peculiar behavior observed when volumes of SF_6 gas at different temperatures are mixed and plays an important role in describing the cooling of an arc and the dispersion of the hot exhaust gas from the interrupter.

An example of this peculiar behavior is found in the hypothetical experiment of mixing two different gases—a gas at T_0, P_0 of volume V_0 and a small amount (ΔV_0) of gas at T_d, P_d. The temperature and pressure of the mixed gas can be obtained from Eqs. (2.1):

$$\boxed{\begin{matrix} V_0 \\ T_0 \\ P_0 \end{matrix}} + \begin{matrix} \Delta V_0 \\ \boxed{\begin{matrix} T_d \\ P_0 \end{matrix}} \end{matrix} \rightarrow \boxed{\begin{matrix} V_0 \\ T_X \\ P_X \end{matrix}}$$

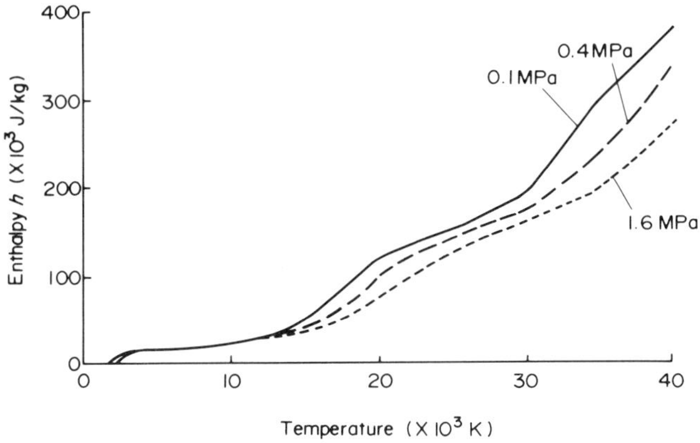

Fig. 2.4 Enthalpy of SF_6 gas at various temperatures. (Frost and Liebermann, 1971. © IEEE.)

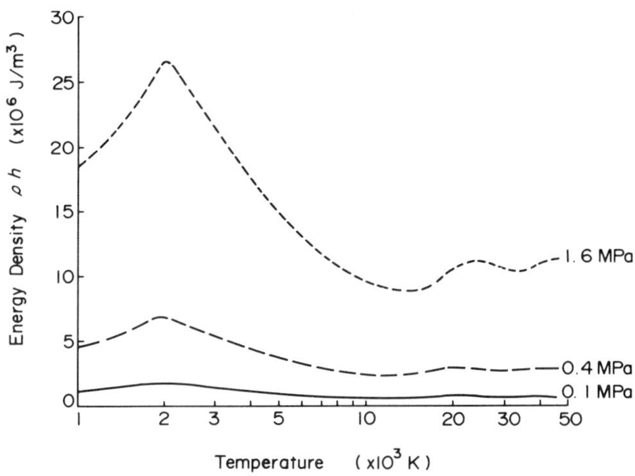

Fig. 2.5 Energy density–temperature (ρh *vs. T*) characteristics of SF_6. (Tanabe et al., 1985. © IEEE.)

$$\rho = \rho_0 + \rho_d \Delta \tag{2.1a}$$

$$h = \frac{\rho_0 h_0 + \rho_d h_d \Delta}{\rho} \tag{2.1b}$$

$$\frac{P(\rho, h)}{P_0} = f(h_0, h_d, \Delta) \tag{2.1c}$$

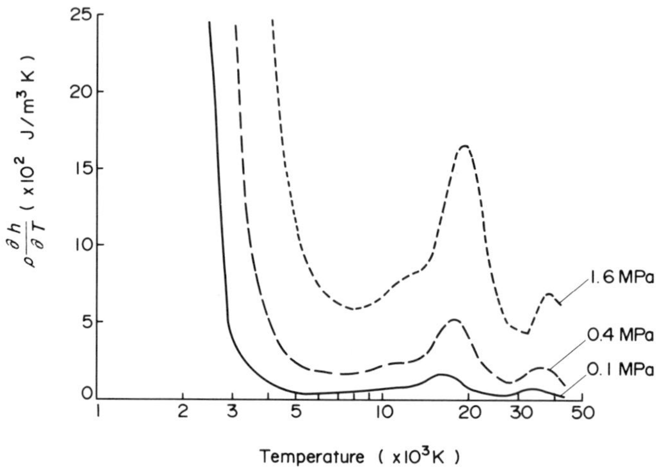

Fig. 2.6 $\rho(\partial h/\partial T)$ of SF_6. In the temperature range where this takes on low-value range. (Tanabe et al., 1985. © IEEE.)

where

Δ = portion of the mixed gas

P_0 = $P(\rho_0, h_0)$, $P_d = P(\rho_d, h_d)$, $P_0 = P_d$

T_0 = $T(h_0)$, $T_d = T(h_d)$ = temperature of the gas

h_0, h_d = enthalpy of the gas

The temperature and pressure are given as functions of T_0 and T_d as shown in Fig. 2.7. When the temperature T_d of the mixed gas of volume ΔV_0 is less than 2000 K, the final temperature of the mixture is close to 2000 K and the pressure is also found to be lower than the initial value. This is understood as a result of the high ρh value of the gas with 2000 K.

The peculiar features of SF_6 are summarized as:

1. A maximum in thermal conductivity is found around 2000 K.
2. The thermal conductivity peak appears before electrical conductivity becomes prominent.
3. Gas at a temperature lower than 2000 K is effective in cooling gas at a temperature above 2000 K because of the high ρh value of SF_6 near 2000 K.

The arc in the interrupter chamber is immersed in this unusual gas, and the *arc structure,* including the temperature distribution, is determined by the characteristics of the gas. The local temperature is determined by the

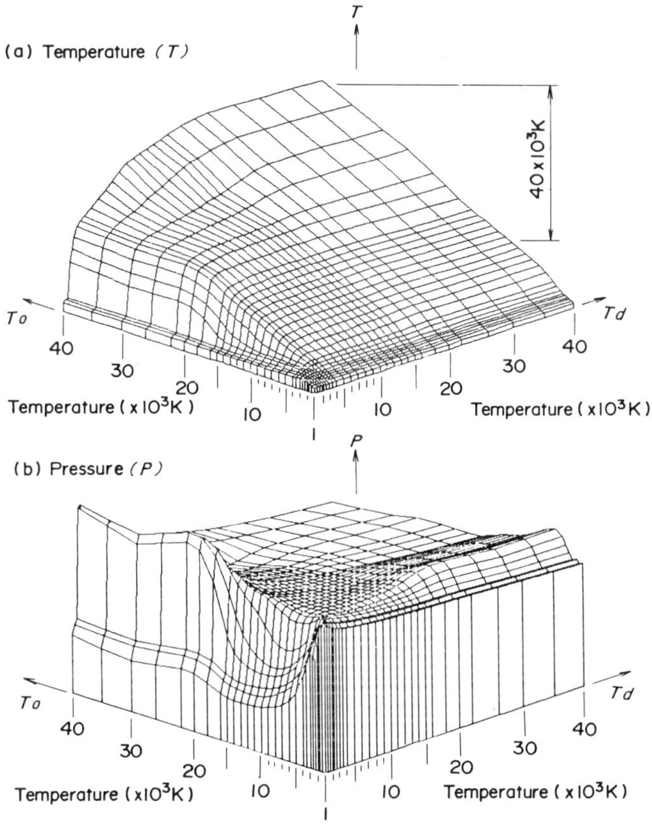

Fig. 2.7 (a) Calculated temperature and (b) pressure obtained as a result of mixing two volumes of SF_6 gas at different temperatures. A small amount of gas (0.1 volume) at temperature T_d is mixed into 1 volume of gas at temperature T_0. When the temperature of the 0.1 vol gas is below 2000 K, the resultant temperature is approximately 2000 K irrespective of T_0. This temperature decrease brought about by the mixing in of low-temperature gas leads to a significant pressure drop. (Tanabe et al., 1985. © IEEE.)

balance of the electrical input energy and the thermal energy flow caused by the temperature gradient in the arc. An example of such a temperature distribution of an arc located in an SF_6 background without flow, which is obtained by solving the classical equation of energy balance for an elementary annulus in the arc [for the Elenbaas–Heller equation, refer to Section 2.3.1(B)] (Elenbaas, 1935; Heller, 1935), is shown in Fig. 2.8. This series of temperature distribution curves shows that the arc radius and arc voltage

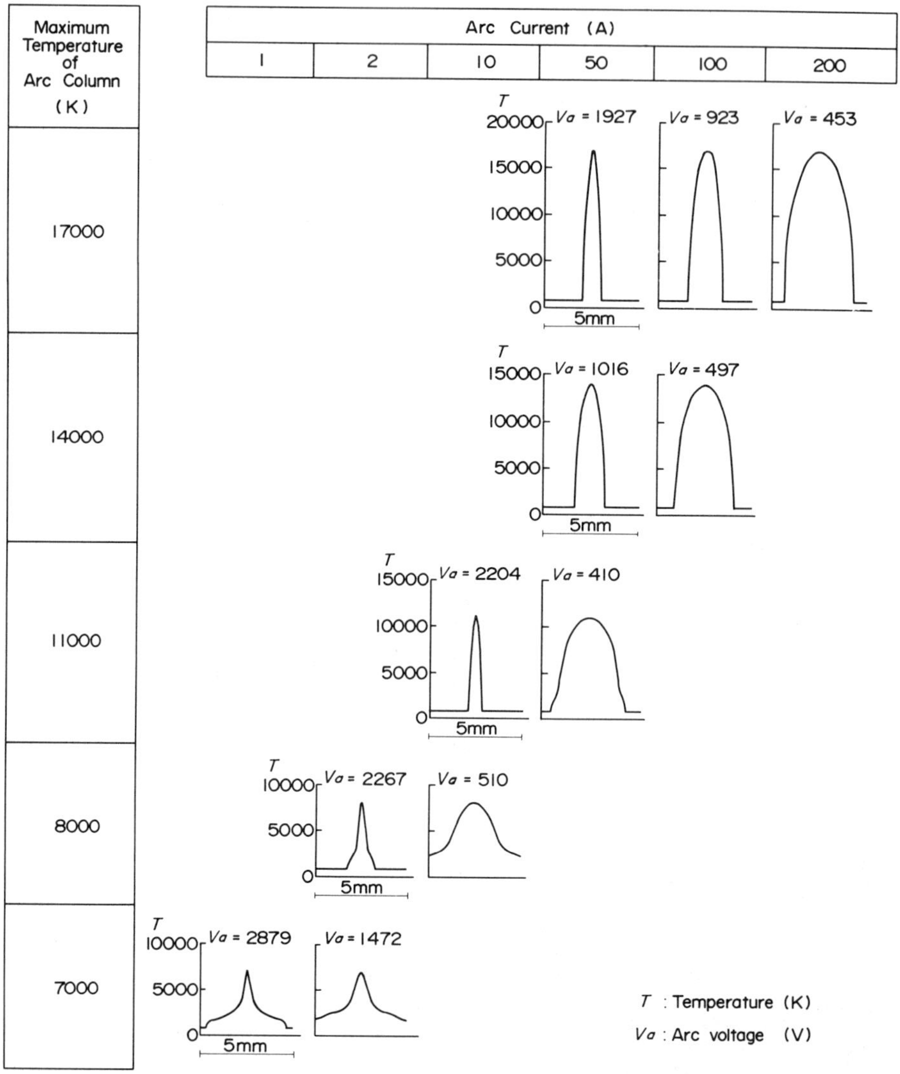

Fig. 2.8 Temperature distribution of SF_6 arc calculated by the Elenbaas–Heller equation for different values of arc current. The arc voltages for arc of 10-cm length are calculated for various values of maximum temperature of the column.

can be determined by the energy carried out from the arc space in a complex way. The heat flow can be given by $\Gamma = -K(dT/dr)$ if it can be assumed to be dependent only on thermal conduction. This means that the high thermal conductivity leads to a smaller temperature gradient, and the gas at 2000 K tends to form a layer with a smaller temperature gradient that surrounds the arc core.

An actual arc of an alternating current in an interrupter is not in the equilibrium state described by the Elenbaas–Heller equation. The input energy and lost energy are not in balance, and the local temperature in the arc is always changing. The arc is usually located in a flow of SF_6 gas, and the actual temperature distribution is still more complex because of the existence of thermal exchange due to turbulence and radiation. Figure 2.9 shows a series of framing camera pictures and the photointensity distribution in the last loop of an interrupted arc. The photointensity distributions in Fig. 2.10 show that the distributions are significantly different in the buildup stage and the decreasing stage of the current (e.g., 0.8 kA in the increasing phase and 1 kA in the decreasing stage) than in the middle stages (15 kA at the midpoint).

The theoretical investigation of the arc structure is also important. Modern computer simulation techniques make it possible to use a complex model based on the fundamental characteristics of the gas. The macroscopic behavior of the interrupter can be simulated if the characteristics of the interruption medium are known. The internal arc structure such as the pressure and temperature distributions, which are difficult to measure, can be visualized with the aid of computer simulation. Figure 2.11 shows an example of such a simulation result, which visualizes the temperature distribution during the interruption of an SF_6 arc.

2.2 ENERGY BALANCE AND THE PHYSICAL ARC MODEL (K. Ibuki, T. Yoshizumi, S. Sakuma)

An actual interrupter using SF_6 gas as the interrupting medium makes a gas flow by some means, which activates thermal exchange with the background. Some examples of the structure that can be thought of are shown in Fig. 2.12. Figure 2.12a illustrates a *plain break;* no explicit gas flow is provided, but the temperature (or density) gradient and gravity cause some gas flow and the flow in turn affects the temperature distribution. The circuit breaker shown in Fig. 2.12b utilizes a *pressure buildup arc* to provide a pressure gradient, and the resultant gas flow to cools the main interrupted arc. In the structure of Fig. 2.12c a high-pressure gas is gener-

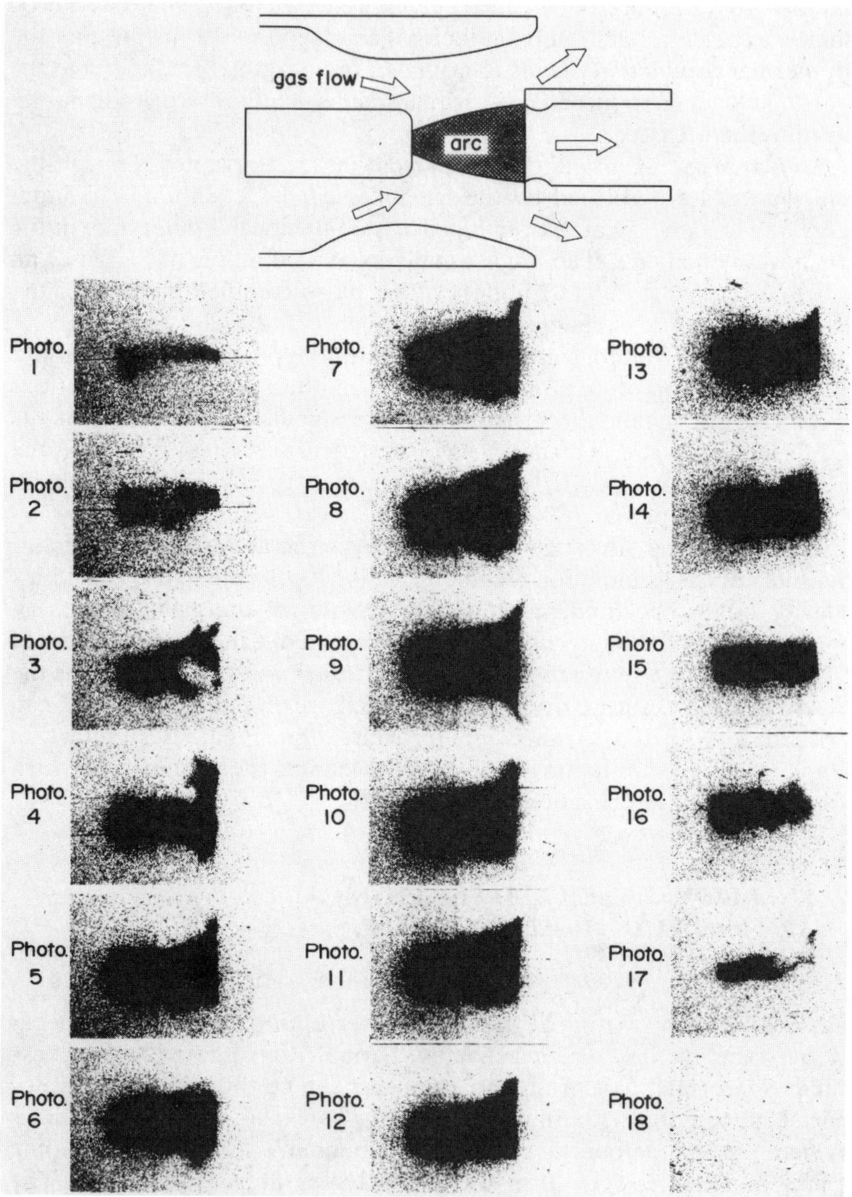

Fig. 2.9 Framing camera pictures watching a fuse-started arc placed in an axial flow of SF_6 gas during one 15-kA loop. The flow guide is made of PTFE walled by perspex plate. (Tominaga et al., 1980. © IEEE.)

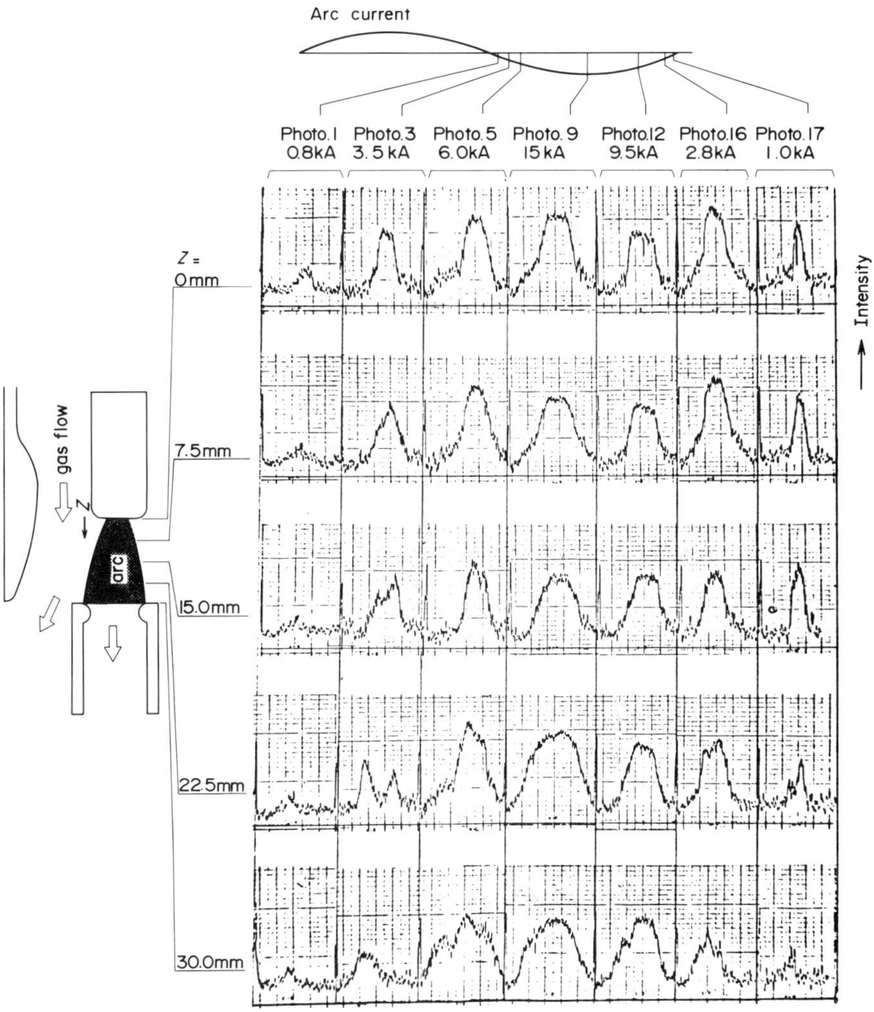

Fig. 2.10 Photointensity distribution of the arc in Fig. 2.9 at various positions along the axis. (Tominaga et al., 1980. © IEEE.)

ated by some separately provided means such as a gas compressor. Figure 2.12d shows an arrangement to generate the gas pressure necessary to build up gas flow by the mechanical motion of contact separation; this is called a *puffer-type* circuit breaker. The pressure buildup in the puffer chamber and the energy exchange in the arc space are the important concepts in understanding interruption. The mechanical motion of the arcing contact, espe-

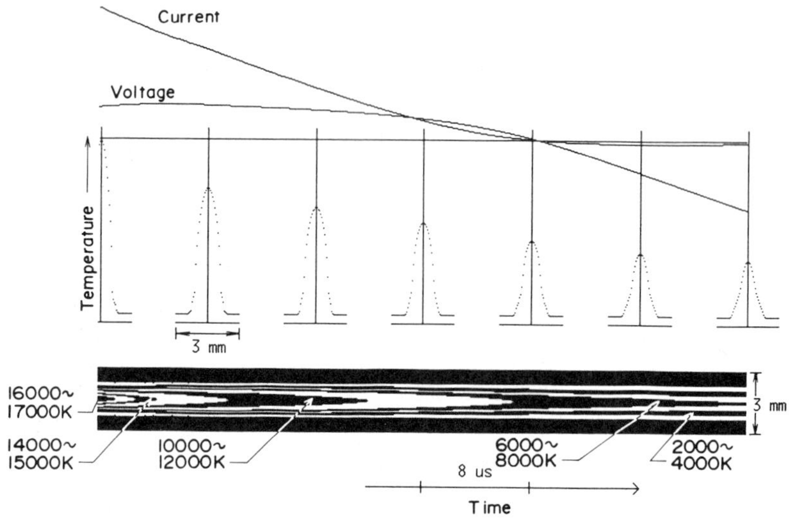

Fig. 2.11 Result of simulating the current interruption of an SF_6 arc. Current and voltage are shown, and the temperature distribution in the radial direction is shown at 0.8-μsec intervals. The lower image shows the temperature change in the arc over time by isothermal lines.

cially in the puffer chamber, is affected by the existence of the arc because of the interference of the pressure buildup in the puffer chamber and the arc current.

The puffer-type arrangement is most widely used in practical circuit breakers, and the gas usually flows in parallel to the arc axis in such an arrangement. The detailed description of this kind of arc requires equations of temperature and flow distributions in both the axial and radial directions. There are three major effects to be included in the description of thermal transfer: (1) the motion of the medium with thermal energy, (2) thermal energy transfer by thermal conduction proportional to the thermal conductivity of the medium, and (3) thermal energy transfer by radiation. The behavior is described by Eqs. (2.2):

$$\begin{aligned}
\frac{\partial \rho}{\partial t} &= -\mathrm{div}(\rho \vec{u}) && \text{(conservation of mass)} \\
\rho \frac{\partial \vec{u}}{\partial t} &= -\mathrm{grad}\, p - \rho\, (\vec{u}\, \mathrm{grad})\vec{u} && \text{(conservation of momentum)} \\
\rho \frac{\partial \vec{h}}{\partial t} &= -\vec{u}\, \mathrm{grad}\, (\rho h) + \mathrm{div}\, (p\vec{u}) + \sigma \vec{E}^2 + \mathrm{div}\, (\kappa\, \mathrm{grad}\, T) - R(T, \rho) \\
&\qquad \text{(conservation of energy)}
\end{aligned} \tag{2.2}$$

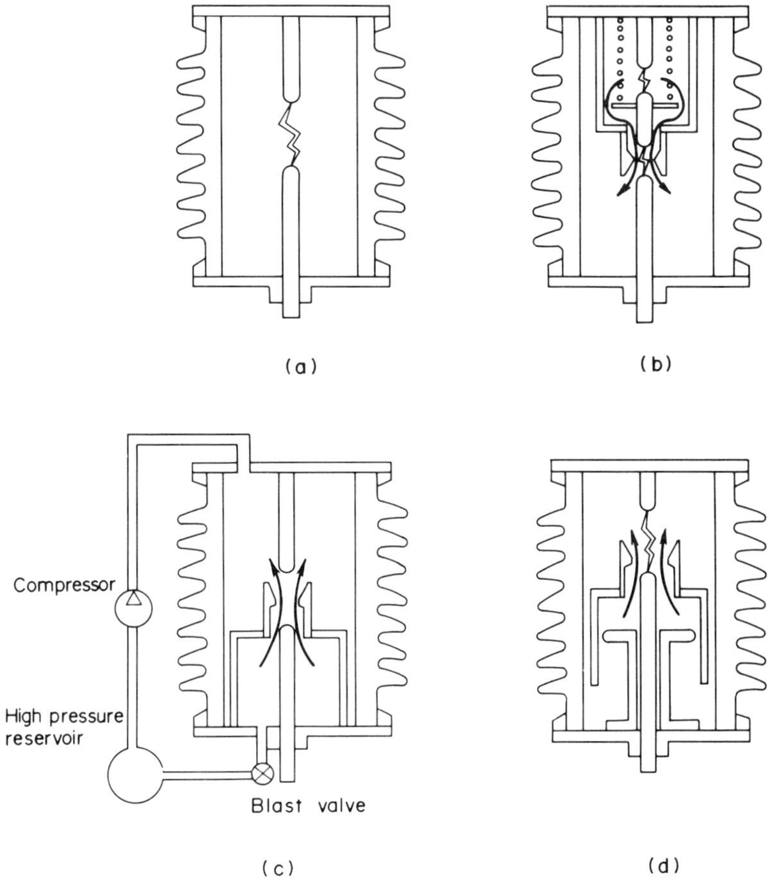

Fig. 2.12 Various arrangement of arc quenching by SF_6 blasting. (a) Plain break; (b) blast by "blasting arc" or self-blast; (c) double-pressure blast; (d) puffer blast.

where

p = pressure
ρ = gas density
$\vec{u}$ = gas flow velocity
h = enthalpy of gas
E = electrical field strength
σ = electrical conductivity
κ = thermal conductivity
T = gas temperature
R = energy loss by radiation

Conservation of Mass

$$\frac{\partial \rho}{\partial t} = -\mathrm{div}(\rho \vec{u})$$

- $\frac{\partial \rho}{\partial t}$: Change rate of density in a unit volume
- $-\mathrm{div}(\rho \vec{u})$: Net mass flow entering a unit volume

Conservation of Momentum

$$\rho \frac{\partial \vec{u}}{\partial t} = -\,\mathrm{grad}\; p - \rho\,(\vec{u}\;\mathrm{grad})\vec{u}$$

- $\rho \frac{\partial \vec{u}}{\partial t}$: Acceleration at a point in the space
- $-\mathrm{grad}\; p$: Accelerating force by pressure distribution
- $-\rho(\vec{u}\;\mathrm{grad})\vec{u}$: Acceleration during the motion along flow lines

Conservation of Energy

$$\rho \frac{\partial h}{\partial t} = -\vec{u}\;\mathrm{grad}(\rho h) + \mathrm{div}(p\vec{u}) + \sigma \vec{E}^2 + \mathrm{div}(\kappa\;\mathrm{grad}\;T) + R(T, \rho)$$

- $\rho \frac{\partial h}{\partial t}$: Change rate of energy in a unit volume
- $-\vec{u}\;\mathrm{grad}(\rho h)$: Net energy input by the mass flow
- $\mathrm{div}(p\vec{u})$: Work perfomed by the flow through pressure gradient
- $\sigma \vec{E}^2$: Heat input by electrical current
- $\mathrm{div}(\kappa\;\mathrm{grad}\;T)$: Net energy lost by the thermal conduction
- $R(T, \rho)$: Energy loss by radiation

The *pressure buildup* in the puffer chamber, when the temperature distribution in the space is negligible, is given by the equations

$$\frac{d\rho}{dt} = -\frac{1}{V}\frac{dV}{dt} - \rho\vec{u}\frac{S}{V} \qquad \text{(density in puffer chamber)}$$

$$\rho\frac{dh}{dt} = \frac{1}{V}f\vec{v} - \vec{u}\;\mathrm{grad}(\rho h) - p\vec{u}\frac{S}{V} \qquad \text{(conservation of energy in puffer chamber)}$$

where

S	=	cross section of the exit
V	=	$V(t)$ = volume of the puffer chamber
f	=	$f(t)$ = driving force
grad (ρh)	=	gradient of ρh at the exit of the puffer chamber
v	=	speed of the compressing pistion

Density in Puffer Chamber

$$\frac{d\rho}{dt} = -\frac{1}{V}\frac{dV}{dt} - \rho\vec{u}\frac{S}{V}$$

- $\frac{d\rho}{dt}$: Change rate of density in the puffer chamber
- $-\frac{1}{V}\frac{dV}{dt}$: Effect of change of the puffer volume or the compression
- $-\rho\vec{u}\frac{S}{V}$: Contribution by total loss of mass by the outward flow

Conservation of Energy in Puffer Chamber

$$\rho\frac{dh}{dt} = \frac{1}{V} f\vec{v} - \vec{u}\,\mathrm{grad}(\rho h) - p\vec{u}\frac{S}{V}$$

- $\rho\frac{dh}{dt}$: Change rate of energy
- $\frac{1}{V} f\vec{v}$: Work performed by the mechanical compression
- $-\vec{u}\,\mathrm{grad}(\rho h)$: Energy carried away by the out flowing gas
- $-p\vec{u}\frac{S}{V}$: Energy carried away by accelerating the out flowing gas

It is not practical to solve these equations by either analytical or numerical methods. Therefore, they must be simplified to some extent for use in any practical situation. *Numerical calculations* are particularly important these days because they can provide a physical image of the phenomena in a handy manner as long as the simplification is adequate both for the capacity of the computers and for yielding the information required for a given purpose of the investigation.

Understanding the insulation characteristic is as important as understanding the interruption problem. An important characteristic of SF_6-insulated equipment such as circuit breakers and *gas-insulated switchgear* (GIS) is that the dielectric characteristic of an interelectrode gap is strongly influenced by the electric field strength, particularly on the surface of the electrodes. This characteristic is found both in the static condition with the gas at room temperature and in the dielectric recovery period after interruption of the current when the temperature is considerably higher than room temperature. Therefore, the understanding of the dielectric characteristics can be helped by obtaining a clear solution to the question of electric field distribution. The electric field with no highly ionized gas can be expressed by Laplace's equation and the necessary boundary conditions.

$$\nabla^2\phi = 0 \qquad \text{(Laplace's equation)}$$

and

$$\phi(r) = U_i$$

where r is the coordinates on the surface of the ith conductor.

Recovery voltage is the voltage that appears between the contacts after the interruption. The interelectrode space where the arc burns during interruption is filled with hot gas, and the temperature in the recovery period decreases, typically, to 2000 K. Ionization in this range of temperature is negligible, and the electric field distribution during this period can be evaluated with the aid of the Laplace equation given above. The actual withstand voltage of the electrode gap is estimated from knowledge of the breakdown characteristics of SF_6 gas.

2.3 COMPUTER-AIDED ANALYSIS (K. Ibuki, T. Yoshizumi, S. Sakuma)

2.3.1 Simplification of the Physical Arc Model

The set of equations should be simplified to highlight the important features of the arc phenomena. The physical parameters that cannot be measured and/or are impractical to measure experimentally can be evaluated by comparing macroscopic experimental results with calculated results. It is useful in practice to separate the whole period of arc into three subperiods: (1) the *high-current period,* (2) the *current-zero period,* and (3) the *dielectric recovery period.* In the dielectric recovery period, analysis of the electric field distribution is as important as analysis of the cooling of the gas in the arc space. Physical models for each period are described in the following sections.

(A) High-Current-Period Regime

The arc in the high-current period can be described as follows:

1. The arc fills the nozzle (or flow guide in some descriptions), and the temperature gradient in the radial direction does not play an important role. This is because the temperature difference between the arc space and the wall is so great and the important heat exchange is found either between the arc space and the wall or between the arc space and the gas flow. This feature allows the simplification of assuming a step-shaped distribution of temperature in the radial direction that will be referred to as *channel shape.*
2. The gas in the arc space is accelerated by the pressure imposed by some external agent, and at the same time the heat input into the arc space causes gas flow in the axial direction.
3. The heat transfer is dominated by gas flow in the axial direction and radiation loss in the radial direction. Heat transfer due to the temperature gradient in the axial direction is less important in this period.

The *enthalpy flow model* (Frost and Liebermann, 1971) and the *one-dimensional model* (Kuwahara et al., 1983) are obtained by simplification based on these features.

(1) Enthalpy Flow Model

A channel-shaped temperature distribution is assumed in this model. The channel is surrounded by cold gas as shown in Fig. 2.13, and the diameter of the channel is decided by the *energy balance* of the input and output energy by the flow in the channel and in the background gas.

1. The arc is represented by a constant-temperature channel where the temperature is independent of either the axial position or nozzle dimension. The temperature outside of the channel is assumed to be equal to that of the upstream gas.
2. All of the arc power is put into the constant-temperature channel.
3. The pressure distribution is assumed to have some simple form (it is usually assumed to be linear) in both the channel and the background gas. The upstream end of the nozzle is connected to some pressure-buildup arrangement such as a puffer chamber. The pressure and density there are assumed to be equal to those in the puffer chamber. Velocities at the exit of the nozzle are assumed to be equal to the sonic velocity at the local temperature.

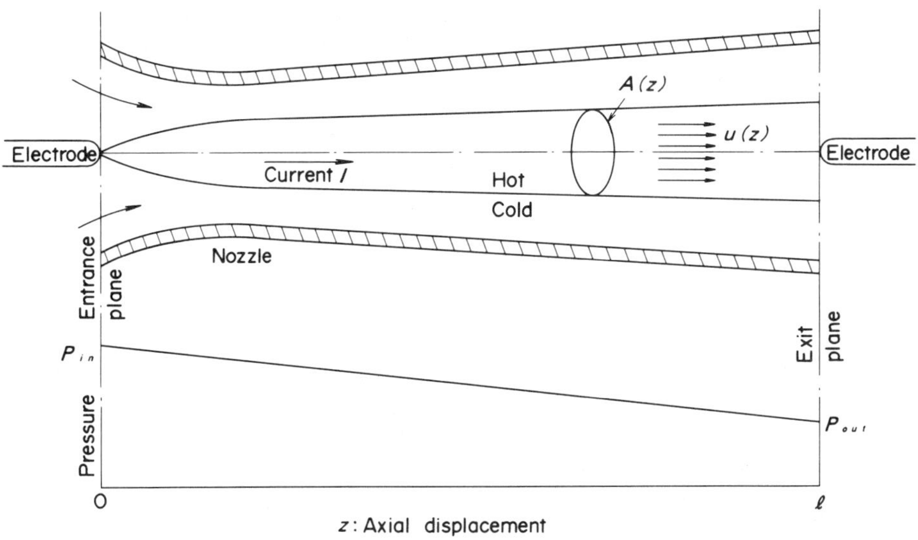

Fig. 2.13 Schematic diagram of the enthalpy flow arc model. Hot channel gas and cold background gas are assumed to be of uniform temperatures. (Frost and Liebermann, 1971. © IEEE.)

The second assumption means that the sum of the energy put in upstream of a given point in the arc must be balanced by the energy flowing through the point in the nozzle. This condition gives the equation

$$IV(z) = h\rho uA$$

where

z	=	postion along the nozzle
I	=	arc current
$V(z)$	=	electrical potential at z
u	=	flow velocity
A	=	cross section of the nozzle at z
h	=	enthalpy
ρ	=	gas density

The third assumption gives the pressure distribution $p(z)$ in the nozzle according to the equation

$$p(z) = \frac{(\ell - z)\, p_{\text{in}} + zp_{\text{out}}}{\ell}$$

where p_{in} is the pressure in the puffer chamber; p_{out}, the pressure outside the nozzle; and ℓ, the length of the nozzle. The assumption that the effect of ionization on the pressure is negligible gives the density as a function of the position in the nozzle as

$$\rho(z) = \frac{(\ell - z)\, \rho_{\text{in}} + z\rho_{\text{out}}}{\ell}$$

where ρ_{in} is the gas density in the puffer chamber and ρ_{out} is the gas density outside the nozzle.

The conservation of momentum, which is given by $dp + u\, du = O$, can be integrated to give

$$\rho u^2 = 2[p_{\text{in}} - p(z)]$$

where u is the flow velocity.

The flow velocity u is obtained from these equations as

$$u^2(z) = \frac{2z(p_{\text{in}} + p_{\text{out}})}{(\ell - z)\, \rho_{\text{in}} + z\rho_{\text{out}}}$$

When the flow velocity at the exit is assumed to be equal to the sonic velocity c, this equation is simplified to

$$u^2 = \frac{\alpha c^2 \eta}{1 - \alpha' \eta}$$

where

$$\alpha = \frac{p_{\text{out}}}{p_{\text{in}}}, \qquad \alpha' = 1 - \alpha, \qquad \eta = \frac{z}{\ell}$$

The equation for the conservation of energy gives expressions for the potential of the arc (V) and the area of the arc (A):

$$V^2(\eta) = \left(\rho_{\text{out}} \frac{ch\ell}{\sigma} \right) (\alpha\alpha'^3)^{1/2} \left[(\alpha'\eta - \frac{1}{2})(\alpha'\eta)^{1/2} (1 - \alpha'\eta)^{1/2} + \frac{1}{4} \sin^{-1} (2\alpha'\eta - 1) + \frac{\pi}{8} \right]$$

and

$$A(\eta) = \frac{I\ell \bar{V}(\eta)}{\sigma V^* [\eta(1 - \alpha'\eta)/\alpha]^{1/2}}$$

where

$$V^* = (\rho_{\text{out}} ch\ell/\sigma)^{1/2} \qquad \text{(characteristic arc voltage)}$$

When the temperature of the arc channel is assumed and the pressure of the puffer chamber and at the nozzle exit are given, the pressure and the flow velocity along the channel axis can be obtained. The puffer-chamber pressure, which is given by the balance of compression and the gas outflow, can be calculated by these parameters. This model was found to be useful because it evaluates the puffer-chamber pressure, which is a convenient parameter because it can be measured experimentally. It was found to be particularly useful for fixing the nozzle size and the mechanical energy necessary for compression by comparing the evaluation with experimental results. In many examples of application of this model, the arc channel temperature was assumed to be 20,000 K, which was justified by experimental temperature measurements of actual arcs.

Research work on the detailed expression of the arc by radial integral showed that the "areas" defined by the integral equation of conservation of momentum, energy, and mass can give a more detailed description of arc behavior by introducing *shape factors* that show the ratios of areas to thermal area. The enthalpy flow model can be understood as the extreme case where the shape factors for these areas are constant (Cowley, 1974; Cowley and Chan, 1974; Chan et al., 1976).

(2) *One-Dimensional Model*

The arc temperature in the enthalpy flow arc model is assumed to be constant, and the diameter of the arc is varied along the axis depending on the electrical energy input from the arc current. Another simplification of practical convenience is found in a model in which the arc temperature is calculated as a function of position in the nozzle by neglecting the radial distribution of temperature. In this model, the gas flow channel is modeled by slicing it into disk-shaped elementary sections of axial length Δx as schematically shown in Fig. 2.14. Each elementary section is assumed to be a disk of uniform diameter, and the gas parameter in this section is assumed to be constant. This model can be used to calculate the influence of the difference in nozzle shape.

The cross section of the elementary sections with axial length Δx are given by $S_1, S_2, \ldots, S_n$ and the gas parameters in these sections are represented by P, ρ, h, and u (pressure, density, enthalpy, and flow velocity) with the number of the section as subscript. The assumption of the uniformity of gas parameters in the radial direction simplifies the basic equations (2.2) into one-dimensional form as

$$\frac{\partial \rho}{\partial t} = -\frac{\partial}{\partial x}(\rho u)$$

$$\rho\frac{\partial u}{\partial t} = -\frac{\partial p}{\partial x} - \rho u\frac{\partial u}{\partial x}$$

$$\rho\frac{\partial h}{\partial t} = -u\frac{\partial(\rho h)}{\partial x} + \frac{\partial(ph)}{\partial x} + \rho\hbar_2$$

where $\hbar_2$ = rate of enthalpy input by arc.

This set of equations and the pressure of the puffer chamber calculated from the balance of the mechanical compression and the amount of gas flowing out from the compression chamber give an expression for the physical behavior of the puffer circuit breaker.

Both the enthalpy flow model and the one-dimensional model could reproduce the behavior of a puffer breaker well. One example of the simulation results obtained with the one-dimensional arc model is given in Fig. 2.15.

(B) Current-Zero-Period Regime

The important aspects of the arc in the *current-zero period* can be summarized as follows:

1. The diameter of the arc in this period is so small that the temperature distribution across the diameter plays an important role.

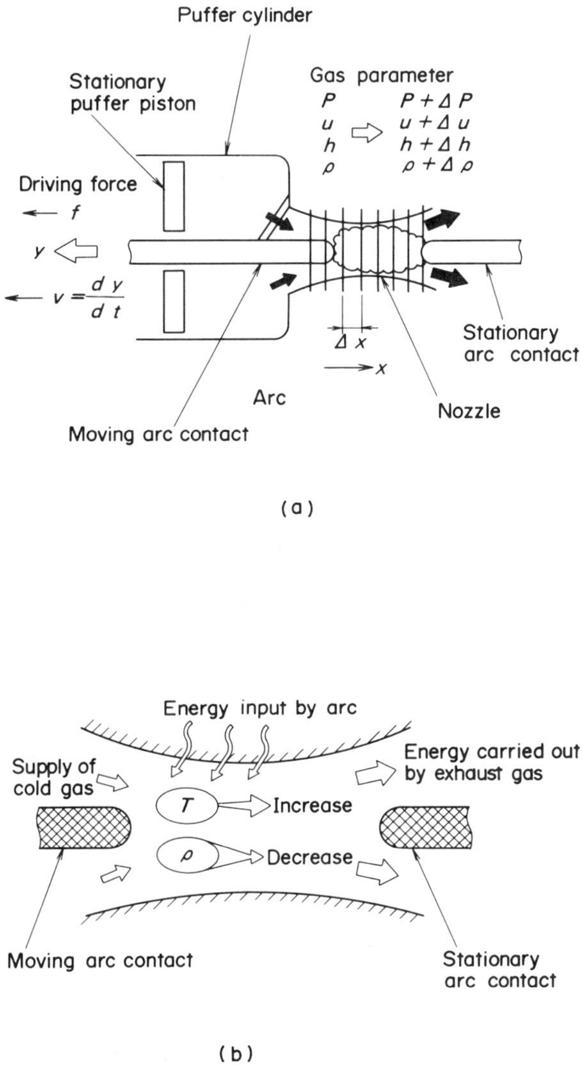

Fig. 2.14 (a) Schematic diagram of one-dimensional model. (b) Nozzle cross section is represented by a series of disks of different diameters and thickness Δx. (Kuwahara et al., 1983. © IEEE.)

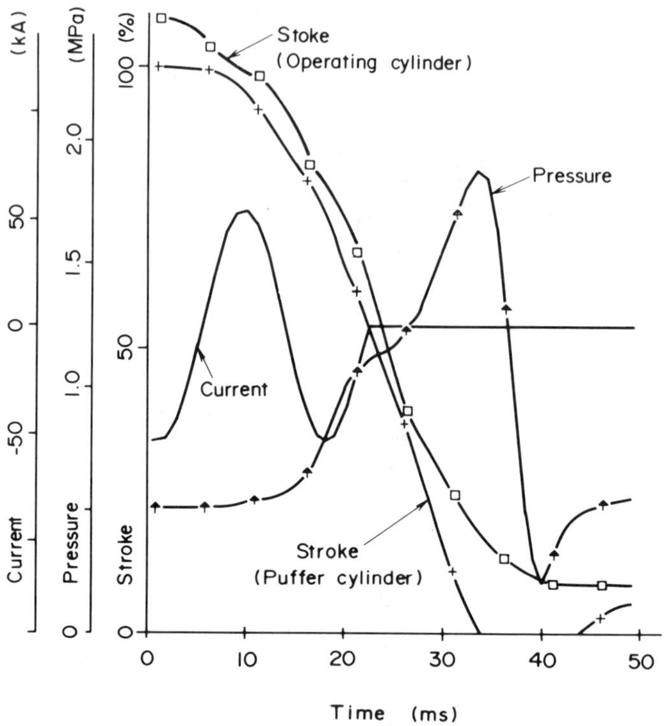

Fig. 2.15 An example of simulation results for 0.9×50 kA by a model breaker with an arcing time of 1 cycle. (Kuwahara et al., 1983. © IEEE.)

2. The transient phenomena in a few microseconds are important, and the displacement of the background gas during this period is typically a few millimeters, up to 10 mm. This small displacement in the axial direction is negligible in comparison with the total arc length.
3. The background gas parameters vary along the axis.

The radial temperature distribution of arcs in oil and air in this period was investigated as early as 1935 (Elenbaas, 1935; Heller, 1935). Balancing the thermal flow across the radius and the electrical energy input gives the following equation as an expression under a steady-state arc:

$$-\frac{1}{r}\frac{\partial}{\partial r}\left(K_r\frac{\partial T}{\partial r}\right) + P_R = \sigma E^2$$

where P_R = radiative power loss and E = electric field intensity. When this equation is solved with assumed boundary conditions it gives the temperature profile of the arc. Knowledge of the temperature profile is assumed in all the current-zero arc models.

Physical arc models of practical use simplify the basic equation by using these features. These models can be classified according to the expression of energy transfer in the radial direction. The number of zones across the diameter of the arc is a convenient index in classifying models.

(1) Single-Zone Models

Single-zone models implicitly assume the temperature distribution in the arc but do not describe the internal structure of the arc and express the net input energy as the difference between the input energy and the energy transmitted to the surrounding space:

$$\frac{dQ}{dt} = EI - N_0$$

The relationship between the arc conductance and the thermal energy in the arc column can be given from a consideration of the arc's internal structure. Mayr (1943) assumed that the temperature distribution was given by a Bessel function and deduced that the arc resistance can be represented by the exponential function

$$R = \text{const} \times \exp\left(-\frac{Q}{Q_0}\right)$$

The time derivative of this equation is given as

$$\frac{dQ}{dt} = -\frac{Q_0}{R}\frac{dR}{dt}$$

Substituting this relationship into the first equation yields

$$\frac{1}{R}\frac{dR}{dt} = \frac{N_0}{Q_0}\left(1 - \frac{EI}{N_0}\right)$$

When θ is defined as $1/\theta = N_0/Q_0$,

$$\frac{1}{R}\frac{dR}{dt} = \frac{1}{\theta}\left(1 - \frac{EI}{N_0}\right)$$

Mayr used this equation to represent the dynamic behavior of the arc.

Cassie (1939) and Cassie and Mason (1956) used the same relationship of energy balance as the basic equation and assumed that the net input energy is given by $dQ/dt = AE^2 - BQ$ and that the arc resistance is mostly dependent on the arc radius as shown in the equation

$$R = \frac{\text{const}}{r^2} \exp\left(-\frac{Q}{r^2 G_0}\right)$$

This expression leads to a different expression of the dynamic behavior with two parameters $\theta = BQG_0/r^2$ and $E_0 = BQ/A$ (characteristic arc voltage) as follows:

$$\frac{1}{R}\frac{dR}{dt} = \frac{1}{\theta}\left(1 - \frac{E^2}{E_0^2}\right)$$

When the conductance of the arc is represented by $G(= 1/R)$, these equations become

$$\frac{1}{G}\frac{dG}{dt} = \frac{1}{\theta}\left(\frac{EI}{N_0} - 1\right)$$

and

$$\frac{1}{G}\frac{dG}{dt} = \frac{1}{\theta}\left(\frac{E^2}{E_0^2} - 1\right)$$

This deduction of the mathematical expression for energy balance under allowable simplifications of the basic model gave convenient models for some practical purposes such as the study of application of circuit breakers and the evaluation of breaker performance. They represent the dynamic behavior of the arc to the necessary extent although they have some limitations in reflecting physical parameters in actual breakers. Because of their convenience, these models are still used for practical purposes with some modifications in spite of the development of more complex models.

One example of the modified Mayr model is found in Urbanek's (1972) work. The arc characteristics are expressed by four parameters θ (arc time constant), E_0 (voltage constant of the arc), P_0 (power loss of the arc), and U_d (breakdown voltage of cold gas):

$$\frac{dG}{dt} = \left\{\frac{IE}{E_0^2} - G - \frac{N_0}{E_0^2}\left[1 - \left(\frac{E}{U_d}\right)^2 - 2\frac{\theta}{u_d^2}E\frac{dE}{dt}\right]\right\}\theta^{-1}$$

(2) Two-Zone Model

The single-zone models can reflect the thermal flow in the radial direction, but they neglect flow in the axial direction and do not take the temperature gradient in the axial direction into account. A model proposed by Ishikawa et al. (1984) handles the flow of energy in a more complex manner by dividing the arc radius into two zones—an arc column region and an arc surrounding zone—as shown in Fig. 2.16. Mass, momentum, and energy conservation equations [Eqs. (a), (b), and (c)] in each zone are used as six basic equations.

Mass conservation:

$$\frac{\partial \rho}{\partial t} + \frac{1}{r}\frac{\partial}{\partial r}(\rho r V_r) + \frac{\partial}{\partial z}(\rho V_z) = 0 \qquad \text{(a)}$$

Momentum conservation:

$$\rho \frac{\partial V_z}{\partial t} + \rho V r \frac{\partial V_z}{\partial r} + \rho V z \frac{\partial V_z}{\partial z} = -\frac{\partial p}{\partial z} + \frac{1}{r}\frac{\partial}{\partial r}\left[(\eta + \eta_t) r \frac{\partial V_z}{\partial r}\right] \qquad \text{(b)}$$

Energy conservation:

$$\frac{\partial}{\partial t}\left[\rho\left(e_i + \frac{V_z^2}{2}\right)\right] + \frac{1}{r}\frac{\partial}{\partial r}\left[\rho r V_r\left(h + \frac{V_z^2}{2}\right)\right] + \frac{\partial}{\partial z}\left[\partial V_z\left(h + \frac{V_z^2}{2}\right)\right]$$
$$= \sigma E^2 - u + \frac{1}{r}\frac{\partial}{\partial r}\left[(\kappa + \kappa\rho_t) r \frac{\partial T}{\partial r}\right] \qquad \text{(c)}$$

where

z = axial coordinate
r = radial coordinate
t = time
V_z = flow velocity in the z direction
V_r = flow velocity in the r direction
ρ = density
p = pressure
η = viscosity
η_t = effect of turbulence on viscosity
κ = heat conductance
κ_t = effect of turbulence on heat conductance
h = enthalpy
e_i = internal energy
σ = electrical conductivity
E = electric field strength
u = radiated energy
T = absolute temperature

The basic equations are integrated in two zones, and the six equations obtained can then be solved as a set of differential equation in the z direction. The numerical calculation gives the distribution of arc voltage and the variation of arc diameter along the axis together with the voltage and current characteristics. The voltage characteristics obtained by numerical calculation with this model reproduce the behavior obtained experimentally as shown in Fig. 2.16.

(3) Three-Zone Model

Hermann and Ragaller (1977) introduced a three-zone model in which three zones with some form of temperature distribution are assumed as shown in Fig. 2.17. In the core zone, a parabolic temperature distribution is assumed, and in the transition zone an exponential distribution is assumed as shown in the following equations. The outer zone is assumed to be at constant temperature.

Zone 1:

$$T(r^2) = T_a - (T_a - T_1)\frac{r^2}{r_1^{\,2}}$$

Zone 2:

$$T(r^2) = T_3 + (T_1 - T_3)e^{-\alpha(r^2 - r_1^2)}$$

where

$$\left(\alpha = \frac{1}{r_2^{\,2} - r_1^{\,2}}\,\ell\eta\,\frac{T_1 - T_3}{T_2 - T_3}\right)$$

Zone 3:

$$T(r) = T_3$$

Here, T_1, T_2, and T_3 are constants and T_a, r_1, and $r_2 2$ are variables. Mass, momentum, and energy conservation in each section and the heat flux flowing through Zone boundaries can give a description of the behavior of variables.

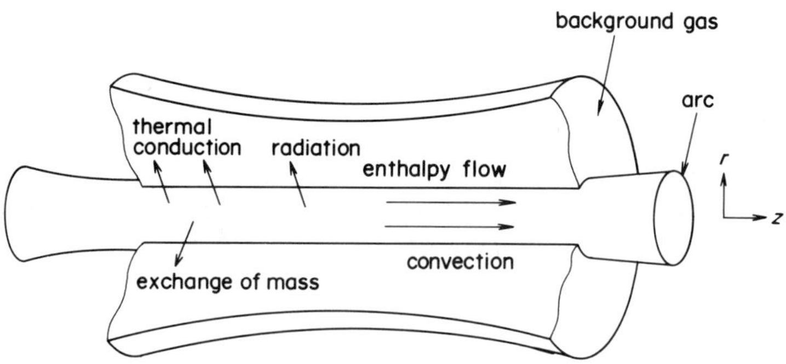

Fig. 2.16 (a) Schematic structure of the two-zone arc model. (b) The voltage characteristics obtained experimentally. (c) The results obtained by numerical calculation with this model. (Ishikawa et al., 1984. © IEEE.)

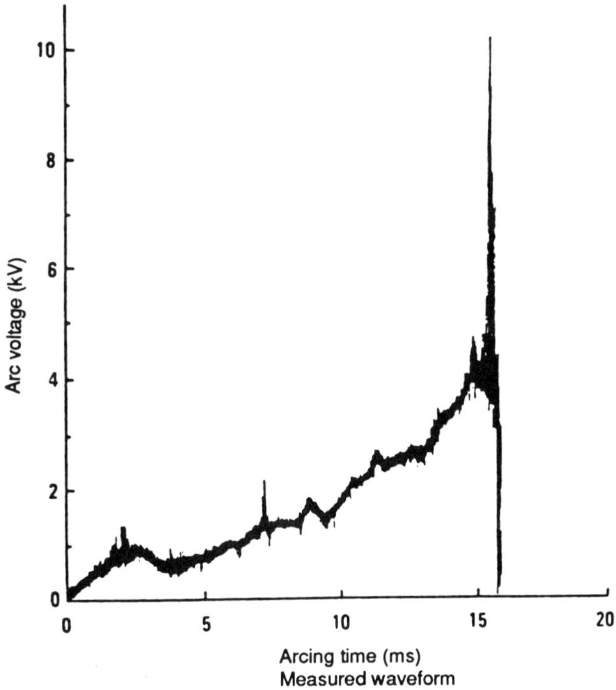
10
8
6
4
2
0
Arc voltage (kV)
0
5
10
15
20
Arcing time (ms)
Measured waveform

(b)

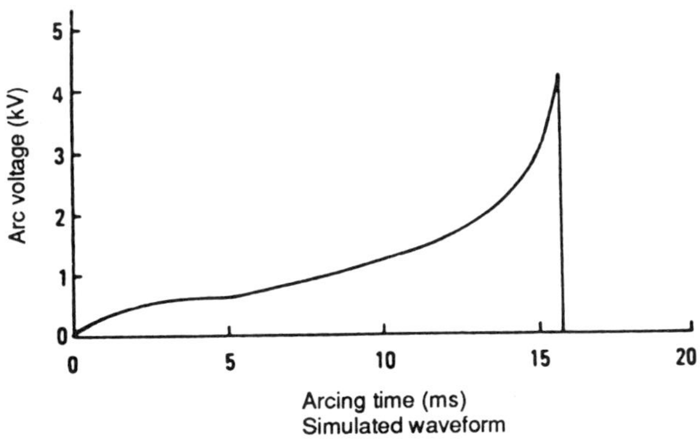
5
4
3
2
1
0
Arc voltage (kV)
0
5
10
15
20
Arcing time (ms)
Simulated waveform

(c)

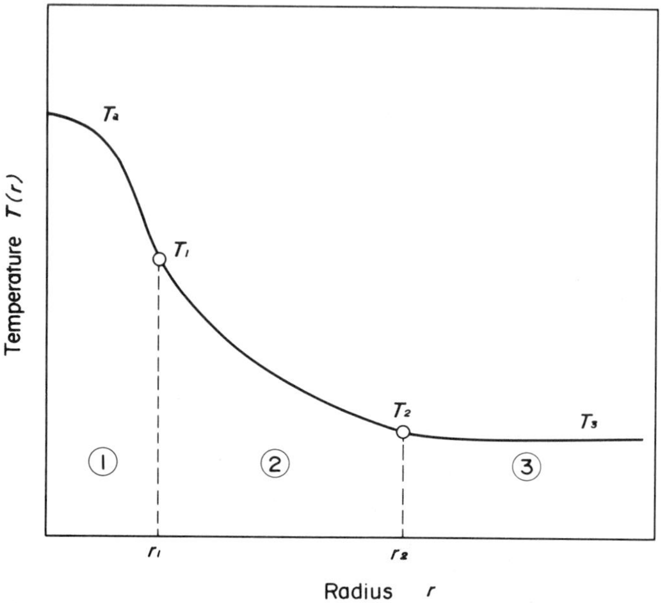

Fig. 2.17 Temperature distribution assumed in integrating energy transfer in the radial direction in the three-zone model. (Hermann and Ragaller, 1977. © IEEE.)

(4) Multiple-Zone Model

The models introduced use a finite number of zones and the concept of heat flux thruogh the boundary. A series of concentric cylinders of thickness *dr* with specified gas parameters can be used instead of integration in the radial direction. A basic idea of this multiple-zone model is shown schematically in Fig. 2.18. The gas parameters such as temperature, pressure (or density), electrical conductivity, and thermal conductivity can be calculated by balancing the input electrical energy and the heat transfer by energy loss caused by the temperature gradient and radiation. The basic equations (2.2) can be simplified by the assumptions that the temperature gradient along the axis is negligible and that the arc is axisymmetric.

$$\rho \frac{\partial h}{\partial t} = -\rho V_r \frac{\partial h}{\partial r} + \frac{1}{r}\frac{\partial}{\partial r}\left(rK^* \frac{\partial T}{\partial r} \right) + \sigma E^2$$

where

$K^* = \kappa + K_r$

κ = thermal conductivity

K_r = equivalent thermal conductivity including the radiation effect

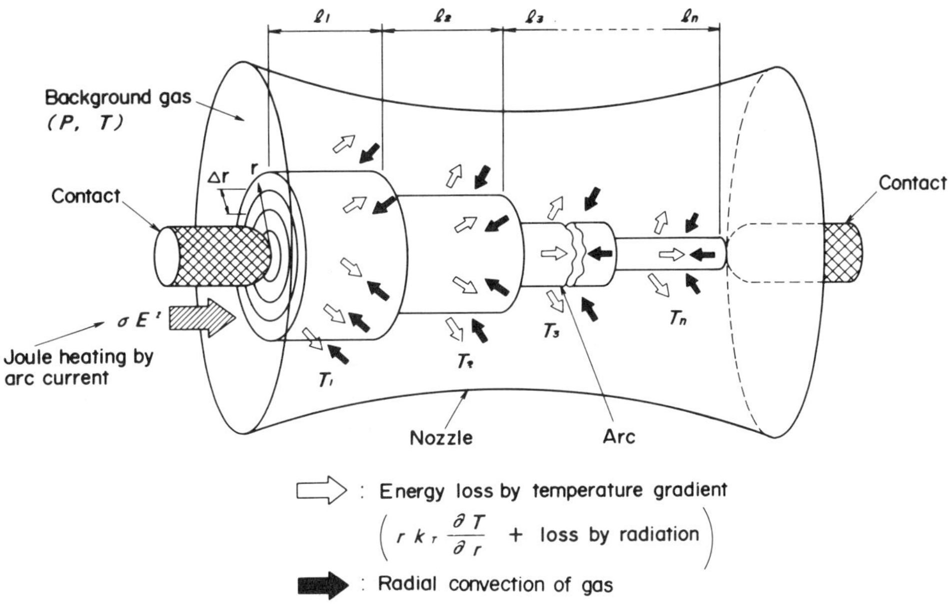

Fig. 2.18 Schematic model of multiple-zone (multiple-layer) model. The temperature in each layer is assumed to be uniform. Arcs with different external temperatures and/or different pressures are connected in series to represent an arc in the flow field of gas. (Kuwahara et al., 1983. © IEEE.)

In actual arcs, the assumption of constant temperature along the axis is not valid, but it has been shown that series connection of this model with background gas of a different temperature can be used as an approximation of actual arc (Kuwahara et al., 1983).

(C) Dielectric Recovery Regime

The electrode gap where the arc was interrupted is exposed to the electric field that is defined by the recovery voltage of the circuit. The period following arc interruption is the period when the arc space is replaced by cool gas and the dielectric withstand capability recovers. Electrical energy input during this period is negligible. Characteristics of this period are summarized as follows.

1. The arc core temperature is cooled to approximately 6000 K, and the arc core diameter is small in comparison with the total arc space.
2. The time span of the phenomenon of interest is of the order of 10 μsec, and gas flow along the axis plays an essential role.

3. The behavior of the exhaust gas outside of the interrupter is also important for the completion of the interruption.

These characteristics require a detailed description of the gas flow inside and outside of the interrupter. The electrical performance is understood by estimating the electrical breakdown characteristics of interelectrode gaps. This is achieved by combining the calculated *distribution of gas density* and the *electric field distribution*.

(1) One-Dimensional Model

The small diameter of the arc space and the relatively quiet transfer of energy along the radius justify application of the *one-dimensional model* for the calculation of energy balance in the high-current period. This model gives an estimation of the gas density at the electrode surface necessary for evaluation of the dielectric withstand. The electric field distribution can be obtained from a separate calculation as will be mentioned in Section 2.3.1(D). The dielectric withstand can be evaluated with the additional assumption that the withstand voltage is determined mostly by the surface field intensity and the density of the gas. This assumption is believed to be valid at temperatures below 2000 K. Kuwahara et al. (1984) gave an example of the calculated time history of the gas density near the cathode and anode. The dielectric recovery estimated from the gas density calculation showed good agreement with the experimentally obtained dielectric recovery (see Figs. 2.19 and 2.20).

(2) Full-Description Model

The fundamental set of equations (2.2) can be solved numerically under axisymmetric two-dimensional coordinates, giving gas temperature and density at the grid points in the calculated region. Ragaller et al. (1982) showed the result of numerical calculation of the decay of the hot-gas channel after current zero of the nozzle-type electrode arc was obtained by transforming the basic equation into isotherms to provide a thermal map of the arc column. The dielectric recovery was obtained by the temperature decay at the stagnation point and showed good agreement with experimental results. An example of the calculated temperature distribution and the dielectric recovery are shown in Figs. 2.21 and 2.22. Calculations of the distribution of the local gas pressure (p) and the local field strength (E_0) indicate the dependence of the breakdown voltage dependent on the position of the electrode (Endo et al., 1989). The theoretically expected breakdown voltage (calculated V_{BD}) deduced from the p and E_0 values is compared with the experimentally obtained breakdown voltage (measured V_{BD}) in Fig. 2.23 (Shade and Ragaller, 1982).

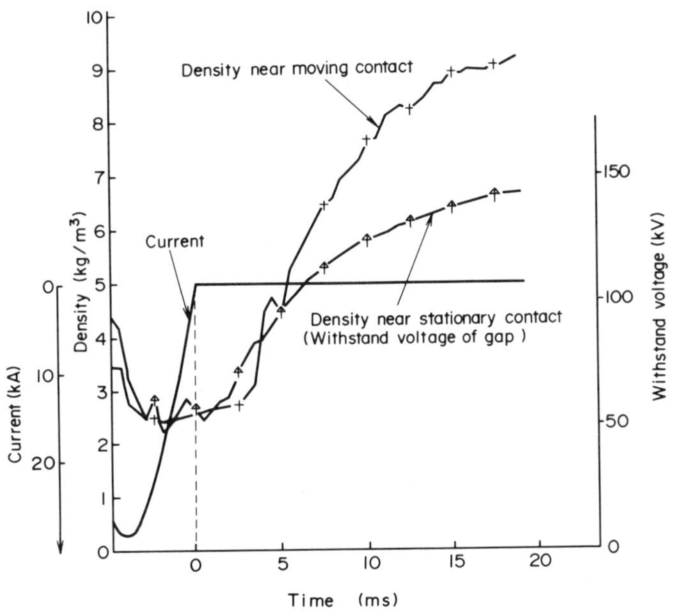

Fig. 2.19 Recovery of the gas density at different places in the interrupter calculated by use of the one-dimensional model. The density near the cathode (in this case the stationary contact) represents the breakdown voltage of the gap. (Kuwahara et al., 1984. © IEEE.)

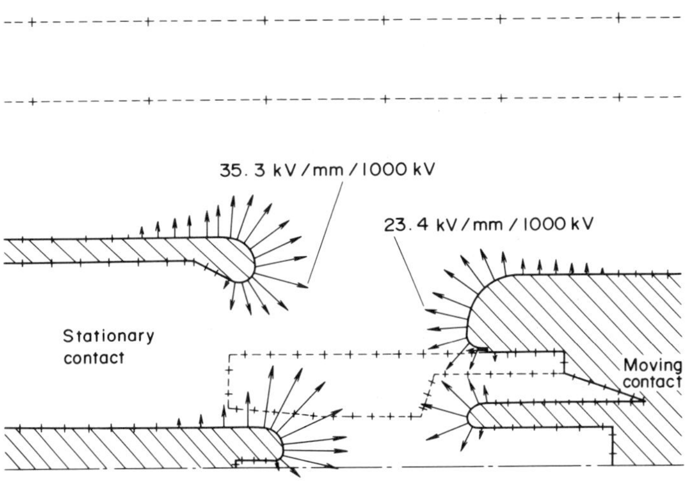

Fig. 2.20 Electric field distribution of the interrupter corresponding to the calculation shown in Fig. 2.19. (Kuwahara et al., 1984. © IEEE.)

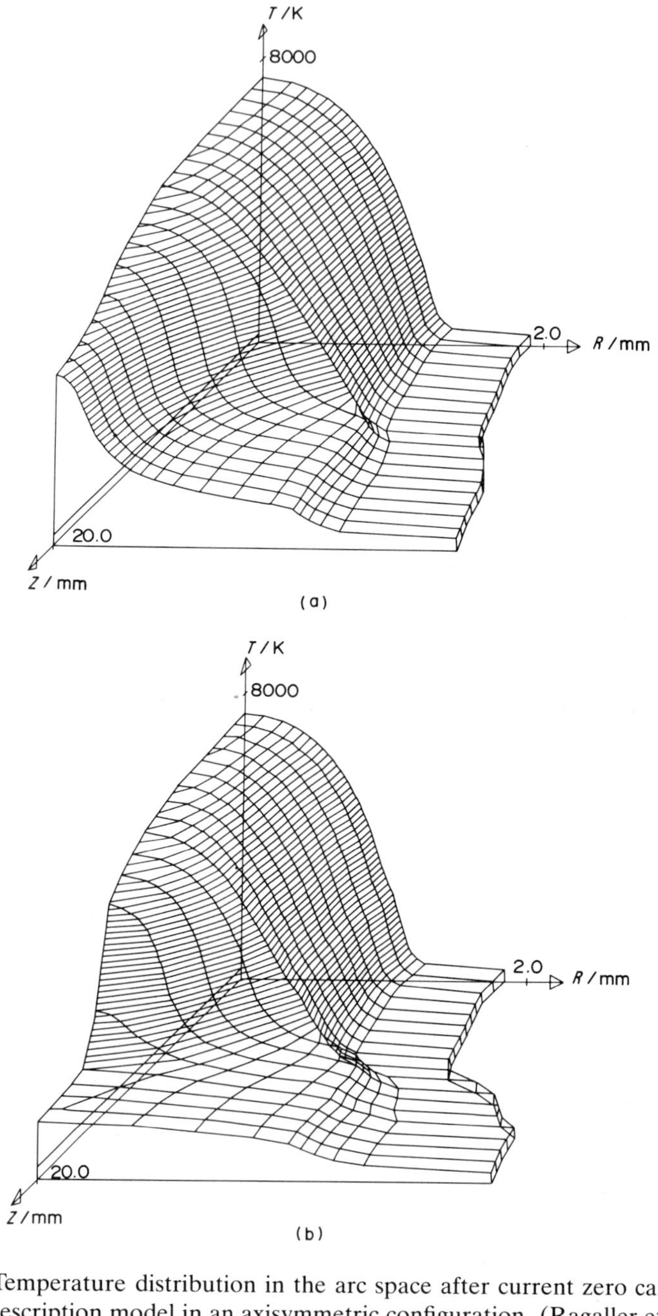

Fig. 2.21 Temperature distribution in the arc space after current zero calculated with a full-description model in an axisymmetric configuration. (Ragaller et al., 1982. © IEEE.)

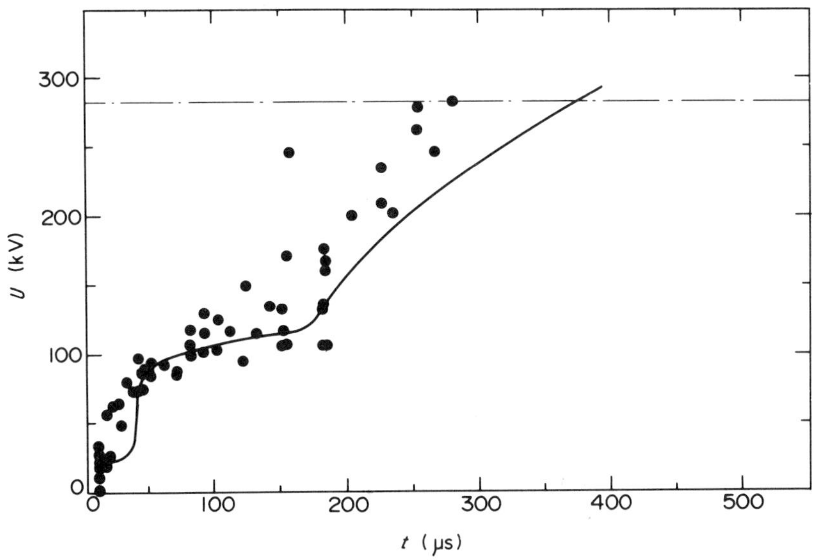

Fig. 2.22 Comparison of calculated dielectric recovery by the full-description model and measured recovery. (Brand et al., 1982. © IEEE.)

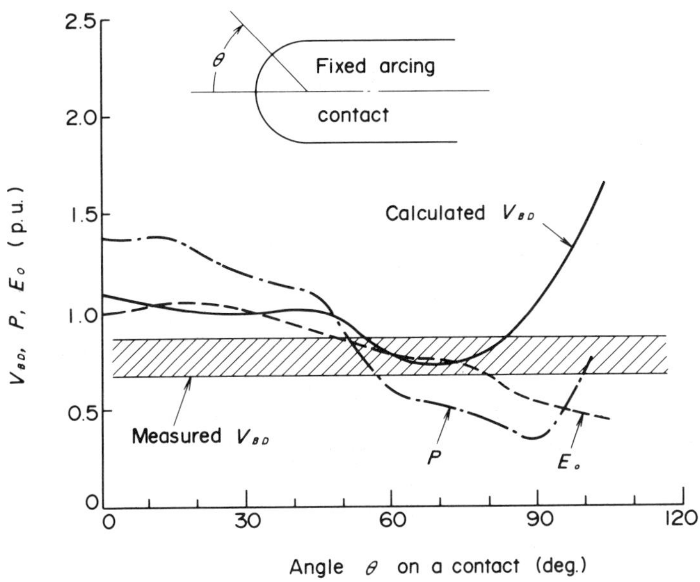

Fig. 2.23 Dependence of the expected breakdown voltage on the position of the electrode calculated by the local density and the electric field intensity shown with the experimentally determined breakdown voltage. (Endo et al., 1989. © IEEE.)

(3) Combined Model

The dielectric recovery of an actual interrupter depends on the distribution of hot gas in the wider region surrounding the current-carrying contacts and the electrical shield as well as the region close to the arcing contents. The hot gas exhausted from the interrupter should not cause any serious lowering of the dielectric characteristics at the interelectrode gaps between phases or between the surrounding shield and ground. The full-description model can be used for this sort of evaluation, but the calculation of the detailed distribution of hot gas in wide space is very time-consuming. As a compromise, the possibility of saving calculation time was shown by Yoshizumi et al. (1988). A full-discretization model is used for the calculation of the distribution in the space of interest, but the arc region that is supplying the hot gas is calculated with a one-dimensional model, and the exhaust gas from the arc region is used as a hot gas source for the surrounding region.

Figure 2.24 shows an example of the hot gas around the interrupter where an arc contact is surrounded by an electrical shield. The exit of the nozzle gives a boundary condition for the calculation in the region of arc contact and the surrounding electrical shield.

Figure 2.25 shows a region still further from the interrupter. This is a calculation outside of the cooling tube of the exhaust gas that is located adjacent to the shield to provide a gas-cooling space. The temperature in the region between the interrupter and the exit of the cooling tube is calculated by a one-dimensional model in this case, and the hot gas flowing out from the cooling tube is assumed to be the input boundary condition for the calculation in the region of interest, which is represented by a full-description model.

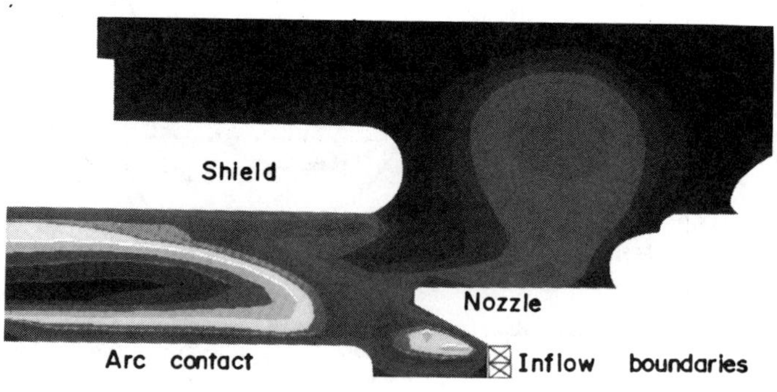

Fig. 2.24 Distribution of the hot exhausted gas from the interrupter. Gas input into the source cell is given by simulation with a one-dimensional model. (Yoshizumi et al., 1989. © IEEE.)

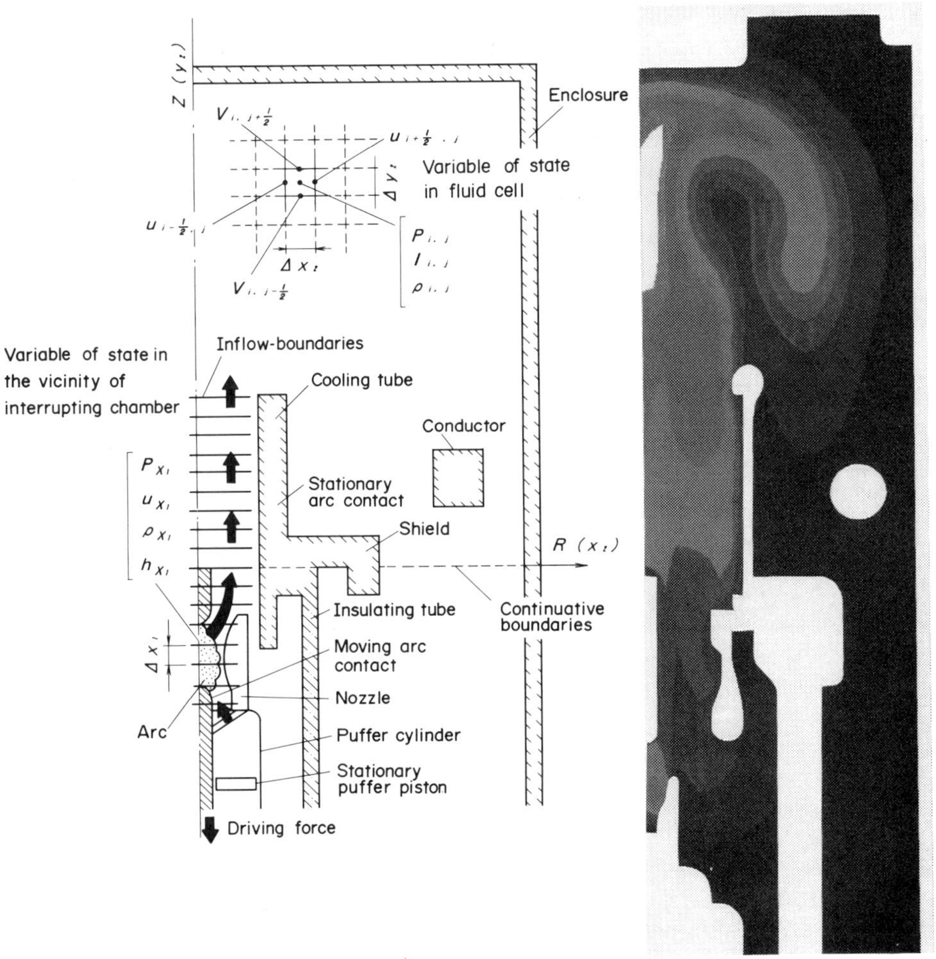

Fig. 2.25 Hot-gas density distribution and the splitting of the space further down the hot gas exhaust for analyzing the effect of the distribution of the flow guide and the electric shield. (Yoshizumi et al., 1989. © IEEE.)

Total calculation time required can be reduced by combining different types of models of sufficient accuracy.

(D) Electric Field Configuration

The electric field distribution is an important factor that determines the dielectric characteristic of an interelectrode gap with a given geometry as pointed out in the preceding sections. The numerical calculation of the field distribution was made possible by various methods. The solution of the electric field distribution is obtained by solving Laplace's equation under the given boundary conditions. Therefore, the question is, How can it be solved most efficiently? in this problem, whereas in solving the temperature distribution in the arc space the question was, How can the complex set of equations be simplified into a form convenient for handling? The examples of practical methods introduced here is based on the same basic equation, and effort is directed to reduce the input time of the boundary condition and/or to improve accuracy.

(1) Finite-Difference Method

The finite-difference method is the most straightforward application of the differential equation. The space of interest is split into a regular grid in orthogonal coordinates. Potentials on the grid points are assigned as ϕ_0, $\phi_1, \ldots, \phi_n$. Potentials $\phi_1, \phi_2, \ldots, \phi_6$, which are adjacent to ϕ_0, can be expressed by Taylor expansion as

$$\phi_1 = \phi_0 + h_x \frac{\partial \phi}{\partial x} + \frac{h_x^{\,2}}{2} \frac{\partial^2 \phi}{\partial x^2}$$

$$\phi_2 = \phi_0 - h_x \frac{\partial \phi}{\partial x} - \frac{h_x^{\,2}}{2} \frac{\partial^2 \phi}{\partial x^2}$$

$$\phi_3 = \phi_0 + h_y \frac{\partial \phi}{\partial y} + \frac{h_y^{\,2}}{2} \frac{\partial^2 \phi}{\partial y^2}$$

$$\vdots$$

$$\phi_6 = \phi_0 - h_z \frac{\partial \phi}{\partial z} - \frac{h_z^{\,2}}{2} \frac{\partial^2 \phi}{\partial z^2}$$

(Refer to Fig. 2.26.) Using these relationships, Laplace's equation $\nabla^2\phi = 0$ can be expressed as

$$2\phi_0\left(\frac{1}{h_x^{\,2}} + \frac{1}{h_y^{\,2}} + \frac{1}{h_z^{\,2}}\right) = \frac{\phi_1 + \phi_2}{h_x^{\,2}} + \frac{\phi_3 + \phi_4}{h_y^{\,2}} + \frac{\phi_5 + \phi_6}{h_z^{\,2}}$$

When this set of equations is solved with the boundary condition $\phi_1 = \phi_2 = \cdots = \phi_n = V_i$, the potentials of all the grid points are obtained. The efficiency of the calculation is greatly affected by the selection of appropriate coordinates and boundary conditions. When the boundary condition is complex, the work necessary to provide the input data should not be overlooked.

(2) Finite-Element Method

The region of interest is split into a set of tetrahedra. The electric field intensity inside the tetrahedron is assumed to be uniform, and the potential

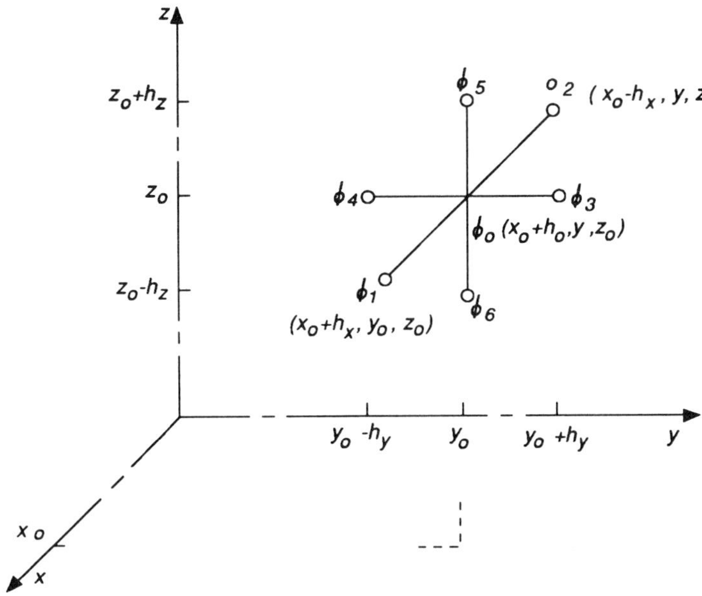

Fig. 2.26 Definition of notations used for the finite-difference method in Cartesian coordinates.

inside of the unit tetrahedron is given by a linear expression for all the coordinate components (refer to Fig. 2.27).

$$\phi = a_1 + a_2x + a_3y + a_4z$$

Therefore, the electrical potential at an arbitrary point in the volume can be expressed by the four potentials, ϕ_1, ϕ_2, ϕ_3, ϕ_4 at the four vertices of the tetrahedron. The total electrostatic energy in the system can then be expressed by $w = \iiint(\varepsilon E^2/2)\, dv$, which can be shown by the equation

$$w = \iiint \frac{\varepsilon}{2}\left[\left(\frac{\partial\phi}{\partial x}\right)^2 + \left(\frac{\partial\phi}{\partial y}\right)^2 + \left(\frac{\partial\phi}{\partial z}\right)^2\right] dx\, dy\, dz$$

The solution in an actual system should make the total energy minimum, and solving the Laplace equation is equivalent to solving the set of ϕ_i's that gives the minimum value of w. $\partial w/\partial\phi_i = 0$ for all the grid points provides a set of equations, and the ϕ_i obtained gives the necessary potential distribution. Preparation of the input data for actual calculations, especially for a three-dimensional calculation, is complex, and not many practical examples are published.

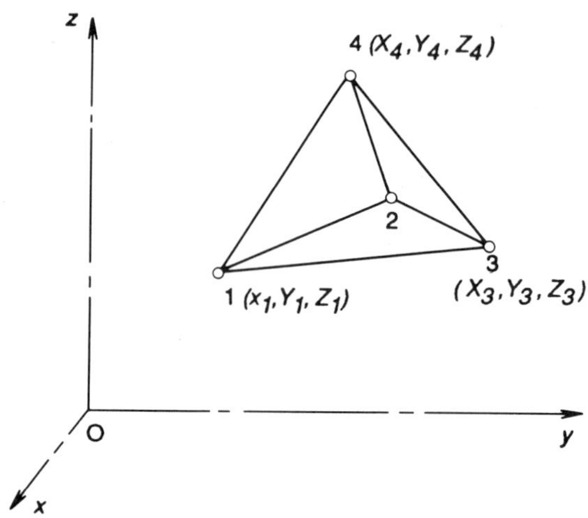

Fig. 2.27 Conceptual diagram of the tetrahedron for the finite-element method.

(3) Charge-Simulation Method

The charge-simulation method assumes a hypothetical set of charges in the system to satisfy the boundary conditions and is applied in practice for axisymmetric or general three-dimensional problems. The hypothetical charges $Q_1, Q_2, \ldots, Q_n$ are assumed to be on the points $A_1, A_2, \ldots, A_n$ inside the electrode C (refer to Fig. 2.28). The electrical potential at a point P can be given by

$$\phi(r) = \sum_{i=1}^{n} \frac{Q_i}{4\pi\varepsilon\,|r - r_1|}$$

The potentials on the points $B_1, B_2, \ldots, B_n$, which are on the surface of electrode C, should each equal U, and this relationship provides the set of equations

$$\begin{bmatrix} P_{11} & \cdots & P_{1n} \\ \vdots & & \vdots \\ P_{n1} & \cdots & P_{nn} \end{bmatrix} \begin{bmatrix} Q_1 \\ \vdots \\ Q_n \end{bmatrix} = \begin{bmatrix} U \\ \vdots \\ U \end{bmatrix}$$

where P_{nn} is the potential coefficient. A computer program for solving this set of equations is relatively simple but the selection of the points to satisfy the boundary condition requires some experience. Because of this complex requirement, solving a problem in general three-dimensional geometry is not a straightforward task.

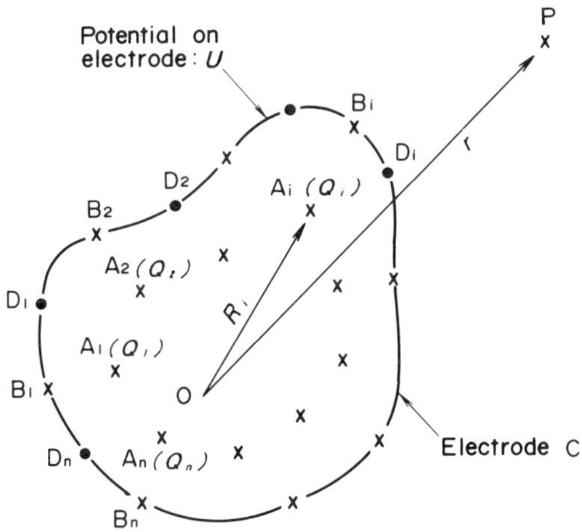

Fig. 2.28 Conceptual diagram of the charge-simulation method.

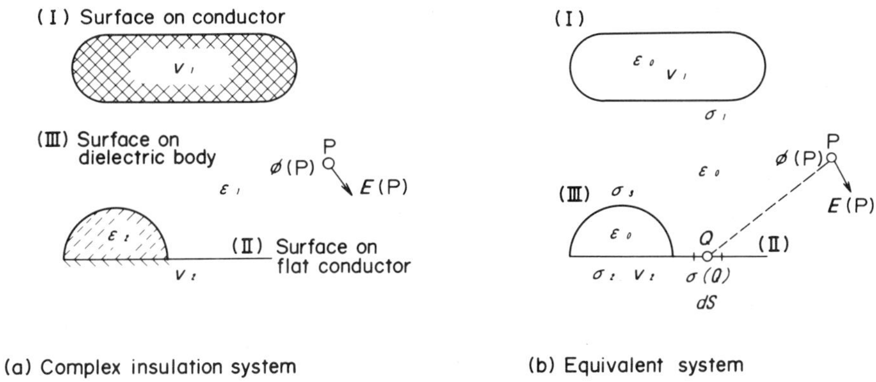

Fig. 2.29 Conceptual diagram of the surface-charge method.

(4) Surface-Charge Method

The surface-charge method assumes a hypothetical electric charge on the boundary of the electrode and insulator bodies. A set of hypothetical electric charges $\sigma_1, \sigma_2, \ldots$ provides a system equivalent to Fig. 2.29a as in Fig. 2.29b. The electrical potential and field intensity at an arbitrary point in Fig. 2.29b can be obtained from the equations

$$\phi(\mathrm{P}) = \int \frac{1}{4\pi\varepsilon_0} \frac{1}{\overline{\mathrm{PQ}}} \sigma(\mathrm{Q})\, dS$$

$$E(\mathrm{P}) = \mathrm{grad}\,\phi(\mathrm{P})$$

The potentials at the point P on the surface of an electrode have a potential $U(\mathrm{P})$, which gives the equation

$$\int \frac{1}{4\pi\varepsilon_0} \frac{1}{\overline{\mathrm{PQ}}} \sigma(\mathrm{Q})\, dS = U(\mathrm{P})$$

When the surface of the insulator has some real electric charge, this is related to other parameters as

$$(\varepsilon_2 - \varepsilon_1)\int \mathrm{grad}_n\, G(\mathrm{P,Q})\sigma(\mathrm{Q})\, dS + \frac{(\varepsilon_1 - \varepsilon_2)}{2\varepsilon_0} \sigma(\mathrm{P}) = \sigma_r(\mathrm{P})$$

$$G = \frac{1}{4\pi\varepsilon_0} \overline{\mathrm{PQ}}$$

where $\sigma_r(\mathrm{P})$ is the real electric charge on the boundary of the insulator. The set of integral equations about $\sigma(\mathrm{Q})$ can eventually be rewritten as a set of first-order equations by splitting the boundary surface into small triangular sections. Figure 3.14 (Chapter 3) shows a example of the splitting of the surface and the obtained distribution of the electric field.

2.3.2 Examples of Analysis by Physical Arc Models

(A) General Behavior of Electric Arc

The *physical arc model* has always been used for improving understanding of the general characteristics of electric arcs. In the early stages of developing the single-zone model, the hysteresis behavior of the arc voltage observed in the actual arc was reproduced by the Mayr arc model. The dependence of the hysteresis behavior on the frequency of the superimposed current was explained, and the voltage behavior caused by superimposing two currents was also reproduced. An example of the voltage hysteresis is shown in Fig. 2.30.

(B) Effect of Nozzle Diameter on Arc Characteristics

The improvement of the physical arc model is closely related to the increases in the capacity of computers. The increases in computer speed and memory capacity made it possible to handle detailed models that can use the physical gas parameters as basic information. These models can relate arc characteristics to physical dimensions and other mechanical parameters. Therefore, these models are more practically used to improve the performance of actual interrupters. The parameters that can be controlled by design are nozzle shape, gas pressure, and driving energy of the puffer chamber, and these parameters can be taken into the set of equations.

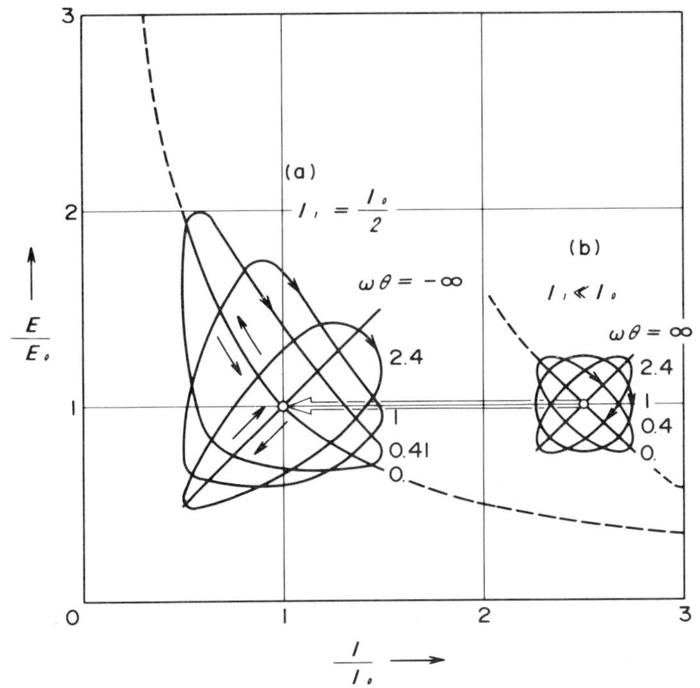

Fig. 2.30 Hysteresis of the voltage at various frequencies of the current reproduced by using the Mayr model. The traces show the voltage behavior with a constant current superimposed with a sinusoidal wave shown as $I = I_0 + I_1 \sin(\omega t)$. (a) $I_1 = I_0/2$; (b) $I_1 << I_0$. (Mayr, 1943. Copyright Springer.)

Figure 2.31 shows an example of the dependence of arc temperature at the nozzle on the nozzle diameter, calculated by using the one dimensional model. This temperature evaluation can then be used for evaluating interruption performance by using the *multiple-zone arc model* to express arc behavior at current zero. The calculated results can be compared with experimental results as shown in Fig. 2.32 (Kuwahara et al., 1983).

A change in the mechanical driving energy affects the temperature and density of the gas in the interrupter at current zero and subsequently changes the estimation of the interruption performance. The electrode shape and shield design can be optimized by applying a model that visualizes the hot gas behavior. These numerical studies are greatly helping to improve and optimize interrupters and their surroundings.

There are examples of successful application of the physical arc model to feasibility studies of circuit breakers in special circuitry. A generator

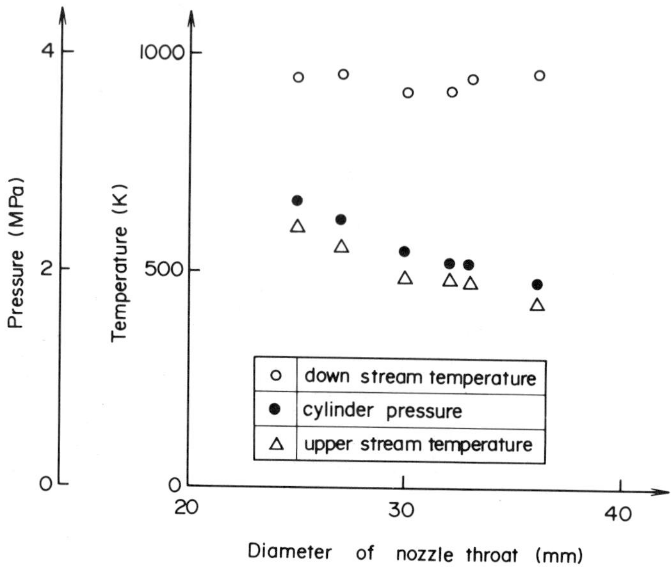

Fig. 2.31 Dependence of the temperature and pressure in a puffer cylinder on the nozzle diameter calculated with a one-dimensional model. (Kuwahara et al., 1983. © IEEE.)

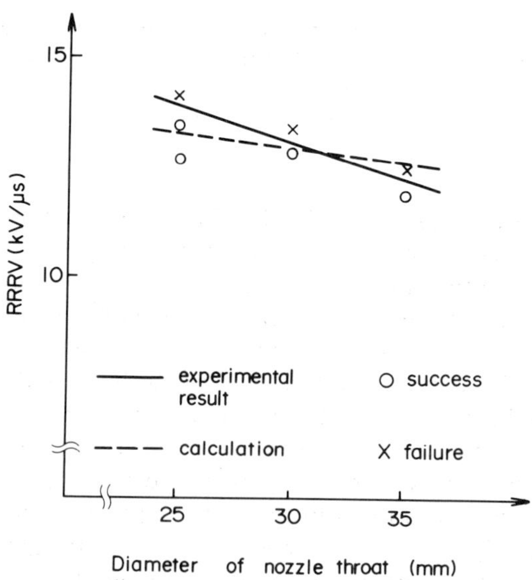

Fig. 2.32 Theoretical and experimental dependence of the interruption performance at various nozzle diameters. (Kuwahara et al., 1983. © IEEE.)

circuit that gives a current-zero missing in an early stage of the fault poses a special condition for a circuit breaker. In Chapter 4, Figs. 4.16 and 4.17 show an example that explains the current zero formed by the arc voltage of the interrupted arc by using a two-zone model.

2.4 ANALYSIS OF TEST RESULTS AND DEDUCTION OF CHARACTERISTIC PARAMETERS (K. Ibuki, T. Yoshizumi, S. Sakuma)

Evaluation of circuit-breaker performance is important for both completed circuit breakers and prototype circuit breakers (Kuwahara 1983). An accurate evaluation of a prototype breaker is particularly important in accelerating the development cycle. Therefore, efforts to assess the performance of a breaker with a smaller number of tests are of general interest to those are working in this field. In many cases, these efforts entail the use of the mathematical arc models described in Chapter 1, often starting with a Mayr arc model and modifying it as needed (Cassie and Rieder, 1964; Haupt and Moeller, 1987).

2.4.1 Evaluation of Characteristic Values

The basic form of the *Mayr model* is shown as $(dR/dt)/R = (1 - ei/N)/\theta$, which is quite simple in handling. However, it is not a simple job to evaluate the two characteristic parameters θ and N appearing in the equation from an actually measured set of current and voltage data. Even when the parameters θ and N are deduced from the measured data with necessary assumptions, they usually vary as a function of time. The *Cassie arc model,* which is another important mathematical arc model, will give similar results in this sense. Therefore, the accurate reproduction of the arc behavior by use of the deduced set of parameters is an important index for checking the practical applicability of the model and the parameter deduction process.

The process of deducing the characteristic parameters can be briefly described as follows. Two sets of data—current $e_1, e_2, \ldots, e_n$ and voltage $i_1, i_2, \ldots, i_n$ of the arc—acquired by sampling with an interval of Δt give a series of data on arc resistance $R_1, R_2, \ldots, R_n$ ($R_j = ej/i_j$). This can then give a series of data $R'_1, R'_2, \ldots, R'_n$ by calculating $\Delta R/\Delta t$. It is not possible to get a unique solution for θ and N from a set of data e_j, i_j, R_j, R'_j at a given instant, as the number of equations is not sufficient to give a unique solution. Therefore, the process of solving this problem can be considered to be equivalent to identifying the points with the same characteristic parameters out of the series of data. Early in the investigation it was recognized that even when we restrict ourselves to think about the current-zero period, arc behavior cannot be expressed adequately with two constant parameters. A number of variations in obtaining the characteristic values have been proposed and used, but no standard method has yet been developed.

(A) Assumption that One Parameter is a Constant

In both the Mayr and Cassie models, $(dR/dt)/R$ (at $i = 0$) $= 1/\theta$, and when the assumption that the arc time constant stays unchanged within the time span of interest is valid, the number of unknown in the equation becomes equal to 1. In this method, a set of data e_j, i_j, R_j, R'_j at a point in time gives the characteristic parameters N or E_0 by

$$N = \frac{ei}{\theta(g'/g) + 1} \quad \text{or} \quad N = \frac{ei}{1 - \theta(R'/R)}$$

for Mayr model or

$$E_0^2 = \frac{e^2}{\theta(g'/g) + 1} \quad \text{or} \quad E_o^2 = \frac{e^2}{1 - \theta(R'/R)}$$

for the Cassie model ($X' \equiv dX/dt$). The characteristic parameters can be investigated by assuming their dependence on the instantaneous value of current or resistance.

(B) Assumption of Points with Identical Parameters

When two different times with identical characteristic parameters can be defined, the parameters can be obtained as follows.

$$N = \frac{g'_1 i_2^2 - g'_2 i_1^2}{g(g'_1 - g'_2)} \qquad \theta = \frac{i_1^2/N - g}{g'_1}$$

where

$$g'_1 = \frac{dg}{dt}\Big|_{t=t_1} \qquad g'_2 = \frac{dg}{dt}\Big|_{t=t_2}$$

$$i_1 = i(t_1) \qquad i_2 = i(t_2)$$

This method should still be classified into two subclasses.

1. *Assumption that the parameters are* $N(|i)$ *and* $\theta(|i)$. One record of unsuccessful interruption can be used to calculate the parameters. Two points with a given absolute value of arc current before and after the current zero are assumed to have the same characteristic values (Hochrainer et al., 1972; Stokes et al., 1984).
2. *Assumption that the parameters are* $N(i)$ *and* $\theta(i)$. More than two records that, for good reasons, are believed to have the same

parameters are used. Two points with the same current according to multiple records are picked for calculation of the characteristic parameters (Haupt, 1987).

A problem common to these models is the question of the adequacy of the assumption that arcs with the same current have the same arc parameters. This question is more exaggerated when one thinks about the fact that the current is externally decided for the arc. Another possible hypothesis is that the same characteristic parameter appears when the same arc resistance is realized. This idea leads to a method to obtain $\theta(g)$ and $N(g)$ instead of $\theta(i)$ and $N(i)$ (Rijento, 1974; Pietch et al., 1975; Amsinck, 1977; Zueckler, 1978; Schmidt, 1982). Experimentally obtained time constants and power losses of air arcs are plotted as a function of g (conductance) in Fig. 2.33.

(C) Introduction of Characteristic Conductance

Consideration of the energy balance can also lead to the expression $(dg/dt)/g = [G(i) - g]/\theta$. The two parameters θ and G (steady-state conductance)

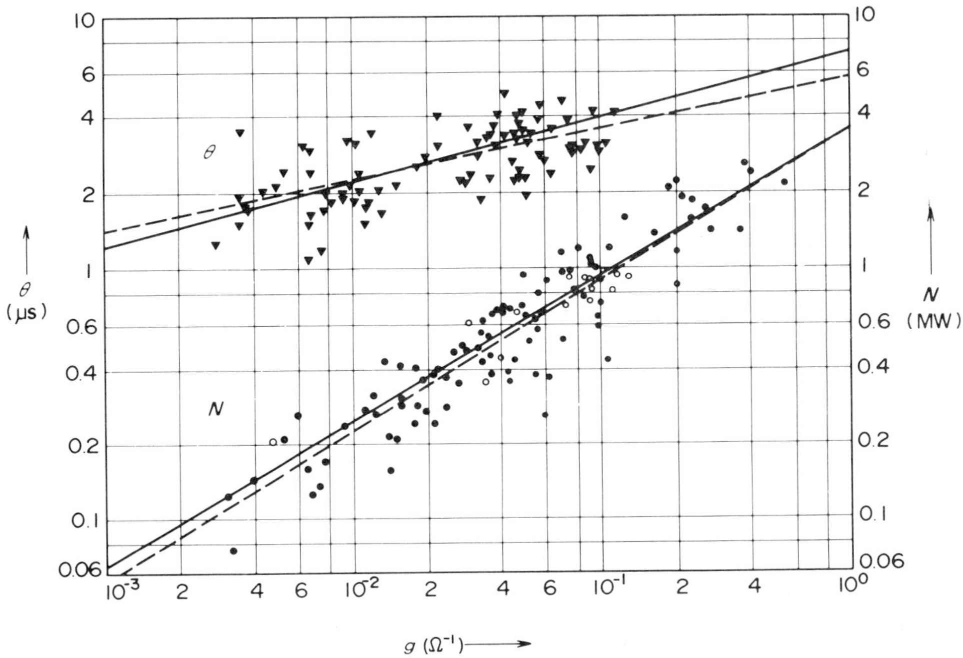

Fig. 2.33 Experimentally evaluated power losses (represented by N) and time constants (represented by θ_T) at various values of g (conductance). (Pietsch et al., 1975. Copyright ETZ.)

instead of θ and N or θ and E can be equally convenient for practical use. The dependence of θ on either i or g classifies the model in the same way as before. The function $G(i)$ and $\theta(i)$ were used in several works (Hochrainer et al., 1972; Stokes et al., 1984), and the expressions $G(i)$ and $\theta(g)$ were used by Matsumura et al. (1985). Two points in time with the same current level out of one or more records of current and voltage provide two sets of data g_1, g_1' and g_2, g_2' for the same value of i, and these data can give θ and G, respectively, by the following equations (Hochrainer et al., 1972):

$$\theta = \frac{g_2 - g_1}{g_2' - g_1'} \qquad G = \frac{g_2 g_1' - g_1 g_2'}{g_1' g_2'}$$

G(i) and $\theta(g)$ can be deduced by an iteration calculation starting from an approximate value of $G(i)$ as an initial value. Iterative calculation of θ and G using the measured data converges into a set of characteristic values (Matsumura et al., 1985).

(D) Use of a Complex Model

The arc parameters obtained in this kind of technique reproduce arc behavior adequately. However, consideration of the physical status of the arc raises a question about the adequacy of the introduced assumption of constant parameters at the same current level or at the same arc resistance. Arcs with a given arc conductance may be realized by different temperature distributions along the diameter. The temperature distribution in the arc near the current zero, in particular, is different from the steady-state distribution and can be dependent on the rate at which current changes and the impedance of the circuit. These considerations suggest the desirability of using a more complex model with more than two parameters (Browne, 1984). However, the number of restrictions in evaluating these parameters from the measured data pose a practical problem. Accuracy of measurement and digitalization noise in the process of evaluation are problems to be overcome because the data require a wide dynamic range and fine resolution during the interruption period. An example of the introduction of a complex model can be found in the application of two Mayr arcs connected in series (Ibuki et al., 1988).

The $(dR/dt)/R$ values obtained from actual measurement, when they are plotted against ei values, can be fitted better by a curve than by a straight line. This was a common observation from the beginning of the experiment (Butler and Whittaker 1972; Karttunen et al., 1972). This curving is also quite clear in the SF_6 gas interruption arc, and the proposed mathematical arc models do not reproduce the behavior for the entire period. However, the 10-μsec period that is of particular importance for the interruption can be represented reasonably well by two Mayr arcs connected in series. The necessary two sets of Mayr parameters were obtained by curve fitting to the curve obtained by plotting $(dR/dt)/R$ with measured values. These parameters can be used to reproduce the interaction of the arc with the circuit. Figure 2.34 shows the behavior of $(dR/dt)/R$ for different arcing times of a model circuit breaker. The two sets of parameters (N_1,θ_1) and $N_2,\theta_2)$ obtained by curve fitting at these arcing times are shown in Fig. 2.35, and the estimated interruption limits are shown together with the experimentally obtained interruption limits.

A similar idea using a Mayr arc and a Cassie arc as series components has also been presented (Habedank, 1988). The characteristic values in this case were obtained by an iterative fitting method.

(E) Evaluation by a Diagrammatic Method

A more straightforward expression of a breaker's interruption capacity is also desirable because of its practical applicability on various occasions. The limitation curve to show the limit of dv/dt versus a given value of di/dt is widely used. This *limit curve* can be obtained both by experimental measurement and by calculation with some sort of arc model. This expression is useful in studying the possibility of application of a breaker to a circuit condition with a different rate of rise or other special interruption condition. The process of estimating this curve for a given circuit breaker

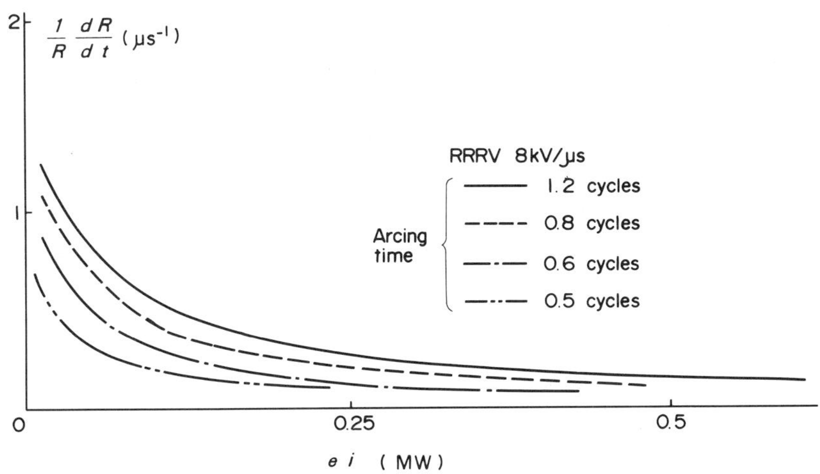

Fig. 2.34 $(dR/dt)/R$ as functions of *ei* (electrical input into arc) at various arcing times of a model circuit breaker. (Ibuki et al., 1988. Copyright IEE.)

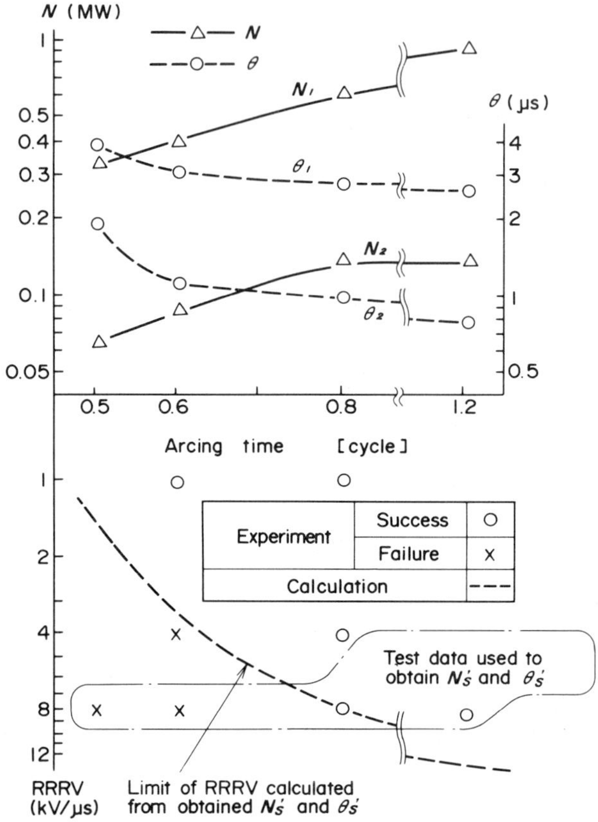

Fig. 2.35 Arc parameters deduced from experiment with various arcing times and the limit of RRRV calculated from the values obtained for the arc parameters. (Ibuki et al., 1988. Copyright IEE.)

requires many interruption tests with different values of *di/dt* and/or *dv/dt* and requires a great deal of work. Figure 2.36 shows the limit curves for various model circuit breakers summarized by Frind (1978).

2.5 CIRCUIT-BREAKER DESIGN (T. Yamauchi)

2.5.1 Electrical Aspects

The most important electrical function of a circuit breaker is to interrupt a fault current and to insulate the electric circuit from the ground potential. Another, equally important, function is to carry the load current when the

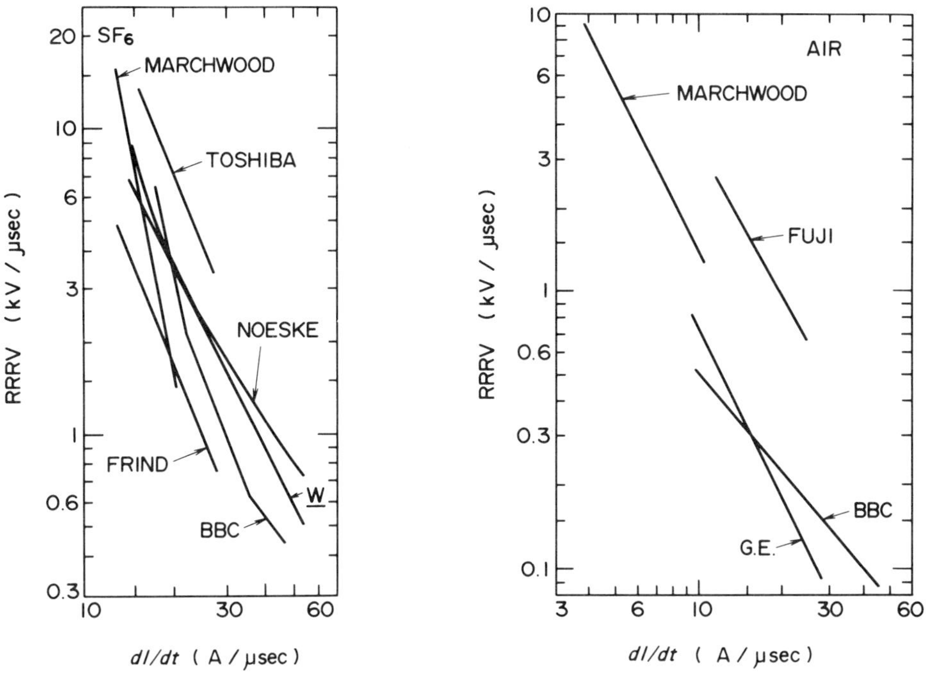

Fig. 2.36 Summary of measured and calculated critical rates of rise dependent on *di/dt* for SF_6 and air. (Frind, 1978. Copyright Plenum Press.)

breaker is closed. The insulation requirement can be satisfied by various arrangements, which can be classified in terms of two major concepts.

The first concept is the use of air as an insulating medium. This is realized by providing a support insulator to hold the interrupter portion at an appropriate level above the ground. The outside insulation of the support structure is limited by the dielectric characteristics of air, and the length of the insulator cannot be less than a specified value. This type of structure is usually called a *live-tank type*. This structure can usually make the breaker lighter than the *dead-tank type* (see below) and is more inexpensive. An example of a circuit breaker employing this concept is shown in Chapter 3 (Fig. 3.7b).

The other idea is to use a special insulating medium for insulation between the ground and the live part. The interrupter portion can be located in an enclosed space at ground potential that is filled with some kind of fluid insulator; in most cases the interruption medium functions as an insulation medium also. This type of structure is called a *dead-tank type;* an example is shown in Chapter 3 in Fig. 3.7a. The connection to the main circuit is made by a bushing on the grounded enclosure, and the insulation outside the bushing is dependent on the capability of the air like the insulation of the live-tank-type design. However, the physical arrangement with the interrupter inside the grounded enclosure located at ground level makes the center of gravity low, and this structure can more efficiently accommodate the current transformers that are needed in applying a breaker to a real system. These differences in the features of the two arrangements decide their fields of application.

The interruption performance and insulation capability between the terminals in the open position of the breaker are important electrical aspects. The breaker must withstand the voltage that appears between the terminals after the interruption of the current. The voltage that can be sustained by a unit interrupter is limited by the structure and dimensions of the unit. Breakers used for higher voltages must withstand high voltages, and when one unit cannot withstand the voltage, two or more units must be connected in series. In most cases, the same interrupter units are used to design the total withstand voltage to meet the system requirement.

An example of a puffer-type interrupter used for a live-tank-type breaker is shown by a schematic drawing in Fig. 2.37. The porcelain that

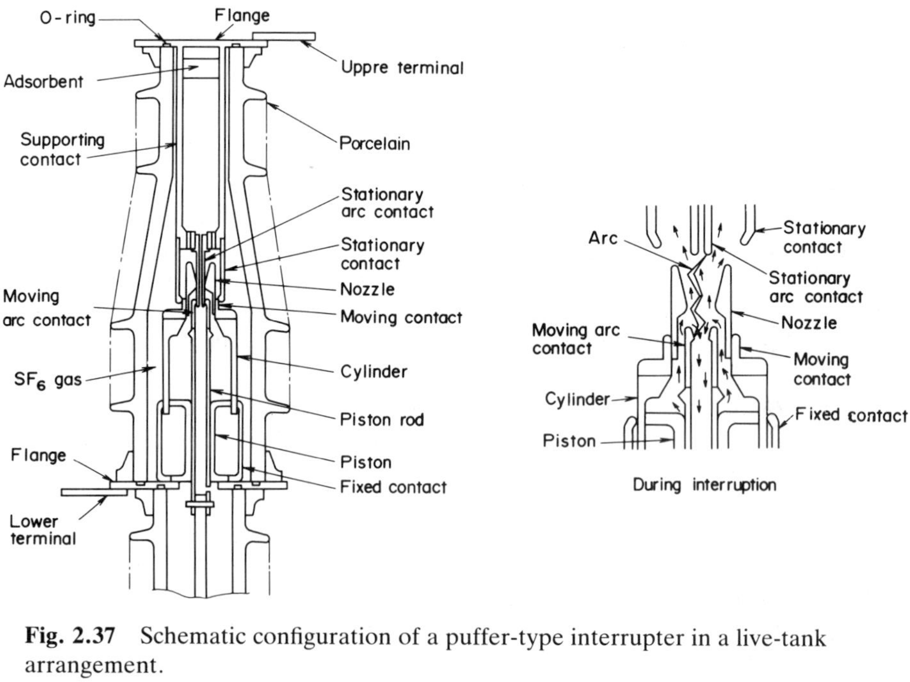

Fig. 2.37 Schematic configuration of a puffer-type interrupter in a live-tank arrangement.

forms an enclosure is filled with SF_6 gas of 0.5–0.6 MPa. Both the main contacts and arcing contacts are engaged, but the major part of the load current goes through the main contact because of the current distribution in the coaxially arranged conductors. When the breaker is driven toward the open position, the main contacts separate first and the current is transferred to the arcing contacts. The separation of the arcing contacts leads to initiation of an electric arc between the arcing contacts. The puffer cylinder, which is formed as one physical unit with moving arcing contact, is driven at the same time and compresses the space between the cylinder and the piston. This compressed gas is led to the arc burning between the arcing contacts. The arc is cooled by the gas flow and is interrupted when the current in the arc reaches zero. The recovery voltage appears between the contacts, and the dielectric strength of the insulation medium filling the gap sustains this voltage.

The history of the increase in the unit voltage of the interrupter was discussed in Chapter 1. The present status of the series number of the interrupter at various voltages is summarized in Fig. 2.38. At voltages below 300 kV, interruption by one break unit is common at present. At voltages of 362–525 kV, interrupters are usually constructed with two break units, and at higher voltages, three or four breaks are used. The necessary number of interrupter units is also dependent on the interruption current.

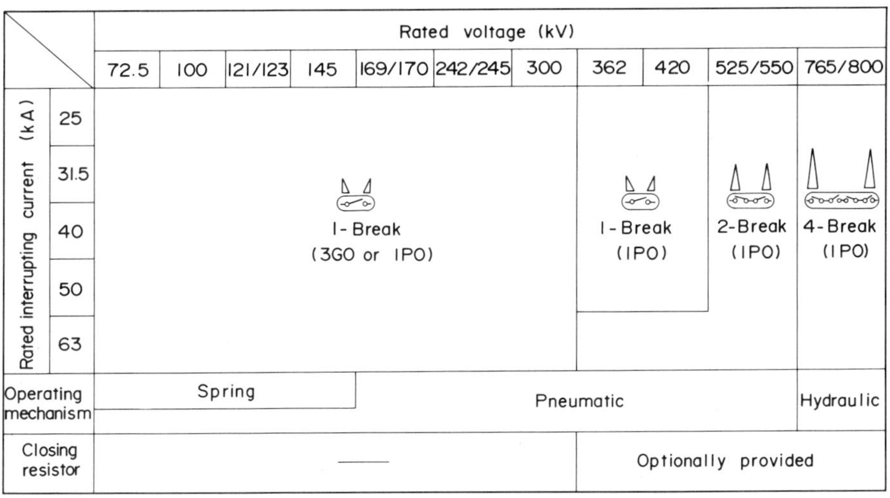

(Note) 3GO: 3-Pole gang operation, IPO: Individual pole operation

Fig. 2.38 An example of the conceptual construction dependent on the ratings in voltage and the interruption current in a dead-tank arrangement.

Another important requirement for a breaker is found in the duty of closing of the system. When an open breaker is closed, it causes some transient phenomena in the power system. Closing of a no-load transmission line is an important case that must be studied with special care. The over voltage caused by the reflection at the open end of the line increases the voltage of the closing surge. Usually, at voltages above 362 kV, some measures must be taken to suppress the closing surge.

2.5.2 Mechanical Aspects

The mechanism that drives the moving part to close and open the breaker is of particular importance as long as the closing/opening is achieved through mechanical motion. The peculiar requirement for circuit-breaker operation poses particular requirements for the operating mechanism. The special requirement is that it will not operate under normal conditions but must operate at very short notice; when the requirement is raised, it must move as quickly as possible. When it is required to operate in another direction, the breaker has to respond as required. The requirements for the circuit-breaker operating mechanism can be summarized as follows:

Close the breaker and keep it in the closed position.
Open the breaker and keep it in the open position.

These commands are spontaneously given to the breaker at irregular times and without any warning.

Another important mechanical aspect is the seismic requirement in the installed condition. Seismic specifications take into account the likely strength of an earthquake and the probable seismic frequency components depending on geographic location.

The necessary characteristics of the operating mechanism and examples of seismic design are explained in the following subsections.

(A) Closing Operation

On receipt of the closing signal, circuit breakers must close fully and remain in the closed position. The most critical requirement for a circuit breaker is a case where the closing causes the maximum fault current to flow. The voltage applied across the terminal causes some prestriking prior to actual metal-to-metal contact. The arc causes wearing of the contact, the current flowing through the breaker imposes mechanical stress, and the electromagnetic force may cause mechanical resistance to the closing torque. The driving force must be designed to overcome these counterforces. The wear caused by the prestrike must be kept small enough to maintain good performance after the closing operation.

(B) Tripping Operation

A circuit breaker must open fully from the closed position on receipt of the tripping signal. It must interrupt a current of specified value by opening. Contact separation causes initiation of an arc between the arcing contacts, and this arc must be extinguished by the motion that follows the contact separation. The current conditions imposed on circuit breakers are of a wide variety, as explained in Chapter 1. The current value can vary from a few amperes to a few tens of kiloamperes, and the recovery voltage can also vary depending on the system condition. Required opening speed depends on the design of the interrupters. Puffer-type SF_6 circuit breakers, which are the most popular in modern technology, compress the interrupting medium by the opening motion.

The mechanical stress caused by acceleration and deceleration of the moving contact is significant, because the opening operation must be completed in a relatively short time (typically 30 msec). A special mechanical stress may be imposed by the immediate tripping operation following the closing of a fault current as mentioned in the preceding subsection. A variety of test conditions of interruption tests are disclosed in Chapter 4, and a circuit breaker must withstand all the stresses experienced during these operations.

(C) High-Speed Reclosing Operation

Closing immediately following the opening operation is also required for circuit breakers to minimize the time during which there is voltage in the power system. When there is no continuing short circuit in the system, clearing of the fault current caused by a flashover through the air gap eliminates the fault, and the closing of the breaker can store normal operation to the power system. When the reclosing time is set less than 1 sec, it is called high-speed reclosing; this adds a special mechanical requirement for breakers.

The general mechanical requirement for the breaker to satisfy this operation is quick resetting of the latch and trigger. Another requirement is the storage of energy, both the energy necessary for the closing operation and the energy that may be required for the tripping operation that might follow the closing. This is achieved either by storing sufficient energy in the form of pressurized fluid or when the energy is stored by springs, in most cases the tripping spring is charged during the closing operation. Therefore, a charged closing spring and a holding latch prepared to hold the breaker in the required position are the necessary conditions for initiation of the closing operation. Most modern circuit breakers are designed to fulfill this requirement.

(D) Seismic Requirements

The requirement about the seismic withstand characteristics is dependent on the geographic location of the installation. The IEEE standard for gas-insulated substations (C37.122) recommends the use of a macrozone map to specify the equipment (IEEE Power Engineering Society, 1981). This map divides the United States into four zones (zones 0–3) of relative seismic activity. In the most severe zone (zone 3), a peak ground motion of 0.5 *G* is specified.

Numerical analysis using a model with a group of concentrated masses connected with a given mechanical rigidity is commonly used to estimate the seismic characteristics of equipment. The resonant frequency and possible mechanical stress are evaluated by assuming the input mechanical vibration into the foundation. An example of the analysis of the vibration mode in a gas circuit breaker is shown in Fig. 2.39. The estimated displacement of various parts and the mechanical stress are shown in Fig. 2.40.

2.5.3 Examples of Operating Mechanisms

The requirements for the operating mechanisms of actual circuit breakers can be fulfilled by various means. They can be classified according to energy source. *Spring-operated mechanisms, compressed air mechanisms,* and *hydraulic mechanisms* are the popular types of operating mechanisms.

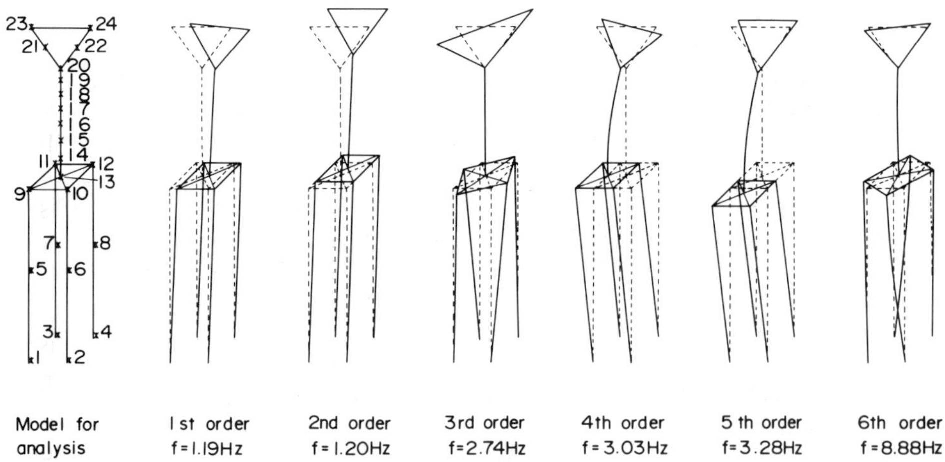

Fig. 2.39 Computer analysis of the response of a live-tank type of gas circuit breaker to a seismic vibration.

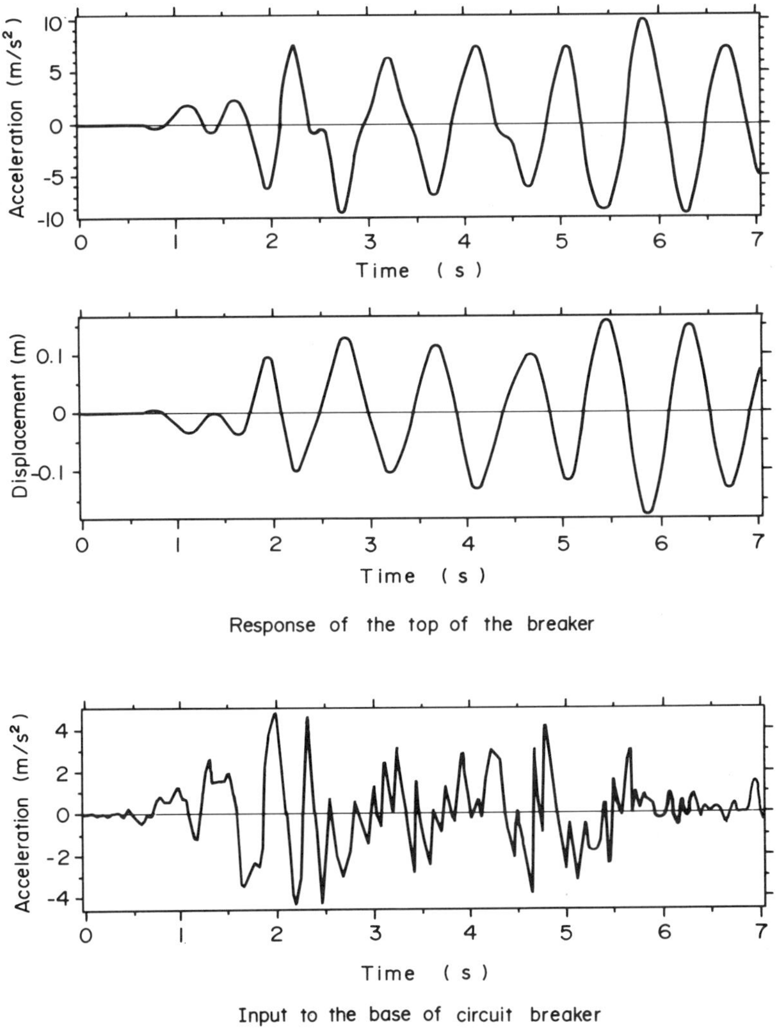

Fig. 2.40 Time history of the displacements and stresses at the top of the breaker shown in Fig. 2.39.

(A) Spring-Operated Mechanisms

A spring-operated mechanism is one driven by the mechanical energy stored in springs. In the most common arrangement the closing spring is mechanically charged by a motor and is held in the charged position by a latch. When the closing signal releases this latch, the spring drives the breaker to close, and in the same motion it charges the trip spring. Another latch holds the breaker in the closed position, and the opening signal releases this latch to open the circuit breaker by using the energy stored in the tripping spring. The closing spring is usually recharged immediately after the closing operation.

Figure 2.41 shows an example of a spring mechanism. The closing spring is charged by rotating the *ratchet wheel* through a pair of ratchets driven by a motor. When the ratchet wheel is driven beyond the dead point, the spring gives a torque in the clockwise direction, and this torque is

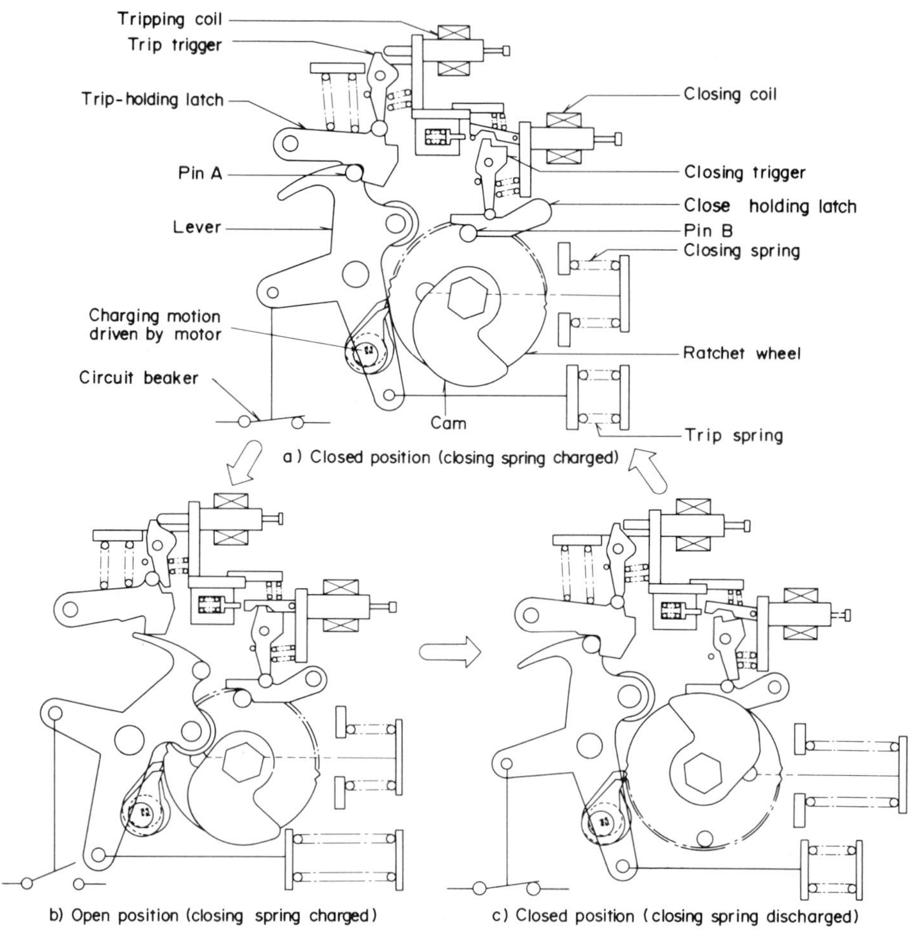

Fig. 2.41 Schematic drawings of a spring-operated mechanism.

sustained by the engagement of *pin B* with the *closing-holding latch.* When this engagement is released, the ratchet wheel is freed and rotates clockwise together with the *closing cam.* The rotation of the closing cam pushes the *lever* counterclockwise. This motion compresses the *trip spring,* and at the end of the tripping motion the lever is held by the engagement of *pin A* and the *trip-holding latch.*

When the energy of the closing spring is fully discharged by the closing operation, the charging motor is switched on, and the charging motion begins without disturbing the position of the output lever. Therefore, starting from the closed position with the charged closing spring, the breaker can complete O-C-O operation (open-close-open) without being recharged by the motor.

This type of mechanism has been used for some extra high-voltage circuit breakers in Europe. However, the energy density stored in a spring is not very good for a large-scale model. This type of mechanism is usually used for relatively low voltage circuit breakers. Improvements in the SF_6 interrupter reduced the amount of energy required for the operation and stimulated the application of this type of mechanism to higher voltages. Application of this mechanism to a higher voltage has been found in Europe, but in Japan, application at voltages of 169/170 kV has become popular.

(B) Pneumatic Operating Mechanism

Pneumatic operation uses compressed air as the energy source for closing and tripping. This has been a widely used mechanism for all voltage classes throughout the history of circuit breakers, because compressed air is a convenient form of stored energy. Output energy can be adjusted to various requirements by changing the size of the driving cylinder, and the application of a spring for one of the operations (close or open) is also quite common.

Figure 2.42 shows an example of a pneumatic operating mechanism. Compressed air is used for opening in this mechanism, and the spring closes the breaker. The *holding latch* holds the breaker in the open position, and at the same time the *cylindrical valve,* which supplies the air to the *driving cylinder,* is held in the shut position by the engagement of the *lever* and the *latch.* A closing signal releases the holding latch, which then releases *pin A;* the *closing spring* is freed, closes the circuit breaker, and then sustains the breaker in the closed position. The opening signal frees the *trigger,* and the cylindrical valve is freed to supply air to the driving cylinder. The compressed air supplied through the *control valve* drives the *piston* to the open position. At the end of the opening stroke, the control valve is mechanically closed by the *cam,* which makes the lever turn in the counterclockwise direction. The number of consecutive operations for this mechanism is

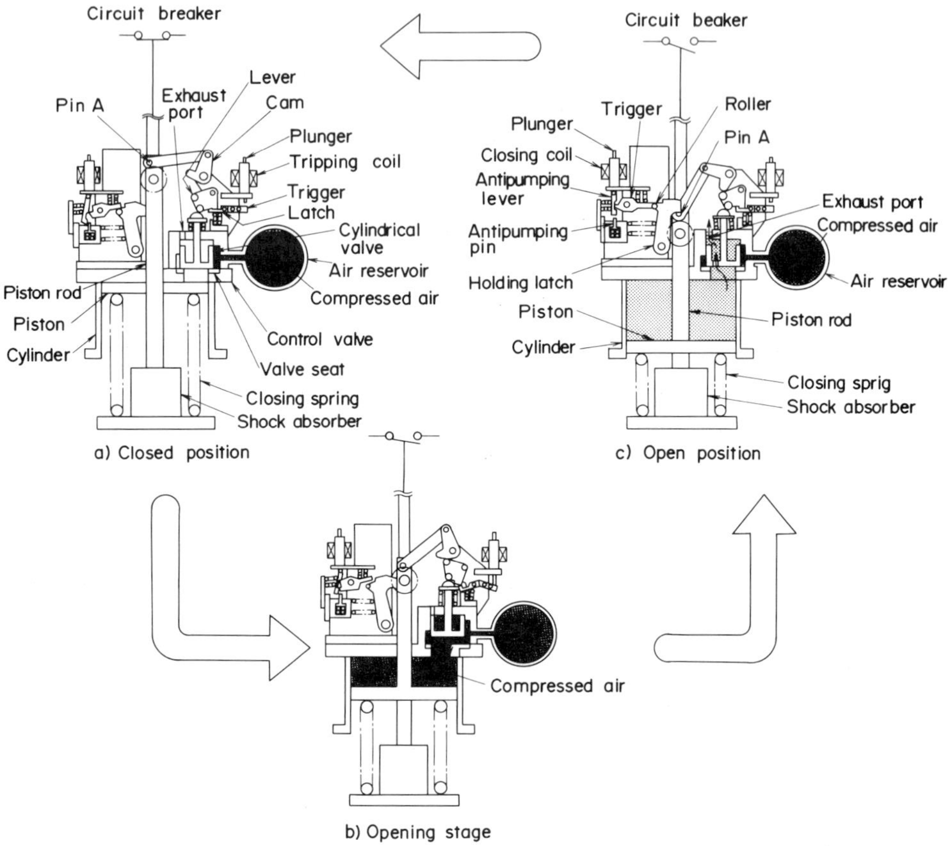

Fig. 2.42 Schematic drawings of a pneumatic operating mechanism.

determined by the size of the *air reservoir.* This is one of the most widely used mechanisms, and many types with numerous variations can be found. These mechanisms have found application in all voltage ranges.

(C) Oil Hydraulic Operating Mechanism

Oil hydraulic operating mechanisms use the energy of pressurized oil in the accumulator; it is common to use the oil energy for both closing and opening operations. The pressure of the oil is of the order of a few tens of megapascals (a few hundred bar), which is much higher than the air pressure of a pneumatic mechanism. Because of this high pressure, the size of the driving cylinder, and consequently the size of the mechanism as a whole, can be less than that of other mechanisms, so that higher driving energy is achieved with a more compact mechanism. This type of operating mechanism is suitable for use in high-voltage breakers.

Figure 2.43 shows an example of a hydraulic operating mechanism. Both the opening and closing energy are supplied by high-pressure oil. The spaces on both sides of the driving piston can be filled with pressurized oil, and the space on the right-hand side of the piston is permanently connected to the accumulator oil. The piston is driven toward the right when both

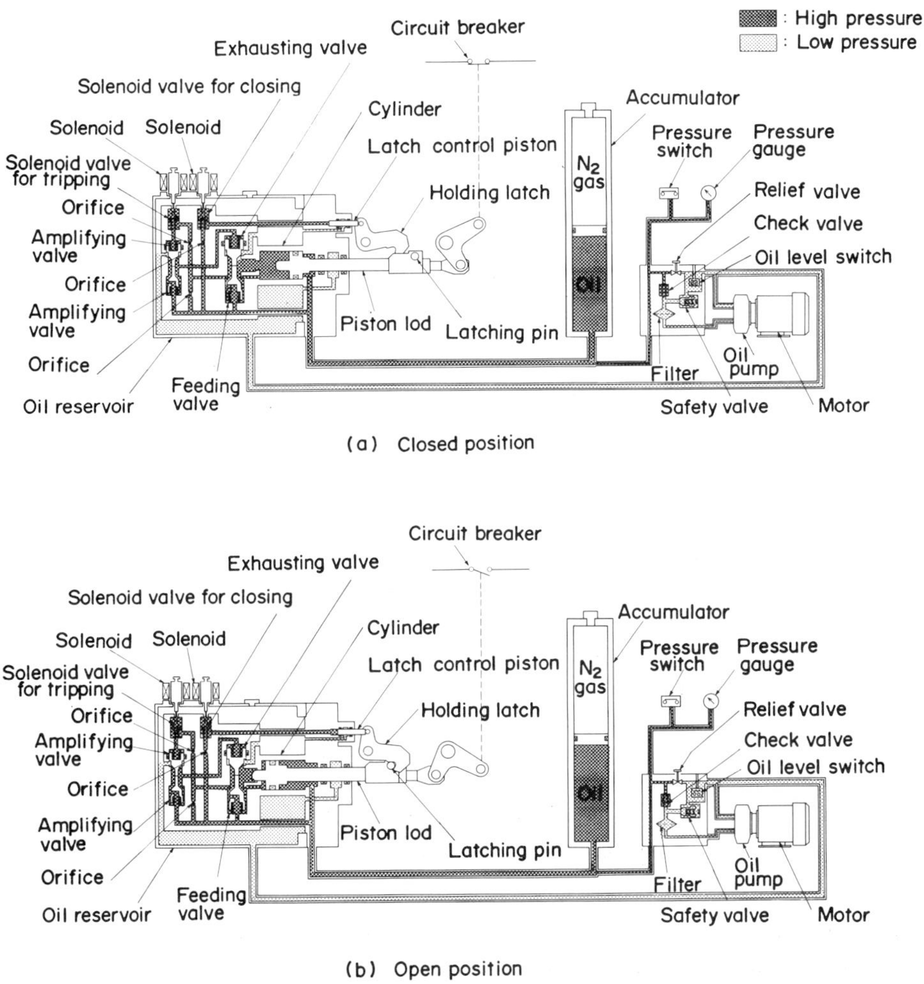

Fig. 2.43 Schematic diagram of an oil hydraulic operating mechanism.

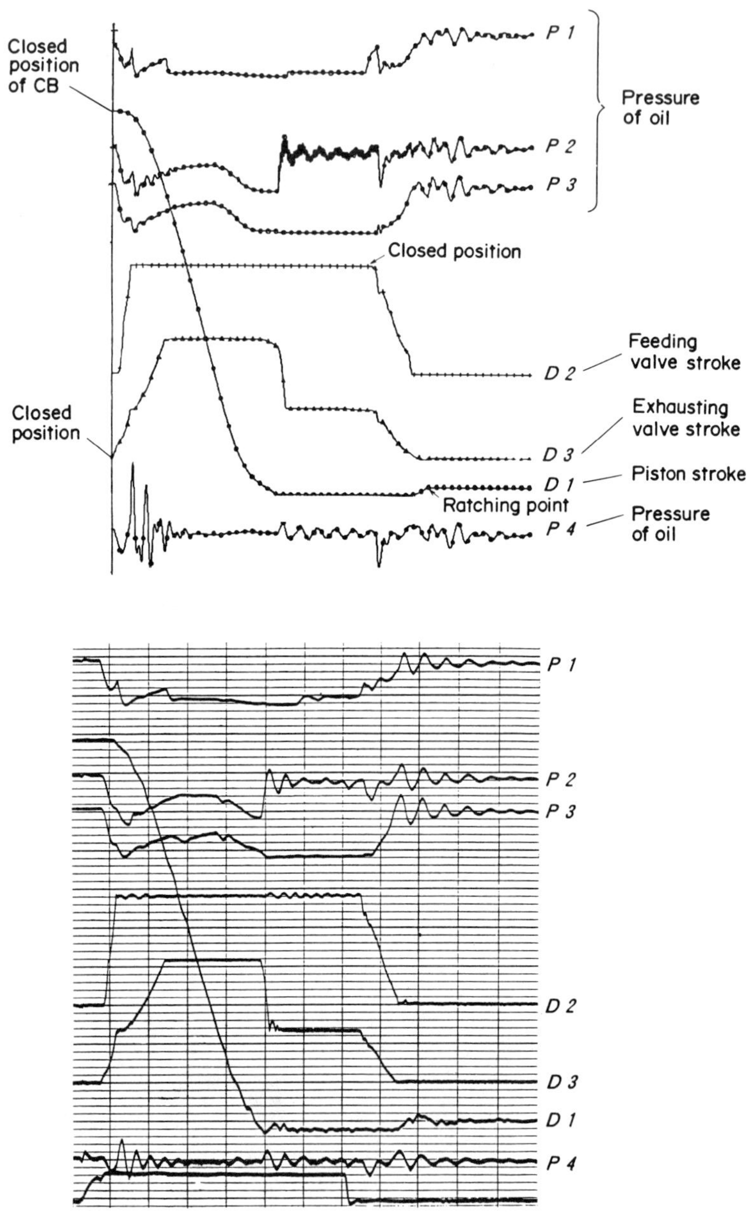

Fig. 2.44 Results of simulated operation of a hydraulic mechanism (top) and operation of the actual mechanism. (Yoshizumi et al., 1988. Copyright Mitsubishi Electric.)

sides of the piston are pressurized, because the pressurized surface area is greater on the left-hand side than on the right-hand side. The piston is driven rightward or leftward by filling or releasing the high-pressure oil in the left-hand space. In the opening operation, the fully open position is detected by a pressure-detecting port that is pressurized in the full open position of the piston. The detection of the full stroke helps to reset the control valves and pushes the *holding latch* through an *auxiliary piston*. The closing operation is initiated by releasing the pressure in the auxiliary piston, which then releases the holding latch. The mechanism is held in the closed position by the oil pressure. The high pressure in the oil could cause some leakage from the joint of the hydraulic circuit components. The high-pressure components are placed in the low-pressure oil tank to minimize the effect of leakage in practical operation.

Components of the hydraulic mechanism need to be made very accurately to achieve consistent operation. However, they are rust free because they are insulated from oxygen in air. High reliability can be achieved when one can utilize this characteristic, and the high available energy widens the application range for circuit breakers. There are various features, and this type of operating mechanism has a long history of application and as many possibilities of detailed variation as the pneumatic mechanism.

Computer analysis plays an important role in the design of operating mechanisms as well as in the design of interrupters. The sizes and the strokes of the valves, for example, affect the resultant movement of the output piston in oil hydraulic mechanisms. Optimization of these parameters is one of the important steps in finalizing a design with a good balance to obtain a fast response. Figure 2.44 shows results of simulated operation of a hydraulic mechanism considering the motion of the valves, along with results of the actual operation of the mechanism.

REFERENCES

Amsinck, R.(1977). Werfahren zur Ermittlung der das Ausschaltverhalten bestimmenden Lichtbogenkenngroessen, *ETZ-A, 98*(8): 566–568.

Brand, K.P., Egli, W., Niemeyer, K., Ragaller, K., and Schade, E. (1982). Dielectric Recovery of Axially Blown SF_6-Arc After Current Zero: Part III. Comparison of Experiment and Theory, *IEEE Trans. Plasma Sci., PS-10*(3): 152–172.

Browne, T.E., Jr.(1984). *Circuit Interruption,* Marcel Dekker, New York.

Butler, T.F., and Whittaker, D.(1972). Distributed Moving-Boundary Circuit-Breaker Model, *PROC. IEE, 119*(9): 1295–1300.

Cassie, A.M. (1939). CIGRE Report 102.

Cassie, A.M., and Mason, F.O.(1956). Post-Arc Conductivity in Gas Blast Circuit Breakers, CIGRE Report 103.

Cassie, A.M., and Rieder, W.(1964). Ancillary Data for Current Zero Recording, *ETZ-A, 85*(10): 312–316.

Chan, S.K., Cowley, M.D., and Fang, M.T.C.(1976). Integral Method of Analysing Electric Arcs: III. Shape-Factor Correlation for Low Radiation and Laminar Flow, *J. Phys. D.: Appl. Phys., 9:* 1085–1099.

Cowley, M.D.(1974). Integral Methods of Analysing Electric Arcs: I. Formation, *J. Phys. D.: Appl. Phys., 7:* 2218–2231.

Cowley, M.D., and Chan, S.K.(1974). Integral Method of Analysing Electric Arcs: II. Examples, *J. Phys. D.: Appl. Phys., 7:* 2232–2245.

Elenbaas, W.(1935). Aehnlichkeitsgesetze der Hochdruckentladung, *Physica, 2:* 169.

Endo, F., Sato, M., Tsukushi, M., Yoshioka, Y., Saito, K., and Hirasawa, K.(1989). Analytical Prediction of Transient Breakdown Characteristics of SF_6 Gas Circuit Breakers, *IEEE Trans. Power Delivery, 4*(3): 1731–1737.

Frind, G.(1978). Experimental Investigation of Limiting Curves for Current Interruption of Gas Blast Breakers, in *Current Interruption in High-Voltage Networks,* K. Ragaller, Ed., Plenum, New York, pp. 67–94.

Frost, L.S., and Liebermann,R.W.(1971). Composition and Transport Properties of SF_6 and Their Use in a Simplified Enthalpy Flow Arc Model, *Proc. IEEE, 59*(4): 474–485.

Habedank, U.(1988). On the Mathematical Description of Arc Behaviour in the Vicinity of Current Zero, *ETZ Arch., 10*(11): 339–343.

Haupt, M.(1987). New Approach for Evaluating Current and Conductance Dependent Arc-Parameters, *ETZ Arch., 9*(6): 195–197.

Haupt, M., and Moeller, K.(1987). Contribution on the Application of Two Pole Arc Models on SF_6-Blast Circuit Breakers, *ETZ Arch. 9*(8): 255–260.

Heller, G.(1935). Dynamical Similarity Laws of the Mercury High Pressure Discharge, *Physica, 6:*385.

Hermann, W., and Ragaller, K.(1977). Theoretical Description of the Current Interruption in HV Gas Breakers, *IEEE Trans. Power Apparatus Syst., PAS-96*(5): 1546–1551.

Hochrainer, A., Gruetz, A., Schwarz, J., and Thiel, H. G. (1972). Study of Arcs in Breakers with the Help of a Cybernetic Model and Under the Influence of Turbulence, CIGRE Paper 13-10.

Ibuki, K., Yoshizumi,T., Yonezawa, T., Sakuma, S., and Kuwahara, H. (1988). An Arc Model Used in Analysis of Interruption Test of GCBs, Proc. 9th International Conference on Gas Discharges and Their Applications.

IEEE Power Engineering Society (1981). Gas-Insulated Substations (GIS) Standard, Recommended Practices and Guides, Publ. C37.122.

Ishikawa, M., Ikeda, H., Yanabu, S., and Yamamoto, M. (1984). Numerical Study of Delayed Current Interruption Phenomena Using Transient Analysis Model for an Arc in SF_6 Flow, *IEEE Trans. Power Apparatus Syst., Pas-103*(12): 3561–3566.

Karttunen, M., Ylostalo, T., and Jussila, J.(1972). Current-Zero Measurements in Circuit-Breaker Development, CIGRE Paper 13-05.

Kovitya, P.(1984). Thermodynamic and Transport Properties of Ablated Vapors of PTEE, Alumina, Perspex and PVC in the Temperature Range 5000–30000 K, *IEEE Trans. Plasma Sci., PS-12:* 38–42.

Kuwahara, H.(1983). Modern Technology for SF_6 Gas Insulated Switchgear Development, CESI Symposium 83.

Kuwahara, H., Tanabe, T., Ibuki, K., Sakai, M., Sato, A., and Sakuma, S.(1983). New Approach to Analysis of Arc Interruption Capability by Simulation Employed in the Development of SF_6 GCB Series with High Capacity Interrupter, *IEEE Trans. Power Apparatus Syst., PAS-102*(7): 2262–2268.

Kuwahara, H., Tanabe, T., Yoshinaga, K., Sakuma, S., Ibuki, K., and Yamada, K. (1984). Investigation of Dielectric Recovery Characteristics of Hot SF_6 Gas After Current Interruption for Developing New 300 kV and 550 kV Gas Circuit Breakers, *IEEE Trans. Power Apparatus Syst., PAS-103*(6): 1371–1376.

Matsumura, T., Oppenlander, W.T., Schmidt-Harns, C.A., and Stokes, A.D. (1985). A Refined Cybernetic Model for Circuit-Breaker Arcs, Proc. 8th International Conference on Gas Discharges and Their Applications, pp. 69–72.

Mayr, O. (1943). Beitraege zur Theorie des statischen und des dynamischen Lichtbogens, *Arch. Electrotech., 37*(12): 588–608.

Pietsch, G., Rijanto, H., and Thiel, G. (1975). Schaltlichtbogen im electrischen Netz, *ETZ-A, 96:* 222–226.

Ragaller, K., Egli, W., and Brand, K.P.(1982). Dielectric Recovery of an Axially Blown SF_6-Arc After Current Zero: Part II. Theoretical Investigations, *IEEE Trans. Plasma Sci., PS-10*(3): 154–162.

Rijanto, H. (1974). Experimentalle Bestimmung der Parameter der Verallgemeinerten Lichtbogengleichung zur Berechung von Schaltvorgaengen, *ETZ-A, 95*(4): 221–223.

Schade, E., and Ragaller, K.(1982). Dielectric Recovery of an Axially Blown SF_6-Arc After Current Zero: Part I. Experimental Investigation, *IEEE Trans. Plasma Sci., PS-10*(3): 141–153.

Schmidt, R.(1982). Erweiterung phaenomenologischer Lichtbogenmodelle fuer einen SF_6-beblasenen Schaltlichtbogen, *ETZ Arch., 4*(11): 341–345.

Standard Handbook for Electrical Engineers, 11th ed., edited by D.G. Fink and H.W. Beaty. McGraw-Hill, New York, 1978.

Stokes, A. D., Manganaro, S., Colombo, E., and Mantovani, E. (1984). Circuit Interruption Prediction Using a Current Dependent Arc Model, CIGRE paper 13-02.

Tanabe, T., Ibuki, K., Sakuma, S., and Yonezawa, T.(1985a). Computer Aided Study of Current Zero Behaviour of SF_6 Gas Arcs, Proc. 8th International Conference on Gas Discharges and Their Applications.

Tanabe, T., Ibuki, K., Sakuma, S., and Yonezawa, T.(1985). Simulation of the SF_6 Arc Behavior by a Cylindrical Arc Model, *IEEE Trans. Power Apparatus Syst., PAS-104*(7): 1903–1909.

Tominaga, S., Kuwahara, H., Tanabe, T., Yoshinaga, K., and Sakuma, S. (1980a). Extremely Short Line Fault Tests of a Puffer-Type Gas Circuit Breaker by Direct and Synthetic Test Methods, *IEEE Trans. Power Apparatus Syst., PAS-100*(2): 563–571.

Tominaga, S., Kuwahara, H., Yoshinaga, K., and Sakuma, S. (1980b). Investigation on Gas Flow of Puffer-Type Gas Circuit Breaker Based on Observation of Arc and Pressure Measurement, *IEEE Trans. Power Apparatus Syst., PAS-99*(6): 2040–2049.

Urbanek, J.(1972). Zur Berechnung des Schaltverhaltens von Leistungsschaltern, ein erweiterte Mayr-Gleichung, *ETZ-A, 93*(7): 381–385.

Yoshizumi, T., Sugiyama, T., Yonezawa, T., Hosomi, M., and Ibuki, K. (1988). Reliability-Improvement Technologies Applied to a New Series of Three-Phase-Enclosure SF_6-Gas Circuit Breakers, *Mitsubishi Denki Giho, 62*(8): 699–704.

Yoshizumi, T., Ibuki, K., Hosomi, M., Yonezawa, T., and Kawasaki, H. (1989). Hot Gas Flow in SF_6 Gas Circuit Breaker During the Short Circuit Interruption, *IEEE Trans. Power Delivery, 4*(3): 1757–1764.

Zueckler, K. (1978). Untersuchungen zum dynamischen Verhalten von Schalterlichtbogen, *ETZ-A, 99:* 546–548.

3
RECENT DEVELOPMENTS IN CIRCUIT BREAKERS

Hiroshi Kuwahara* and Takao Yamauchi Mitsubishi Electric Corporation, Amagasaki, Japan

Yoshio Yoshioka, Yukio Kurosawa, and Kunio Hirasawa Hitachi Ltd., Hitachi, Japan

Satoru Yanabu and Eiji Kaneko Toshiba Corporation, Tokyo, Japan

Sulfur hexafluoride (SF_6) circuit breakers, especially puffer-type breakers, are becoming the main equipment in switchgear for distribution voltage through the ultrahigh voltage (UHV) class. A review of recent trends with some background history will be an aid toward a better understanding of the technology.

3.1 HISTORY OF CIRCUIT BREAKERS (H. Kuwahara)

The first arc-extinguishing medium in the history of electric power switching was air, and experience with mineral insulating oil is almost as old as that with air. The first 40 years of experience of circuit switching of high power was based on oil, which satisfied the requirement of uprating in both current and voltage during this time. An example of the early application of compressed air for blasting the arc can be found in the 1920s (Whitney and Wedmore, 1926). In the 1940s, the investigation on the use of com-

**Current affiliation:* SPC Electronics Corporation, Tokyo, Japan

pressed SF_6 gas as an insulating medium gave successful results (Cooper, 1940). Fundamental research on SF_6 as an interrupting medium also was begun in the 1950s, and in 1959 the first SF_6 breaker for practical use was introduced by Westinghouse in the United States (Friedrich and Yeckley, 1959). SF_6 circuit breakers have extended the field of application since that time supported by various improvements and developments. At present, SF_6 puffer circuit breakers are the most important type of breakers.

The transition in switching equipment can be seen by looking at the statistics on breakers installed since 1955. Figure 3.1 shows 1955–1985 data for Japan, where the transition from oil circuit breakers (OCBs) to air blast breakers (ABBs) and further to gas circuit breakers (GCBs) was thought to be most rapid. The principles of these circuit breakers are reviewed in the following sections.

3.1.1 Oil Circuit Breakers

The history of oil circuit breakers can be traced back to the experience of putting an air switch into mineral insulation oil in 1900 to cope with an increase in voltage and current (Kleman, 1907). Oil generates hydrogen and some carbon compound with low molecular weight and cools down the arc space. The cooling effect can be used to increase the electrical resis-

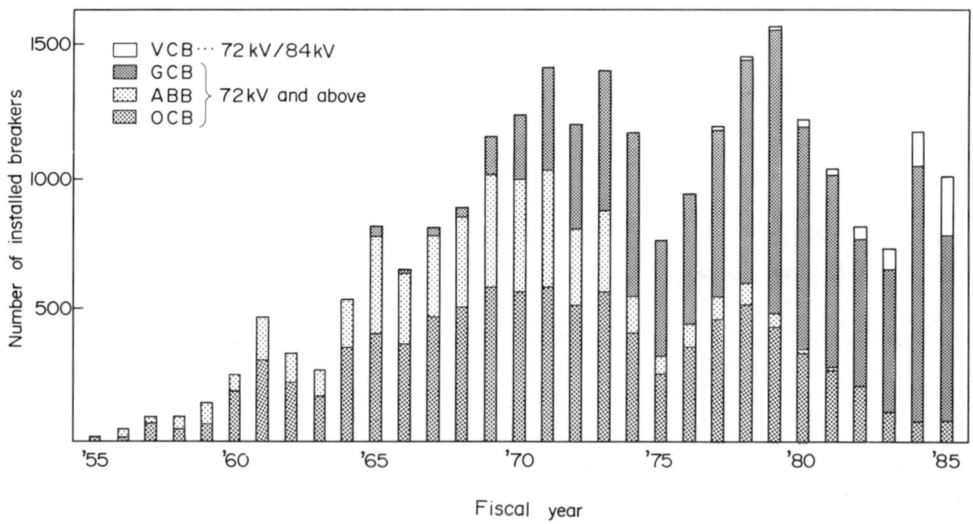

Fig. 3.1 Number of yearly installations of circuit breakers rated at 72 kV and above in Japan from the 1950s. (Denkigakkai, 1989. Copyright IEE of Japan.)

tance of the arc and facilitates interruption of the arc (Biermanns, 1929, 1936). The control to supply fresh oil continuously during the interruption makes the cooling more efficient. This is the principle of using oil as an interruption medium, and the plain break, in which the contacts are simply separated without the provision of any special oil flow, is the most primitive form of oil circuit breaker. Series-connected interrupting contacts in this plain-break arrangement were also commonly used without providing any special means to control the voltage distribution. Although the arrangement was simple, these breakers gave good results, which is understandable because relatively high leakage current after the interruption worked to ease the recovery voltage, and the voltage distribution in the transient was controlled automatically even without applying grading resistors.

In the 1930s, various ideas for increasing the interruption capacity were pursued (Whitney and Wedmore, 1930). A metal explosion pot filled with oil and fitted with an insulation nozzle through which the moving contacts are withdrawn to increase the pressure of the arc's surroundings was proposed by General Electric in the United States (Prince and Skeats, 1931). Application of a deionization grid made of fiber plates (which was effective in pushing the arc into a narrow path) was proposed by Westinghouse (Baker and Wilcox, 1930). Figure 3.2 shows various ideas proposed to increase the effectiveness of the cooling capacity of the oil, and the use of a blasting arc or of pressure buildup by the interrupted arc itself are important ideas that influenced the history of circuit breakers from that time on. In actual circuit breakers, some of these units were connected in series to

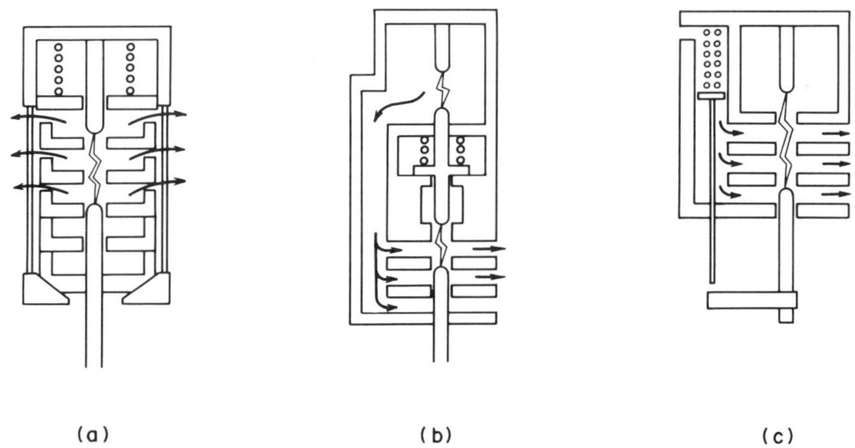

Fig. 3.2 Various arrangements of cooling arcs with mineral insulation oil for interrupting current.

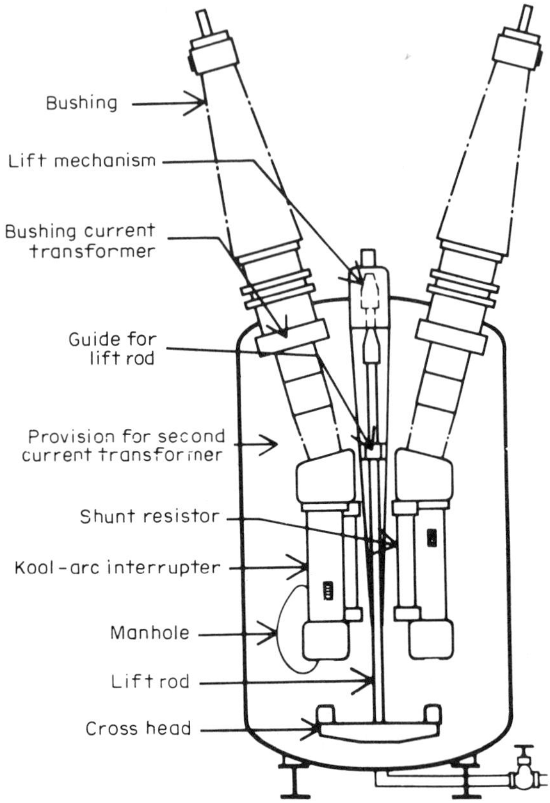

Fig. 3.3 Outline of a typical dead-tank outdoor oil circuit breaker.

provide the required current and voltage capacity. Figure 3.3 shows an example of a dead-tank type of arrangement in which the interrupting units are submerged in oil. The idea of putting the interrupter units in porcelain insulators to reduce the amount of oil needed became popular in Europe. The voltage per interrupter unit, which was restricted by the characteristics of the insulating oil, was around 36 kV; thus a 170-kV breaker can be made with four units in series, and a 300-kV breaker with eight units in series.

Another example of a noncompressible interrupting medium is water (H_2O), and some practical circuit breakers were constructed to use it (Leeds, 1941). The idea of the interrupter unit was based on the same philosophy, and the interruption medium was replaced with an inflammable substitute.

3.1.2 Air-Blast Circuit Breakers

It is obvious that air is a convenient insulating medium, and most switching equipment for distribution voltage is basically insulated by air. However, the use of air as an interruption medium is restricted as long as it is used at a pressure of 1 bar. Greater pressure or another arrangement was required for higher voltages. Application of air to high-voltage switchgear was first proposed at the British Electrical Research Association (Whitney and Wedmore, 1926). The first idea was what is classified as an *axial blast type,* and further investigations carried out in Germany and Switzerland played an important role in this field (Walty, 1935).

An idea for a *deionization grid* for a 15-kV circuit breaker was proposed by a Westinghouse team (Slepian, 1929). Metal plates arranged with a gap between them were located so that an arc can stay between the plates. The arc burning between the plates cannot be sustained when the applied voltage is less than the sum of the cathode drops of the series of arcs, and this characteristic was utilized in interruption. The dielectric strength increases when the air is compressed, and the compressed air can also be used as the driving energy source. Because of these basic characteristics, this idea was also widely used in distribution voltage switchgear.

The interruption principle of air-blast circuit breakers can be classified into two major groups—the *axial blast* arrangement and the *cross-blast* arrangement—as shown in Fig. 3.4. The cross-blast arrangement gives a blast from the vertical direction of the arc, and the arc is expanded into the

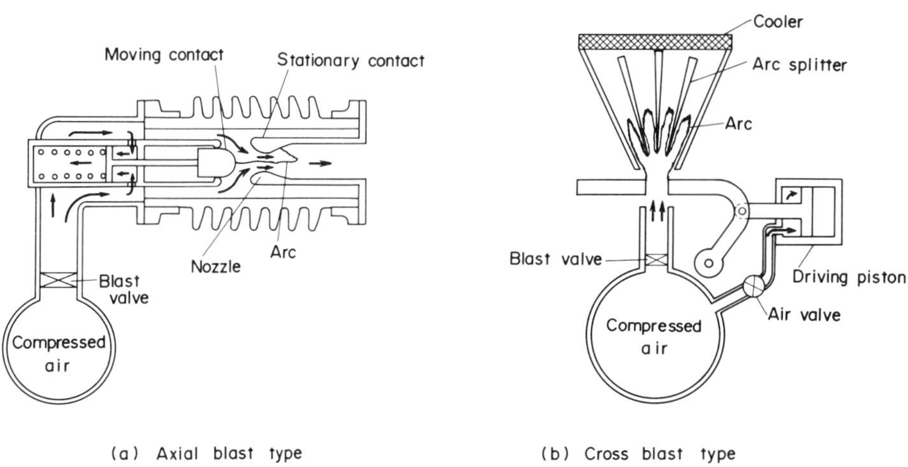

Fig. 3.4 Two principal types of air-blast circuit breakers.

arc chute. This arc chute arrangement places some restriction on how small the breaker can be but the current-carrying contact can be designed rather independently of the interruption arrangement. These characteristics are particularly appreciated at voltages below 36 kV. The axial-blast system can further be classified into the *insulator nozzle* type like the AEG design, and the *metal nozzle* type employed in most other designs. The unit voltage of the insulator nozzle types can be extended to 110 kV, whereas that of the metal nozzle types is around 35 kV. In both types, the interruption capability depends strongly on the geometry of the electrode. The optimum separation of nozzles is on the order of their diameter, and this gap is usually not sufficient for dielectric withstand voltage. These restrictions on the interruption characteristics require complex arrangements. One example is the series connection of interrupting contact and isolating contact that disconnect the circuit after the interruption is completed. Another example is the second opening motion of the interrupting contact itself after it completes the interruption. A parallel resistor to control the transient recovery voltage across the terminal is required in some cases. Many variations in design details and in overall design are found in air-blast circuit breakers.

The critical rate of rise of the recovery voltage decreases as the interrupted current increases. The *short-line fault* condition becomes most crucial at around 60–75% in most cases. The exploding noise caused by releasing the high-pressure air into free space requires various additional arrangements to suppress it.

3.1.3 Vacuum Circuit Breakers

A high dielectric characteristic was realized as early as the 1920s, and the first investigation of vacuum as an interruption medium is found in the report of 40-kV interruption published by Sorensen and Mendenhall (1926). The dielectric breakdown in this high vacuum can be better understood when it is considered as a metal surface phenomenon rather than a phenomenon in the insulating medium. In this sense, the investigation on vacuum circuit breakers differed greatly from that of other breakers. A truly practical vacuum interrupter was realized when developments in metallurgy in the 1960s made the supply of gas-free electrode materials possible. Practical applications began at distribution voltage and expanded to higher voltages. Because of the low arc voltage, contact wear is small and contacts are durable in long-term operation.

No mechanical blast or means to help cooling of the arc can be applied to vacuum circuit interrupters; the only physical process by which the space can be restored to insulating space is diffusion of the arc itself. A *magnetic field* is the only means left to have some interaction with the arc. The geometry to use the magnetic field generated by the interrupted current itself was investigated, and many practical configurations were proposed. Geometries to generate some azimuthal component of the current in the electrode were of particular interest. Slits in the electrode to make the azimuthal current flow vary in detail (an electrode may be spiral or contrate). An arrangement to generate an axial magnetic field by self-current is of great importance in expanding the field of application, especially in high-capacity interrupters.

In the 1970s, 36 kV was accepted as general unit voltage, and in the 1980s a 72-kV unit was developed. The possibility of still higher unit voltage such as 120 kV was also reported (Noda et al., 1987; Shioiri et al., 1983). In a vacuum circuit breaker, both closing and opening operations are accomplished by means of a moving contact, and no blast valve or compression of the medium is needed. Therefore the energy required for operation is less than for other types of breakers, and the spring-operated mechanism is most conveniently used with this interrupter.

There are some problems such as the relatively low unit voltage and the difficulty of making a unit with a high continuous current. However, the peculiar feature, that the interrupter unit is completely sealed offers ease of maintenance, which is appreciated as it reduces total maintenance work.

3.1.4 SF_6 Gas Circuit Breakers

The superior dielectric characteristics of SF_6 were recognized as early as 1920, but investigation into its interruption capability (by Westinghouse in the United States) was not begun until the 1940s. Some patents related to the use of SF_6 gas as an insulating medium appeared in 1940 (Cooper, 1940). The fundamental research in the 1940s was mostly directed to dielectric applications, and no prominent work is recognized before the 1950s in the application of SF_6 in interrupters (Lingal et al., 1953).

Industrial production of SF_6 gas started in the United States in 1947, and it was first used for switchgear in 1953 (Lingal and Owens, 1953). Westinghouse first used this gas for power circuit breakers in 1959 (Friedrich and Yeckley, 1959); these breakers operated on the same principle as air circuit breakers. The interruption arc was cooled by a blast from a compressed gas reservoir of 1.5 MPa (15 bar) into a gas space of 0.3 MPa (3 bar); this was later classified as a *double-pressure type* interrupter. When the pressure in the high-pressure reservoir decreased, a gas compressor pumped up gas from the low-pressure reservoir. The interrupter arrangement was of the axial-blast type, and this also was quite similar to the situation in air-blast breakers. Figure 3.5 shows a schematic cross-sectional drawing of the breaker, which had three units in series for 230 kV.

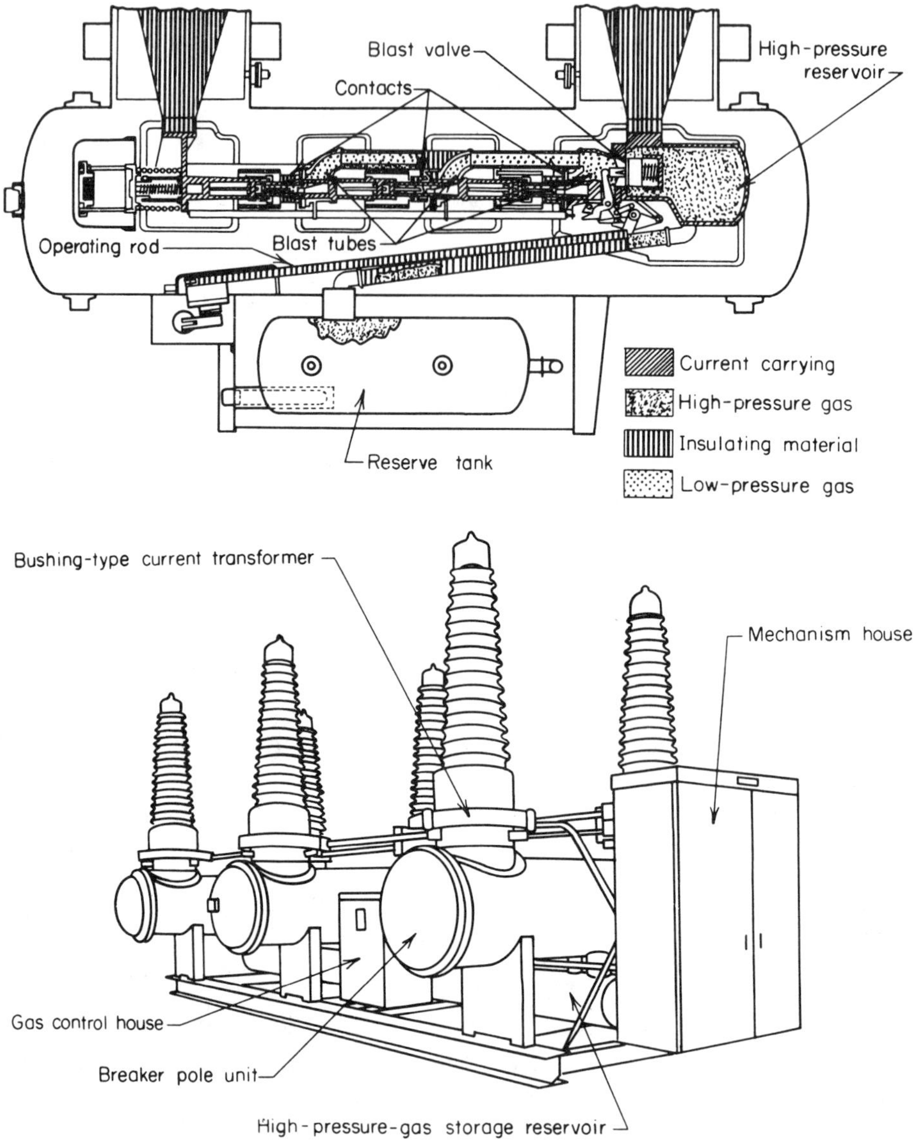

Fig. 3.5 A schematic cross section of a 230-kV SF_6 circuit breaker operating on the double-pressure principle. The rated interruption current was 40 kA and there were three interrupters in series. (Westinghouse Electric Corporation.)

A live-tank type of breaker in which the interrupter is supported by porcelain was immediately developed with this "double-pressure" arrangement. The investigative work that followed the first step of the development achieved an increase in interruption capacity and at the same time indicated the possibility of eliminating the gas suppressor. This elimination of the gas compressor was achieved by using the gas compressed during the opening stroke of the moving contact. The idea of a *puffer-type* interrupter was proposed, and models of 115 kV, 1000 MVA (Cromer and Friedrich, 1956) and of 46 kV, 250 MVA (Yeckley and Cunningham, 1958) were first completed in 1955 (Easley and Telford, 1964). This model is accepted as the basic form of today's buffer interrupter (Friedrich and Yeckley, 1959). A Polytetrafluoroethylene (PTFE) nozzle and 4–6-bar background gas pressure are most commonly used. (Refer to Chapter 2, Fig. 2.37.)

Development work in Europe was going on in parallel, and a double-pressure live-tank type of interrupter was developed by Siemens in 1964 (Einsele, 1964; Biersack et al., 1966). This had a rating of 220 kV, 15,000 MVA with four breaks in series. Workers in Japan also succeeded in installing an 84-kV, 5000-MVA circuit breaker for a practical power system in 1965.

Development efforts became particularly active in 1969 when the patent on the principle of GCBs (Lingal et al., 1956) expired in many countries. Switchgear manufacturers completed in developing various ideas for better puffer-type circuit breakers. Flow guide design and the optimization of compressed gas volume were the major subjects investigated. The main objects of the competition were the uprating in interruption current and the uprating of unit voltage, which can lead to a reduction in circuit-breaker size. The increases in interruption current and unit voltage over the past years are summarized in Fig. 3.6. The interruption current expanded from 20 kA to 63 kA (or 130 kA in 24 kV), and unit voltage, which was 72 kV in the first breakers is now 360 or 420 kV. A 550-kV breaker, which required four breaks in series in the 1960s, can now be made up with only two units in either dead-tank or live-tank type (refer to Fig. 3.7), and it is said that, from a purely technical point of view, a 550-kV breaker with a simple break is possible.

In some places 245-kV, 80-kA breakers are practically applied and a special application of circuit breakers to generator main circuits developed breakers with high interrupting ratings, such as 110–130 kA. Figure 3.8 shows an example of breakers for a generator main circuit that is conveniently fitted to isolated phase busbars by making the general arrangement coaxial to the grounded sheath.

The possible advantage of using a GCB connected in series with a vacuum interrupter has been studied (Natsui et al., 1988). This kind of complex application may stimulate expanded use of SF_6 gas circuit breakers.

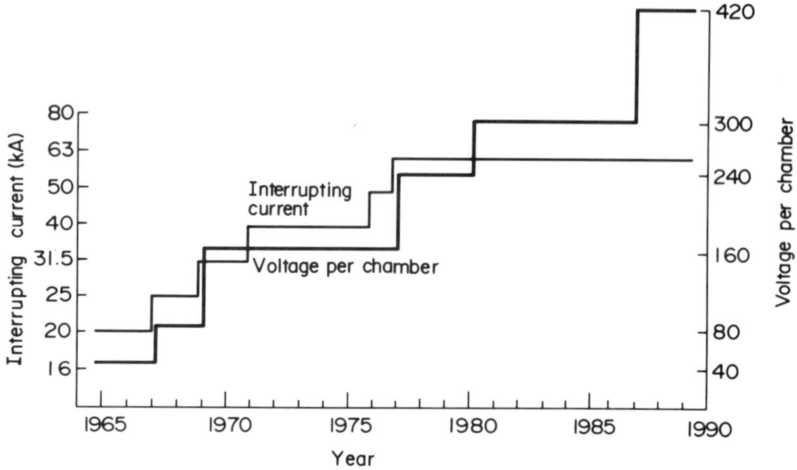

Fig. 3.6 The history of increases in the unit voltage and the interruption current of SF_6 interrupters.

Fig. 3.7a Dead-tank 550 kV circuit breakers.

Fig. 3.7b Live-tank tw-break 550 kV circuit beakers.

The history of SF_6 gas circuit breakers can be divided into three major periods:

Double-pressure units (1960) → puffer-type units (1970s) → high-voltage units (1980s)

3.2 RECENT DEVELOPMENT OF PUFFER-TYPE GAS CIRCUIT BREAKERS (T. Yamauchi)

The developmental work and fundamental research achieved 360 kV/420 kV, 40 kA/50 kA circuit breakers (e.g., Fig. 3.9) (Yoshizumi et al., 1989).

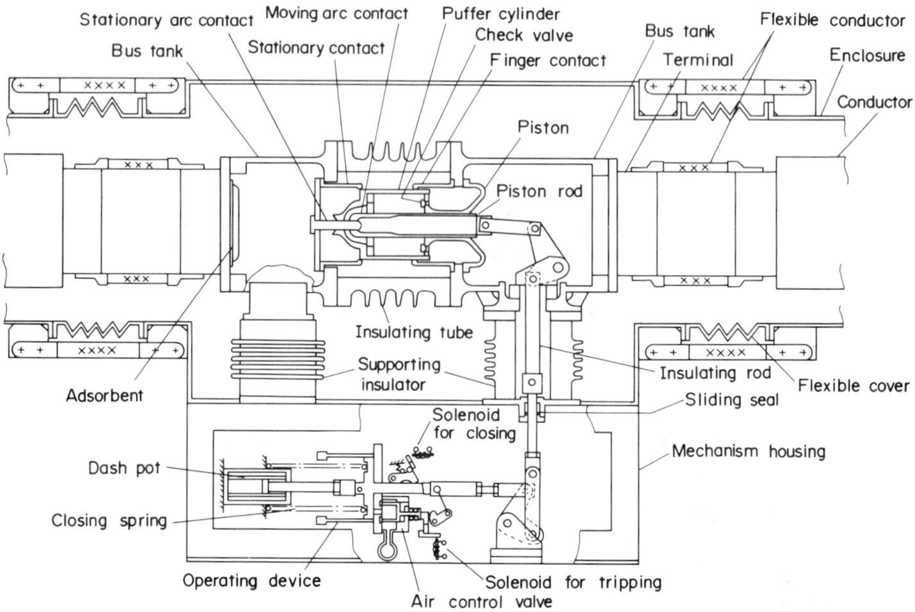

Fig. 3.8 A cross-sectional view of an SF_6 generator main circuit breaker installed in series with isolated phase busbars.

Improved understanding of arc interruption and thermodynamic characteristics stimulated efforts to reduce the mechanical energy required in interruption by using the thermal energy of the arc itself. The technique that realized the energy saving materialized in the design to build up the pressure necessary for blasting the arc with the help of the arc itself. This concept is quite similar to that of self-extinction in oil circuit breakers. The thermal energy is somehow fed back to the pressure buildup space to help the pressure buildup by mechanical drive. Figure 3.10 shows an example of this idea. A part of the heated gas from the arc is fed back to the puffer cylinder, increasing the pressure in this space. The puffer cylinder is compressed further by the time of the actual interruption. The increase in the cooling capacity of the blasted gas flow caused by the increase in pressure exceeds the decrease in the cooling power supplied by the temperature increase in the blasted gas in this arrangement. The interruption capacity is improved by 20% in short-line fault (SLF) conditions as shown in Fig. 3.11.

Improvements in the mechanical structure significantly decreased the mechanical energy requirements for interruption. A circuit breaker for 72 kV, 31.5 kA required mechanical energy of 3 tons × 0.14 m in 1967, and a modern puffer circuit breaker with the same capacity produced by the same

Fig. 3.9 A one-break 362/420-kV 50-kA circuit breaker.

manufacturer requires approximately 40% of that. The arrangement to feed back the hot gas to the compression chamber in the early stage of development varies widely, but this is one of the most interesting subjects in the present-day competition.

3.2.1 Application for Gas-Insulated Switchgear

The application of SF_6 gas as an interruption medium is an important application, and a still more important application is *gas-insulated switch*

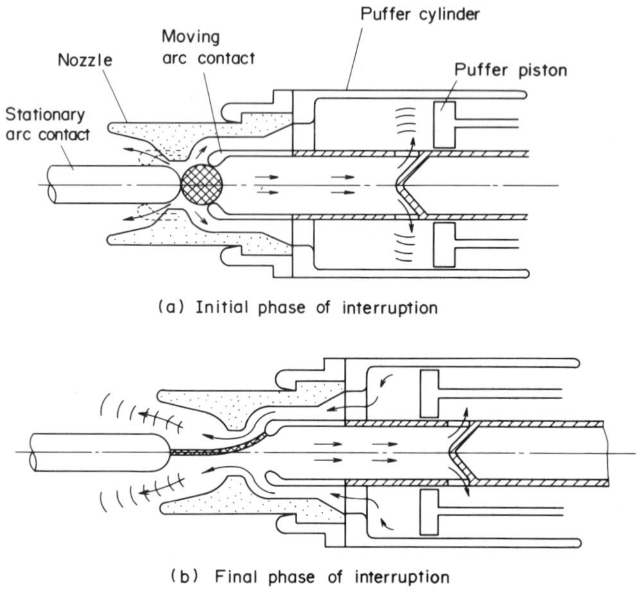

Fig. 3.10 Schematic sketch to show the principle of pressure buildup with the help of arc energy (hybrid puffer). (Ozaki et al., 1988. Copyright Toshiba.)

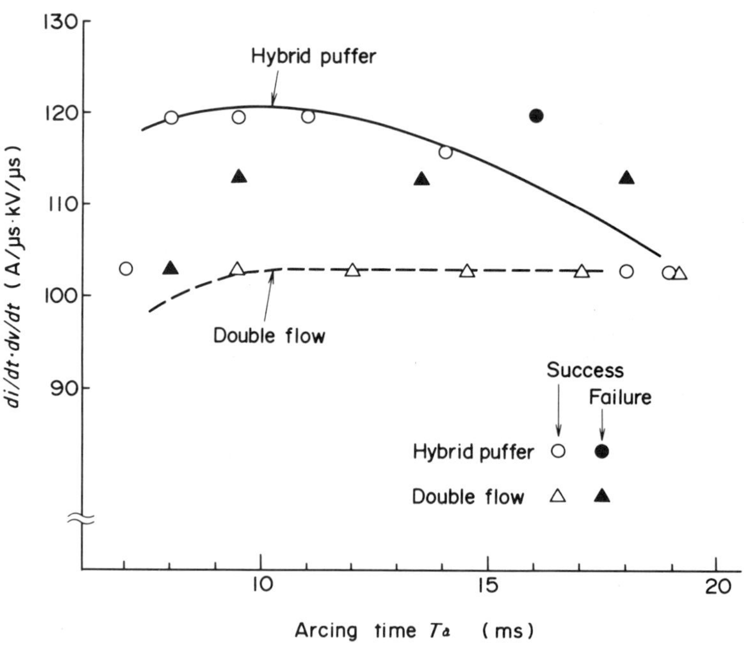

Fig. 3.11 Comparison of the interruption performance of conventional puffer and hybrid puffer interrupters in SLF interruption. (Ozaki et al., 1988). Copyright Toshiba.)

gear (GIS), which uses the dielectric capability of the gas to reduce the size of the switchgear drastically. The basic idea of GIS is to insulate all the equipment connected to the busbars of substations with SF_6 gas in a metal enclosure that contains all the equipment. Disconnecting switches, potential transformers, and current transformers can be insulated with the gas, and any type of circuit breaker can also be put into the enclosure. In most practical applications, SF_6 circuit breakers are used, and SF_6 gas is commonly used for interruption and insulation.

Practical GIS came into use in the late 1960s in Japan and Europe. One major advantage of this equipment is the reduction achieved in size, but the important features are the minimization of exposure of the insulators to the atmosphere and the convenience of satisfying environmental requirements by exploiting the low profile nature of the total system. Endurance under seismic stress is another important feature of GIS. The application of GIS is expanding in densely populated areas and highly salt contaminated areas in Japan and Europe, and the application is becoming popular in the United States, mostly in urban areas.

The preparation of the sealed gas compartment (gas zoning) developed from fine segregation to rough segregation as experience accumulated. In the first stage the three phases were isolated in separate enclosures. This was referred to as a *single-phase-types* GIS.

Figure 3.12 shows an example where the three phase busbars are together in one enclosure but other components are enclosed in isolated phase enclosures. Figure 3.13 shows an example where all the equipment is enclosed in three-phase common enclosures.

A three-phase circuit breaker with isolated phase enclosures is a simple extension of the dead-tank design of SF_6 circuit breakers, because the only important difference is the connection through insulator spacers instead of through bushings. Therefore, the important technology was the application of insulating spacers in SF_6 gas and the insulation design between phases especially in the presence of hot exhaust gas. The ratings that could be covered by the *three-phase-enclosure* type of arrangement expanded in parallel with the expansion of dead-tank circuit breakers. However, active works in this area was stimulated by the introduction of three-phase common enclosures for circuit-breaker units, especially for EHV gas-insulated switchgear. The example in Fig. 3.14 shows a complex nonaxisymmetric enclosure of GIS components with an insulating rod. The presence of insulators increases the complexity of the problem, and analysis by numerical calculation by computers became essential.

A combined analysis of the electric field and the hot gas dynamics of the exhausted gas was required to establish the design. In chapter 2, the examples shown in Figs. 2.20, 2.18, Fig. 2.24 are combined with numerical

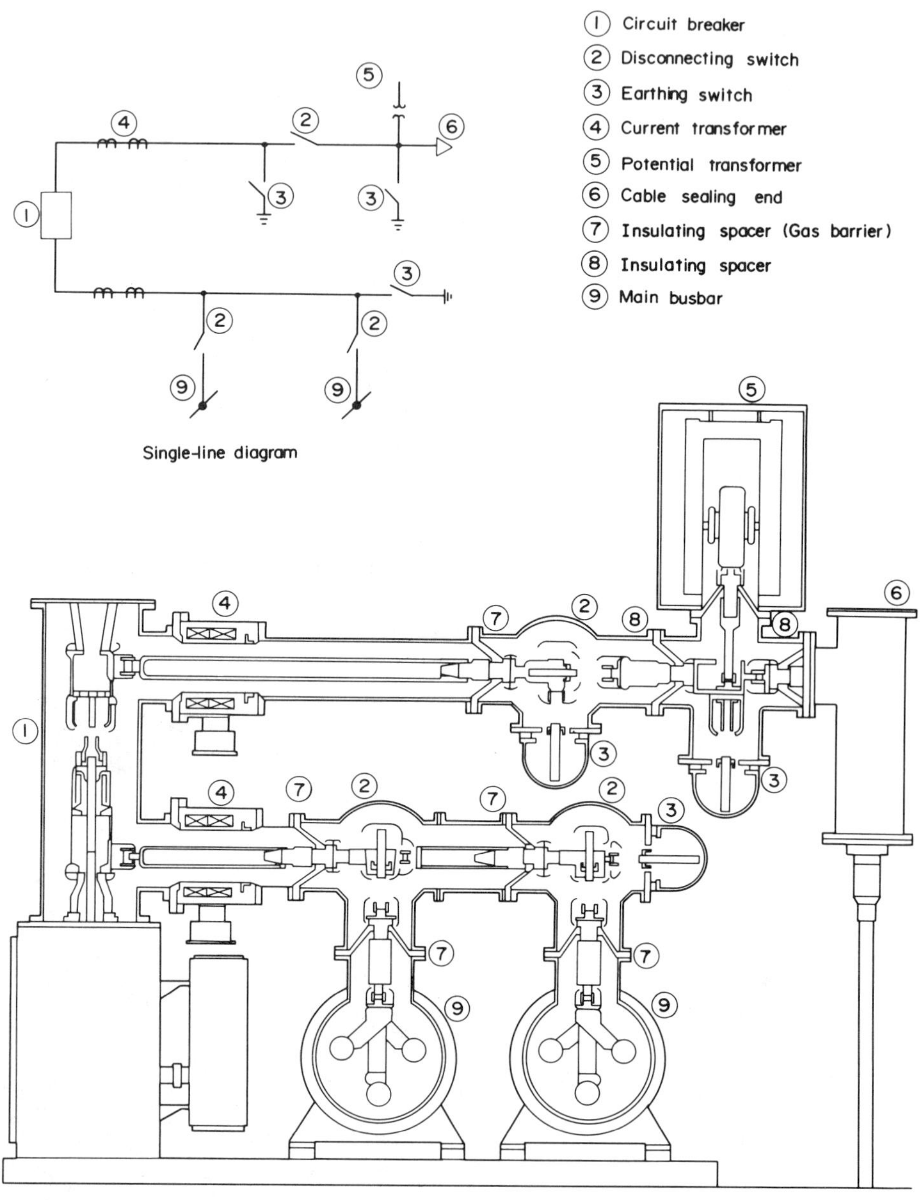

Fig. 3.12 Typical arrangement of a GIS with double busbar with three-phase common enclosure.

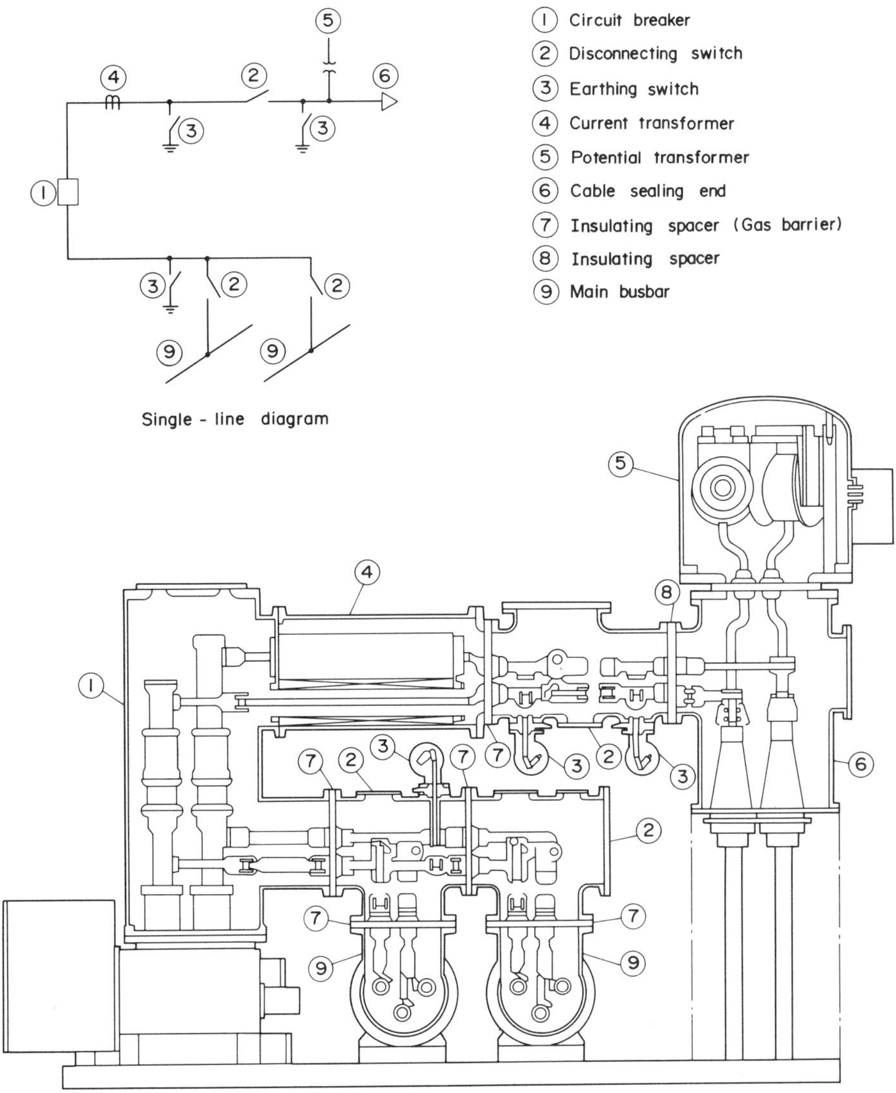

Fig. 3.13 Typical arrangement of a GIS with double busbar. All the components including the circuit-breaker unit are enclosed in three-phase common enclosures.

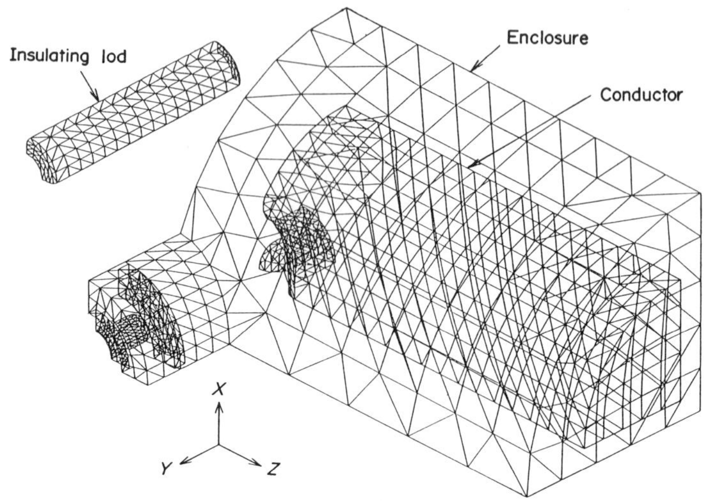

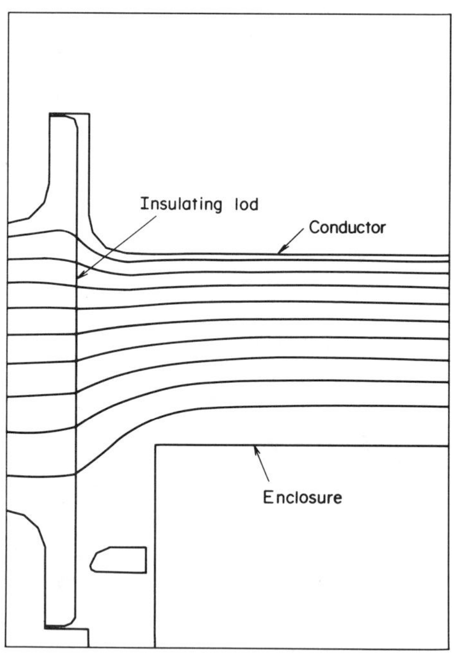

Fig. 3.14 (Top) An example of a mesh diagram for calculation of electric field intensity by the surface-charge method for an insulator that is not axisymmetric. (Bottom) Results of the field intensity calculation.

calculation of electric field distribution, and the expected dielectric characteristics are evaluated.

3.3 GAS-INSULATED SWITCHGEAR (S. Yanabu)

In gas-insulated switchgear GIS, almost all components of a substation are housed in metallic enclosures filled with high-pressure (4–5 bar) SF_6 gas, as previously described. The excellent dielectric characteristics of SF_6gas enable GIS to be designed much smaller than conventional substation equipment, thus finding additional applications in substations in wide areas. Also, totally earthed and free from the external environment, GIS is very safe and reliable. For the reasons mentioned above, GIS recently has been put into wide use in densely populated areas of North America, Europe, and Japan, as well as in areas with severe environments such as the Arabian countries and southeast Asia.

Since its first application to a 66-kV system at the end of the 1960s, improvements in GIS have significantly extended its range of applicable voltages: to 500 kV in the latter half of the 1970s and 800 kV in the middle of the 1980s (Yanabu et al., 1985).

The configuration of GIS largely depends on circuit-breaker structure. The realization of more compact GIS has been promoted by reducing the number of contacts. The diagram of Fig. 3.15 compares GIS with a one-break GCB with GIS with a two-break GCB. By employing a one-break GCB, the effects listed in Table 3.1 are obtained (Ikeda et al., 1984).

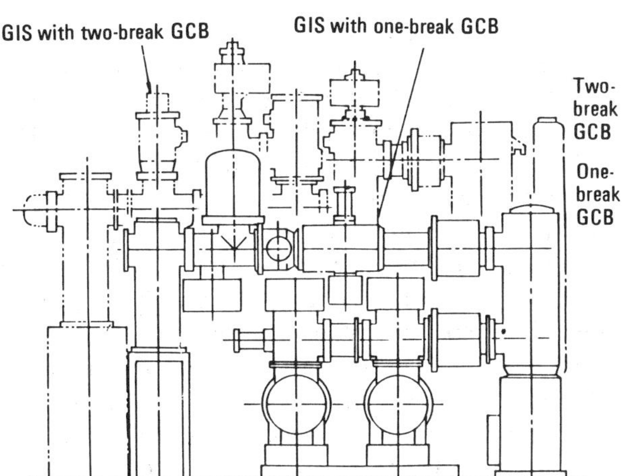

Fig. 3.15 Gas-insulated switchgear employing one-break and two-break GCBs (rated voltage: 300 kV) (Ikeda et al., 1984. © IEEE.)

Table 3.1 Comparison of Major Dimensions of One-Break and Two-Break Gas Circuit Breakers

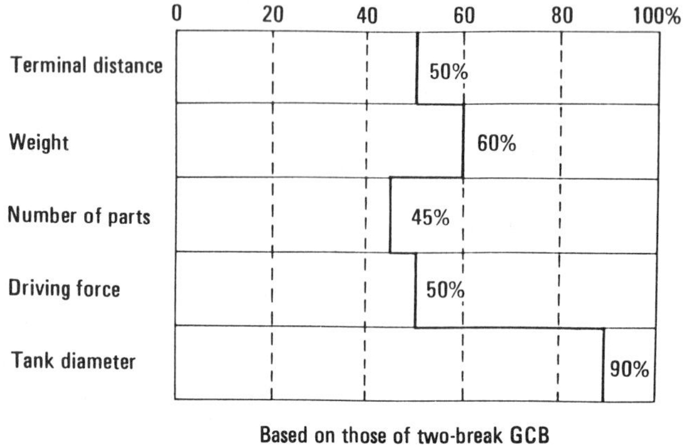

Source: Ikeda et al. (1984). © IEEE.

Since GIS has switching components that include circuit breakers and disconnecting switches sealed in metallic enclosures, special consideration must be given to its design and manufacture in comparison with conventional substation components.

3.3.1 Circuit breakers

Puffer-type SF_6 gas circuit breakers are predominantly used in GIS, and their interrupting capacity has recently been increased. The largest of the products in use today are 420-kV, 50-kA one-break models. Puffer-type SF_6 gas circuit breakers are designed with the following features. The interrupting chamber is housed in an SF_6 gas-filled metallic enclosure installed at ground. As to a multibreak circuit breaker for GIS, the distance between the interrupting chamber and the enclosure is so short that capacitance to ground becomes large. Therefore, to average the voltage distribution on each contact, it is necessary to connect a relatively large capacitance in parallel with the interrupting chamber. A circuit breaker at a voltage level of 362 kV or above usually requires a *closing resistor* and must be housed together with its closing resistor and a *resistor switch* in an enclosure.

Also, when a circuit breaker interrupts a current, *hot gas* generated by arcing pervades the space between the interrupter unit and the metallic enclosure, possibly causing a decrease in dielectric performance. Therefore, it is necessary to ensure dielectric performance to ground at interruption, and special caution should be taken in designing the configuration of the shield that will be exposed to hot gas.

In addition to the interruption tests described in Chapter 4, analytical studies of dielectric performance are conducted. Figure 3.16 shows an example of the calculation of hot-gas distribution near an interrupting chamber during current interruption. This calculation used an axisymmetric, three-dimensional *compressive gas* analysis program. The product of arcing voltage and current is input as thermal energy. Gas temperatures and densities at related portions corresponding to time variations in the opening process are calculated, and the results are compared with measured data, showing good agreement (Suzuki et al., 1989).

To realize even more compact GIS, *three-phase-in-one-tank* type gas circuit breakers with three-phase interrupter units sealed in an enclosure have been developed; their rated voltage ranges up to 300 kV. These interrupter units and phase conductors are positioned close to each other. The shield configuration between phases must be carefully designed to avoid high field strength in an area where hot gas will exist.

3.3.2 Disconnecting Switches

Disconnecting switches are opened at lower speeds (2–3 m/sec) than circuit breakers. Thus, opening a short GIS bus by a disconnecting switch causes a charging current of about 1 A to be interrupted. Here, as shown in Fig.

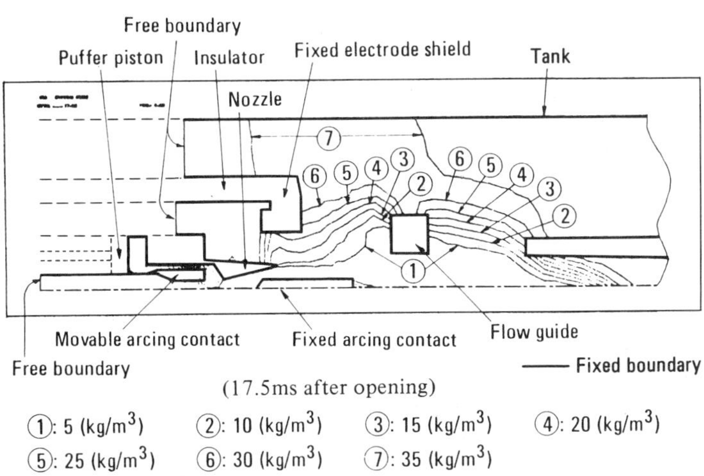

Fig. 3.16 Density distribution after contact opening. (Suzuki et al., 1989. © IEEE.)

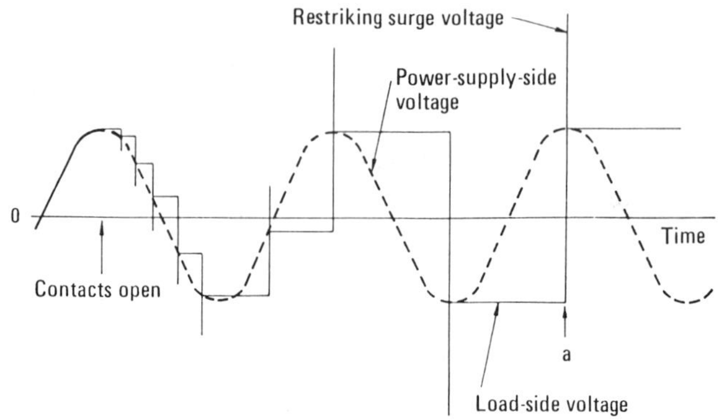

Fig. 3.17 Mechanism of disconnecting switch switching surge. (Nishiwaki et al., 1983. © IEEE.)

3.17, interruption and restriking are repeated before interruption is completed. This is because it takes some time before interelectrode dielectric performance recovers completely, since dielectric recovery performance of the disconnecting switch is slower than variations in the ac voltage applied.

Restriking causes a surge voltage in the circuit. An equivalent circuit for testing this surge voltage is shown in Fig. 3.18. Basically, this is a transient phenomenon of contact closing, with dc voltage remaining in the bus and ac voltage on the power-source side. Since GIS is an ideal coaxial line and the bus on the load side is short, surge voltage is a double-frequency voltage containing a high-frequency component at around 1–10 MHz. Although surge voltages can be calculated in relatively good agreement with measured values by simulating GIS components as accurately as possible, for the sake of higher accuracy, consideration must be given to time for breakdown to occur between contacts (*collapse time*) and accurate modeling of components (Yanabu et al., 1988).

This surge voltage is applied to the spark column generated between electrodes in restriking and to GIS components. If the distance between contacts of the disconnecting switch is relatively long, the surge voltage is high, as shown in Fig. 3.17. Here, on the surface of the spark column between electrodes, there appears a very high nonuniform electric field distribution toward the enclosure, as if it were a thin needle. In an extreme case, it has been proved experimentally that flashover occurs come from the spark column to the enclosure. Figure 3.19 shows an example (Nishiwaki et al., 1983). Although it may be considered that with a nonuniform electric field distribution the breakdown voltage is increased by the field relaxation effect caused by corona discharges, it is recognized that with a

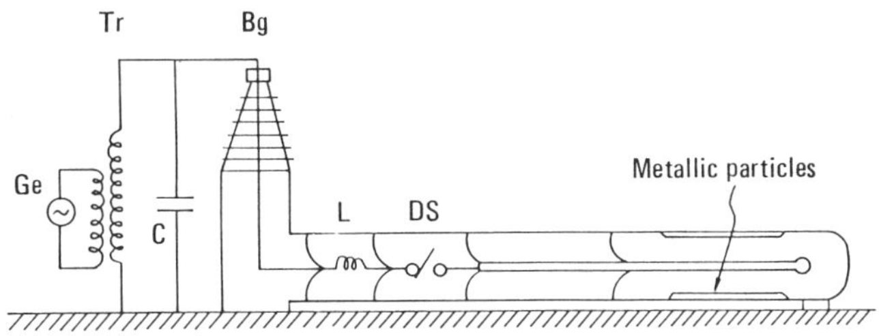

DS: Disconnector; L: Reactor; Bg: Bushing; C: Capacitor, 60 000pF;
Tr: Transformer; Ge: Short-circuit generator

(a) Equivalent circuit

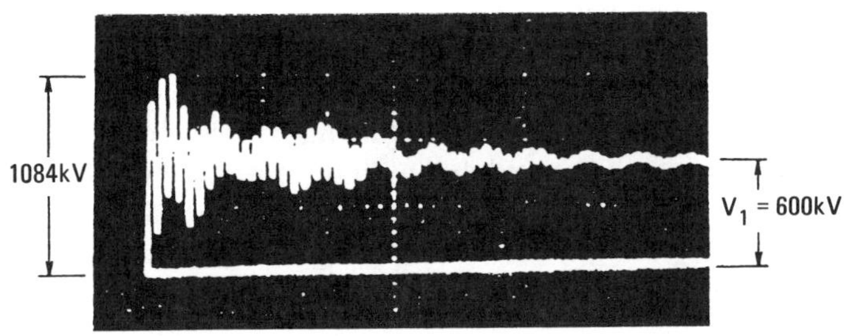

(b) Surge waveform (360 kV/div, 2 μs/div) measured at 500-kV GIS (Enlargement of phenomena in point "a" shown in Fig. 3.17)

Fig. 3.18 Typical waveform and test circuit of disconnecting switch switching surge.

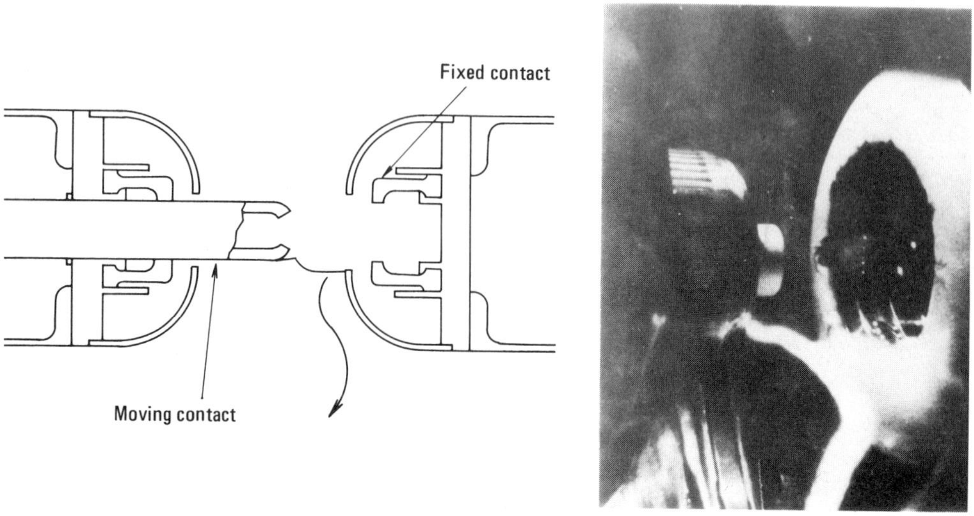

Fig. 3.19 Flashover to enclosure from the spark under DS switching operation. (Nishiwaki et al., 1983. © IEEE.)

steep waveform such as that of disconnecting-switch surges, a streamer proceeds to strengthen before the relaxation effect appears.

Since it is very desirable for such breakdowns to occur, factory proof tests must be conducted. Figure 3.20 shows two test circuits. In (a) a no-load bus is used to operate the disconnecting switch. In (b), operation of the disconnecting switch is simulated with out-of-phase voltage applied via a capacitor connected in parallel with a circuit breaker. For example, a voltage 180° different in phase is applied from the power source side to the load side via a *grading capacitor,* which is equipped for equalizing voltage distribution between breaks. There are also other test methods under study worldwide (CIGRE Working Group 33/13-09, 1988).

Surge voltages spread over various parts of the GIS when the disconnecting switch is operated. Figure 3.21 shows levels of surge voltages and a frequency distribution on a 500-kV GIS (Ogawa et al., 1983). The levels of surge voltages are not abnormally high but are sufficiently high to cause dielectric breakdowns in the presence of a jumping *metallic particle* or nonuniform electric field electrodes, because there is no electric field relaxation effect caused by corona discharges.

Figure 3.22 shows an example of a breakdown voltage with a coaxial cylindrical electrode. In this test, with metallic particles of various lengths attached to the central conductor, breakdown voltages were obtained while the duration of the positive impulse wave front was varied. As the duration

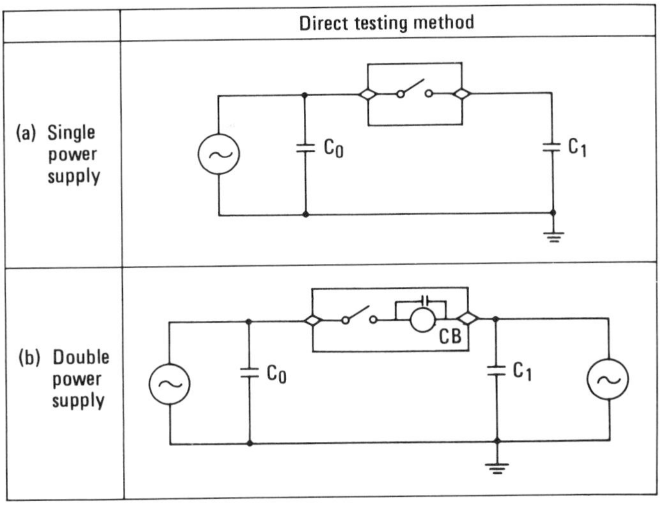

Fig. 3.20 Proposed test circuits for disconnecting switch switching surge.

of the wave front decreases, the breakdown voltage also decreases. Figure 3.23 shows an example of breakdown voltages obtained by including metallic particles of various lengths in GIS equipped with a disconnecting switch (Kobayashi et al., 1987). Breakdown voltages were obtained under various conditions: (1) under clean conditions, (2) with a metallic particle and an independent ac voltage applied, (3) with a surge caused by operating the disconnecting switch while applying an ac voltage, (4) with an impulse and ac voltage superimposed with a metallic particle crossing from enclosure to high-potential conductor, and (5) with an impulse and ac voltage superimposed without a metallic particle crossing. It is evident, when a high-frequency surge such as a disconnecting-switch surge is applied in the presence of a metallic particle crossing, the breakdown voltage decreases extremely. These phenomena occur because high-frequency disconnecting-switch surges are generated in the GIS. Studies on them are being energetically pursued worldwide along with basic studies on the process of dielectric breakdowns. Extensive investigations are also being made of the effect of disconnecting-switch surges applied to GIS components. One of the examples already known is as follows. If a disconnecting-switch surge propagates into a metallic enclosure through an electrically discontinuous part such as a bushing, this leaked surge will enter the control circuit, causing faults such as malfunctioning of electronic devices.

One of the special requirements of GIS disconnecting switches is *closed-loop current switching;* that is, switching the load current of the

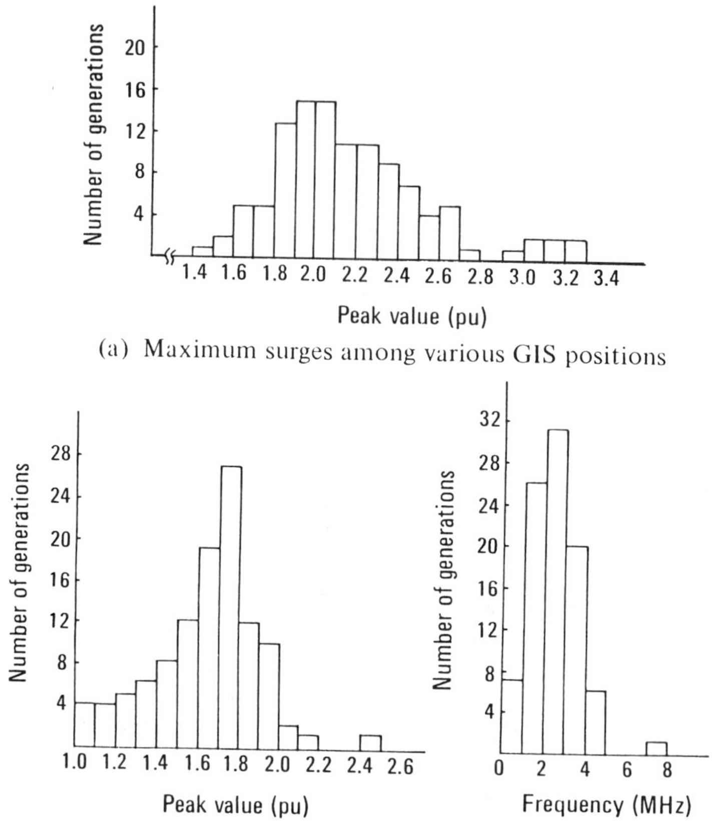

(a) Maximum surges among various GIS positions

(b) Surges at operated disconnector terminals

(c) Frequencies of surges at disconnector terminals

Fig. 3.21 Disconnector restriking surge calculation results (calculations on 111 cases on 10 GISs). (c) Results on 91 cases. (Ogawa et al., 1983. © IEEE.)

double-bus circuit as shown in Fig. 3.24. While the recovery voltage is as low as around 100 V, the interrupting current is comparable to the rated current, a severe level for disconnecting switches. To cope with this requirement, various measures are taken, such as adding a puffer-type interrupting chamber as is done with circuit breakers, or a coil to extinguish the arc by rotating it with the force of a magnetic field (Yanabu et al., 1984; Suzuki et al., 1984). Arcing time is minimized so that dielectric performance will not be strongly affected by *by-product materials* generated at the interruption of a large current. Figure 3.25 shows several examples of interrupting

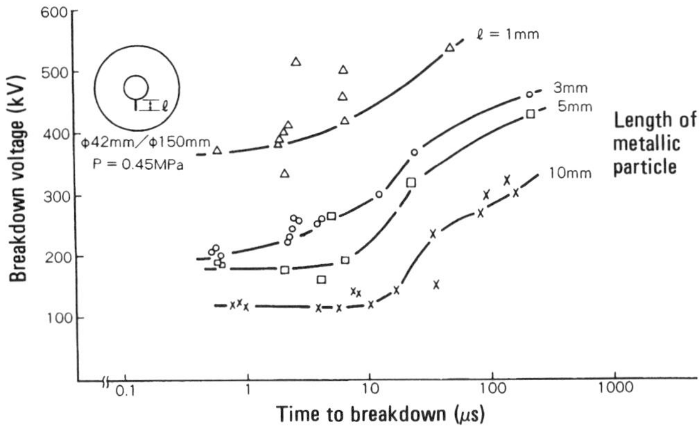

Fig. 3.22 Voltage–time (V–t) characteristic of metallic particles under positive impulse voltage. (Kobayashi et al., 1987. © IEEE.)

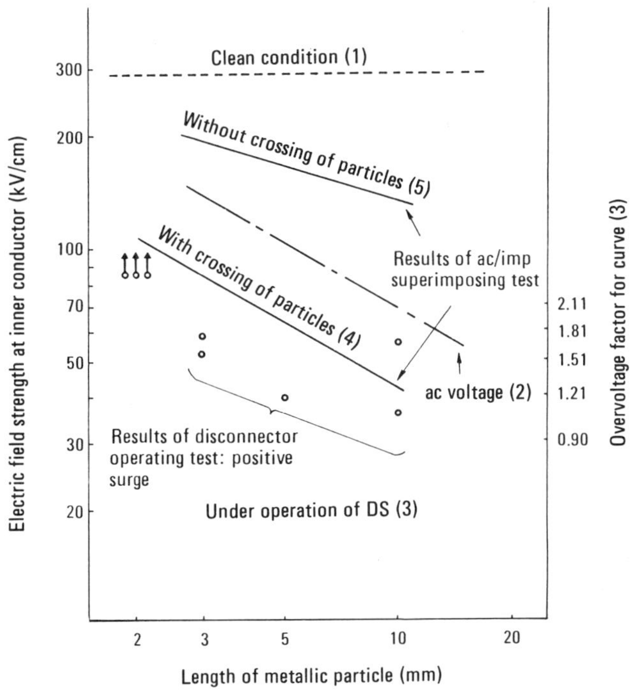

Fig. 3.23 Effect of DS switching surge on metallic particles in 500-kV GIS. (Kobayashi et al., 1987. © IEEE.)

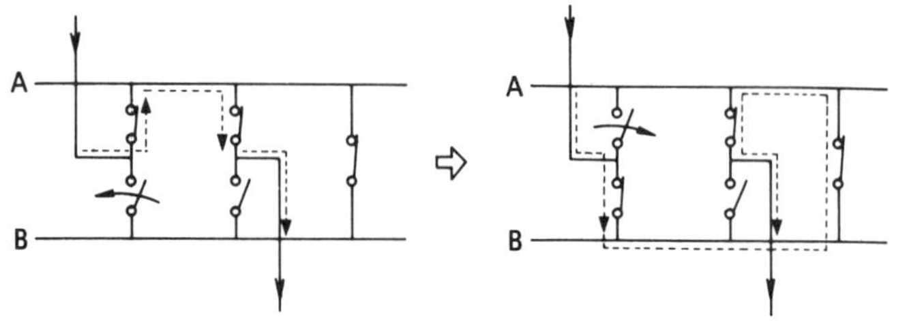

Fig. 3.24 Interruption of closed-loop current. (Suzuki et al., 1984. © IEEE.)

chambers and relations between arcing time and opening speed obtained with them.

Both compact and in excellent harmony with the environment, GIS is prevailing in many countries. Since GIS incorporates switching elements such as circuit breakers and disconnecting switches in a tank, particular consideration must be given to transient phenomena at the time of arc interruption or making. In this respect, important factors include the dielectric performance of the hot gas generated by arcing and the high-frequency surges caused by disconnecting switches. Studies on these phenomena are being rapidly pursued.

3.4 VACUUM CIRCUIT BREAKERS (Y. Kurosawa, E. Kaneko)

3.4.1 Background

A vacuum is such an excellent insulating medium that for a long time various attempts have been made to apply it to switching equipment. Back in the 1890s, a study was made on vacuum arcs, and tests on vacuum switches conducted by Sorenson and coworkers (Sorenson and Mendenhall, 1926) in the 1920s are well known. Despite the long history, it was not until the 1950s that the first practical vacuum circuit breakers were produced. Lack of industrial technology for mass production was the reason for the delay in their practical development. During this period, however, there were various industrial technological innovations such as the invention of vacuum leakage detectors and advances in metal outgassing techniques, techniques for hermetic joining between metal and glass or ce-

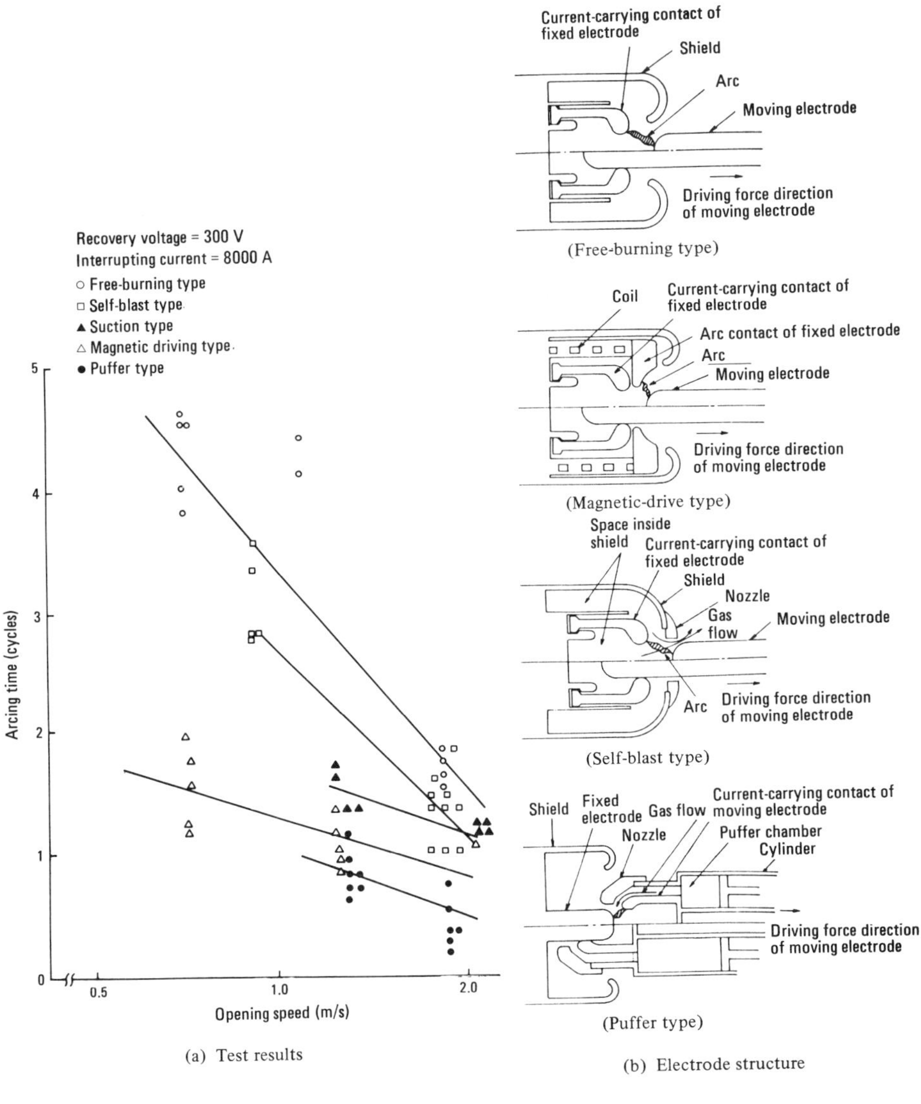

Fig. 3.25 Comparison of arcing time by extinction system for closed-loop current-interrupting tests. (Yanabu et al., 1984; Suzuki et al., 1984. © IEEE.)

ramics, and bellows for maintaining the airtightness of moving parts. Also, as a result of advances in materials technology, electrode materials resistant to large currents and suitable for vacuum circuit breakers were developed, and these have been successively applied in products up to the present.

A cross-sectional drawing of a typical vacuum interrupter is shown in Fig. 3.26. The vacuum interrupter is composed of a pair of electrodes installed in a glass or ceramic insulating enclosure. The movable electrode is driven by an external driving mechanism. Both electrodes are surrounded by an arc shield to protect interior surfaces of the insulating enclosure from metal vapor.

Vacuum circuit breakers (VCBs), in contrast to other circuit breakers, possess the following features:

1. They provide high interrupting performance, reliability, and safety.
2. They are pollution-free, offering low noise and no fire hazards.
3. They are compact, lightweight, and easy to carry and require only small installation space.
4. They require less labor for maintenance and inspection than other types of breakers.

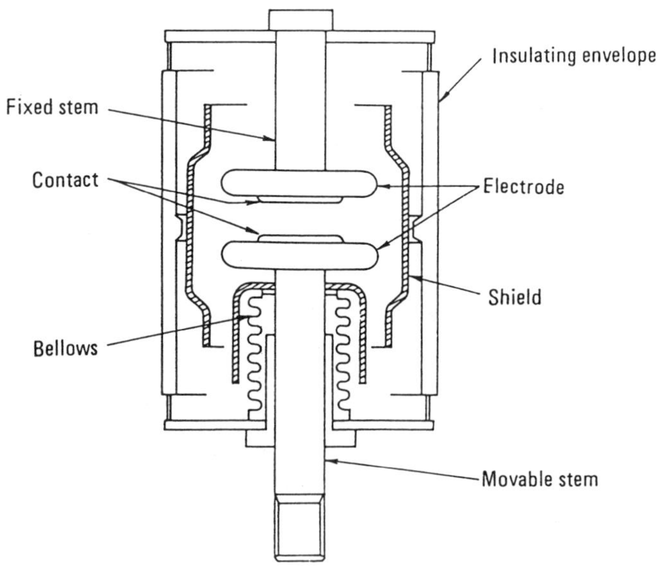

Fig. 3.26 A cross-sectional drawing of a typical vacuum interrupter. (Okumura and Kaneko, 1982.)

As these excellent features have been gradually recognized, VCBs have come into wide use in industry including electric power systems. In terms of rating, VCBs are used mainly with voltages of 3.6–36 kV and interrupting currents of 8–63 kA.

Figure 3.27 shows an example of vacuum circuit breakers for distribution voltage.

3.4.2 Principles of Interruption in Vacuum

The VCB interrupts an electric circuit by separating a pair of electrodes installed together in a vacuum enclosure. When the electrodes are separated, the vacuum arc that is generated is maintained by metal plasma supplied from the cathode spot. At current zero, the metal plasma rapidly diffuses and disappears. This has been recognized as the reason VCBs possess high current-interrupting capabilities. As shown by the high-speed photograph in Fig.3.28, a vacuum arc is generally composed of small arcs. With copper electrodes, the current per cathode spot is about 100 A. It is

Fig. 3.27 A vacuum circuit breaker for distribution voltage.

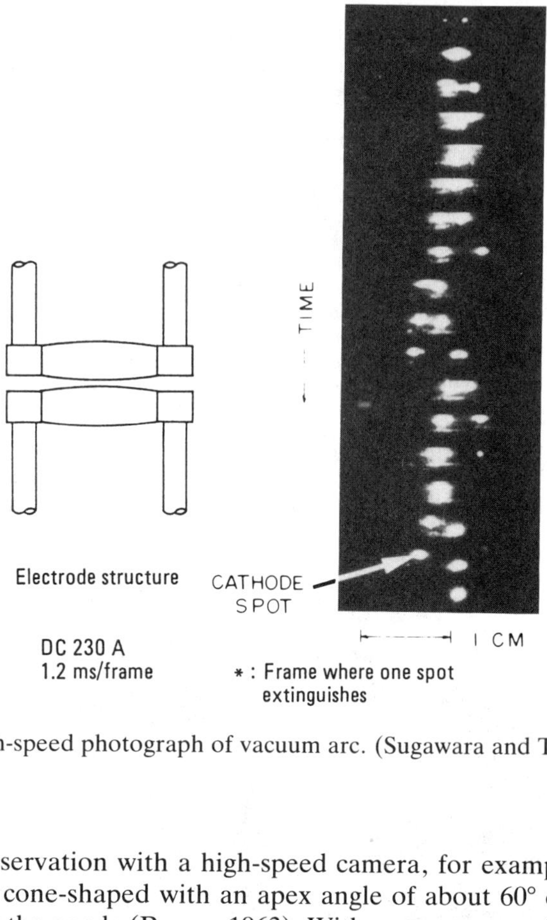

Fig. 3.28 High-speed photograph of vacuum arc. (Sugawara and Takasuna, 1972. © IEEE.)

known, by observation with a high-speed camera, for example, that each arc column is cone-shaped with an apex angle of about 60° directed from the cathode to the anode (Reece, 1963). With a vacuum arc of up to several thousand amperes, several cathode spots are dispersed over the cathode surface. One such unit consists of arc columns ignited parallel with the cathode spots and is said to be in the *diffused mode.* With such arcs in a diffused mode, arc voltages are low; with copper electrodes, they are about 20 V irrespective of electrode spacing.

As the current increases, several arcs gradually gather into a single thick arc (shown in Fig.3.29), and arc voltage may rise to more than 200 V. In such a situation, bright points appear on the anode where it receives most of the arc energy, and melting occurs as shown in Fig.3.30. The arc in this condition is said to be in the *constricted mode.* In this situation, it is assumed that metal vapor is generated in excess of diffusion, forming a high atmospheric pressure arc. Once this condition obtains, the dielectric

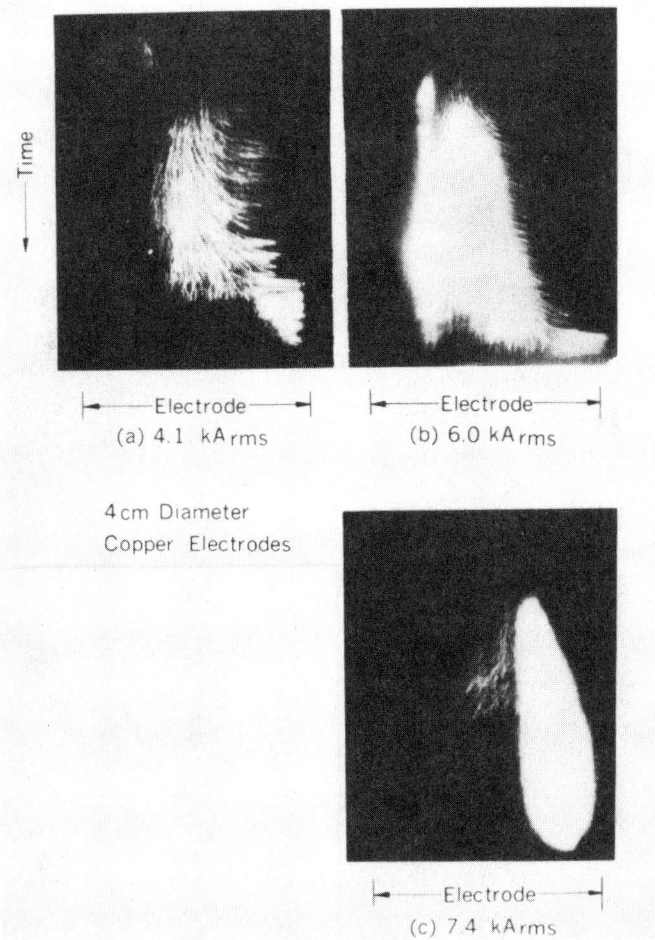

Fig. 3.29 High-speed photograph of vacuum arc. (Sugawara and Takasuna, 1972. © IEEE.)

recovery capacity is reduced by the emission of voluminous metal vapor, causing the interrupting performance to be greatly reduced. For VCBs, therefore, several measures are taken to increase the interrupting performance by preventing constricted arcs. One of the well-known methods is to prevent the anode from melting by continuously moving the anode spots by applying a transverse magnetic field to the arc. Figure 3.31 illustrates arc behavior with a high-speed photograph of the arc with a transverse mag-

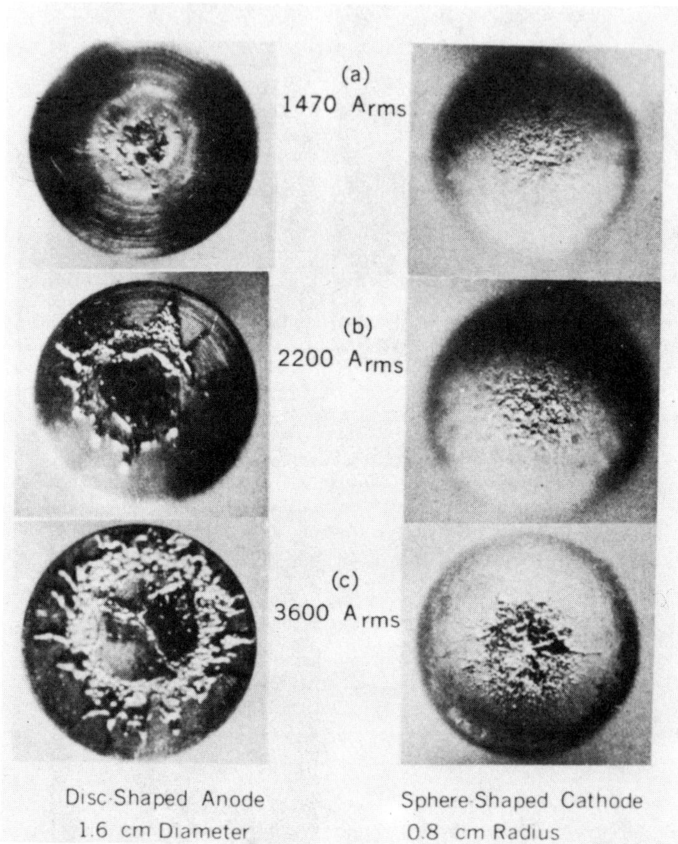

Fig. 3.30 Melting after current interruption. (Sugawara and Takasuna, 1972. © IEEE.)

netic field applied, revealing good movement of the arc. Another method is to reduce the density of the incident energy to the anode by spreading the arc over the full area of the electrode through application of an axial magnetic field. Figures 3.32 and 3.33 show high-speed photos of arc with an axial magnetic field applied. Even with a large current, the vacuum arc is distributed as stringy, thin arcs over the full area of the electrode. In all cases, rises in arc voltages are well controlled (Sugawara and Takasuna 1972). Figures 3.34 and 3.35 show arc voltage characteristics with transverse and axial magnetic fields, respectively, applied. With an axial mag-

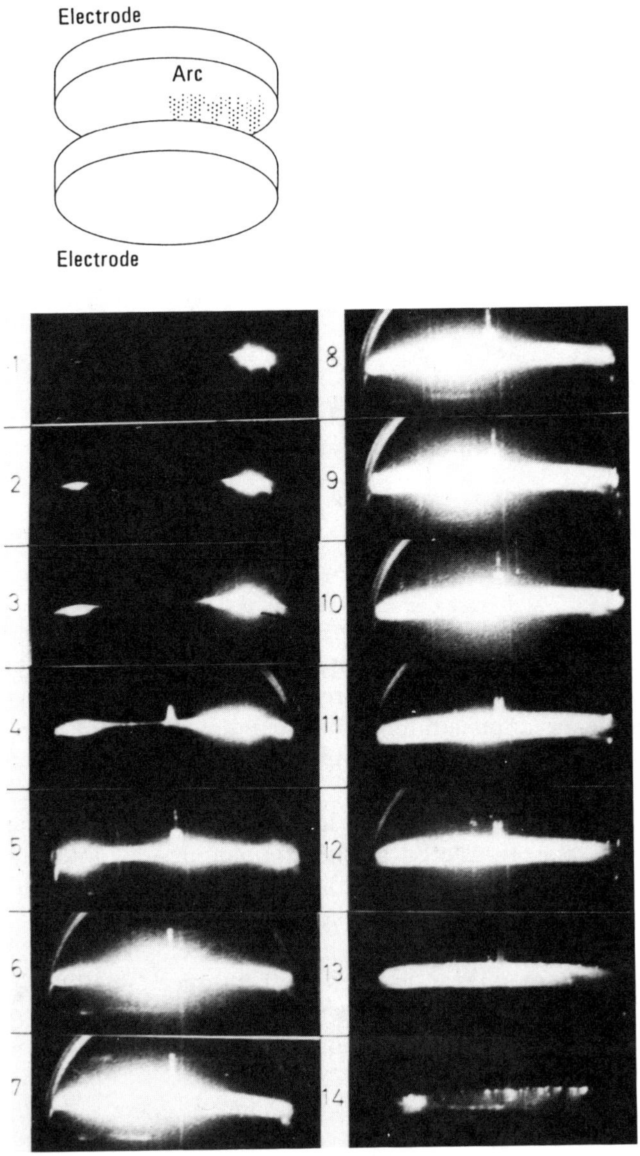

Fig. 3.31 Arc behavior with transverse magnetic field applied. (Kurosawa et al., 1980. © IEEE.)

Fig. 3.32 Arc behavior with axial magnetic field applied. (Yanabu et al., 1981. © IEEE.)

netic field applied, not only does the arc voltage show smooth characteristics up to large current ranges, but also the current flowing between the electrode and the shield (shield current) is very small, as shown in Fig.3.36, permitting the arc to be contained between the electrodes.

Directly after current interruption, charged particles remain between the electrodes, and when a recovery voltage rises between the electrodes, a post-arc current flows as shown in Fig.3.37. With GCBs, for example, post-arc current is closely related to the completeness of interruption. With VCBs, in general, post-arc current is larger than with gas circuit breakers, but there are no signs of so-called thermal interruption failure as a result of plasma reheating due to the flow of post-arc current. This may be because a large volume of metal vapor is emitted by anode melting, causing the vacuum pressure to rise almost up to the minimum point of the Paschen curve. The apparent impedance changes obtained by division of the voltage between electrodes by the post-arc current increase nearly linearly on a semilog graph as shown in Fig.3.38, some tending toward saturation (Hakamata et al., 1988).

As shown in Fig.3.39, the post-arc current first rises slowly with the interrupting current, but at some point it suddenly turns upward. This point of inflection, which varies according to the electrode material, seems to nearly agree with the start of anode brightening (Yanabu et al., 1985).

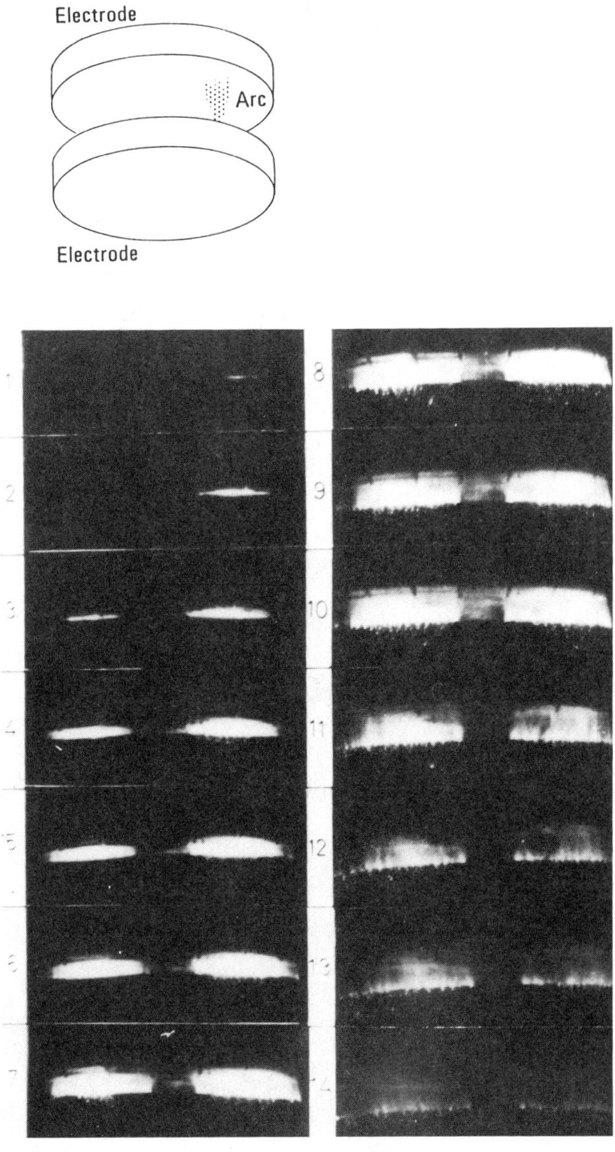

Fig. 3.33 Arc behavior with axial magnetic field applied. (Kurosawa et al., 1980. © IEEE.)

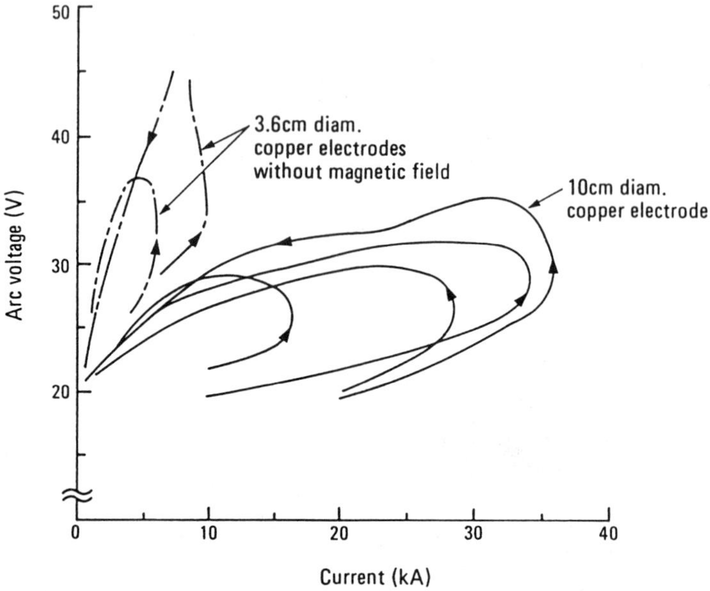

Fig. 3.34 Arc voltage characteristics with transverse magnetic field applied. (Data from Sugawara and Takasuna, 1972. © IEEE.)

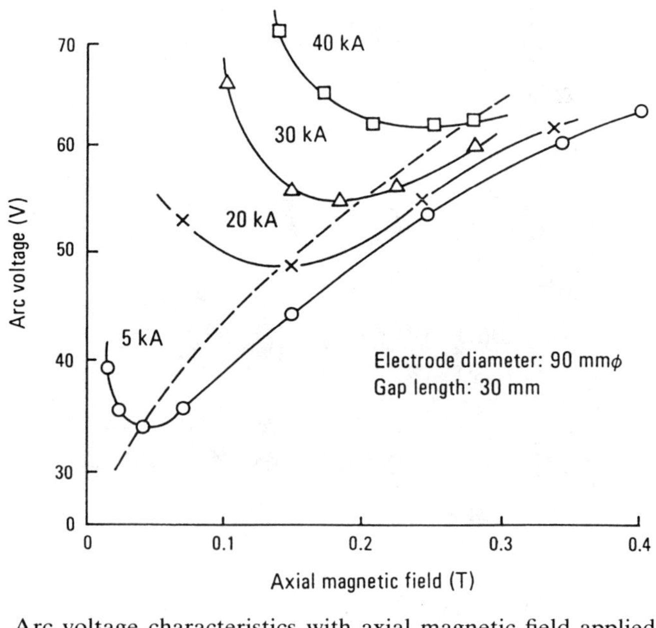

Fig. 3.35 Arc voltage characteristics with axial magnetic field applied. (Yanabu et al., 1981. © IEEE.)

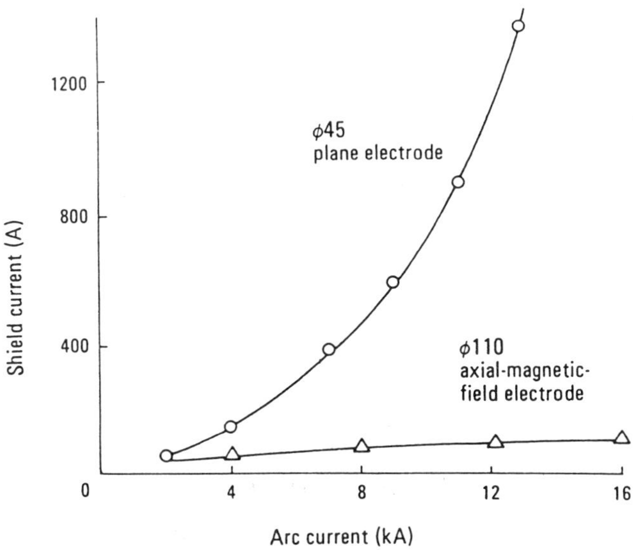

Fig. 3.36 Measurement of shield current. (Yanabu et al., 1981. © IEEE.)

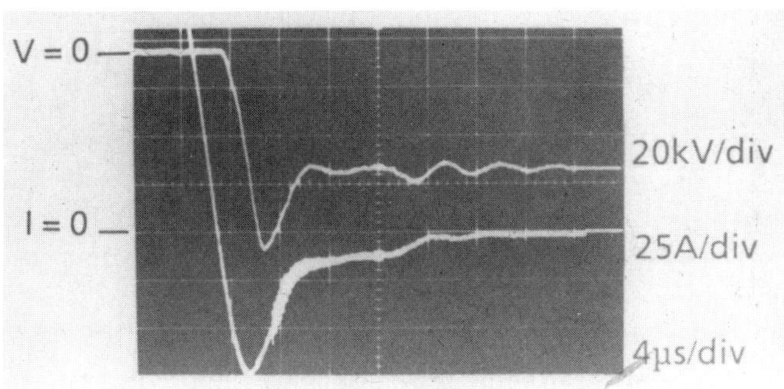

Fig. 3.37 Measured post-arc currents. (Hakamata et al., 1988. © IEEE.)

3.4.3 Practical VCBs and Electrode Configuration

The ordinary structure of vacuum interrupters as VCB interrupting chambers has already been shown in Fig.3.26. Electrode construction varies according to the required performance specifications. Electrode structures used for VCBs can be classified into three types:

1. Plane electrodes
2. Driving magnetic field electrodes
3. Axial magnetic field electrodes

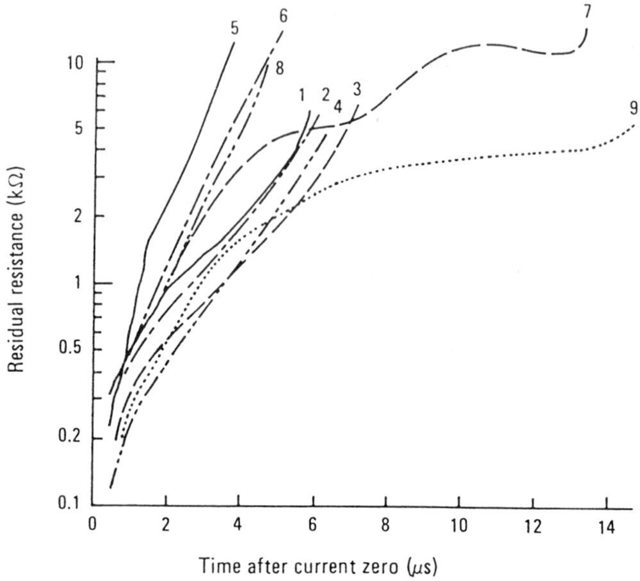

Fig. 3.38 Impedance changes. (Hakamata et al., 1988. © IEEE.)

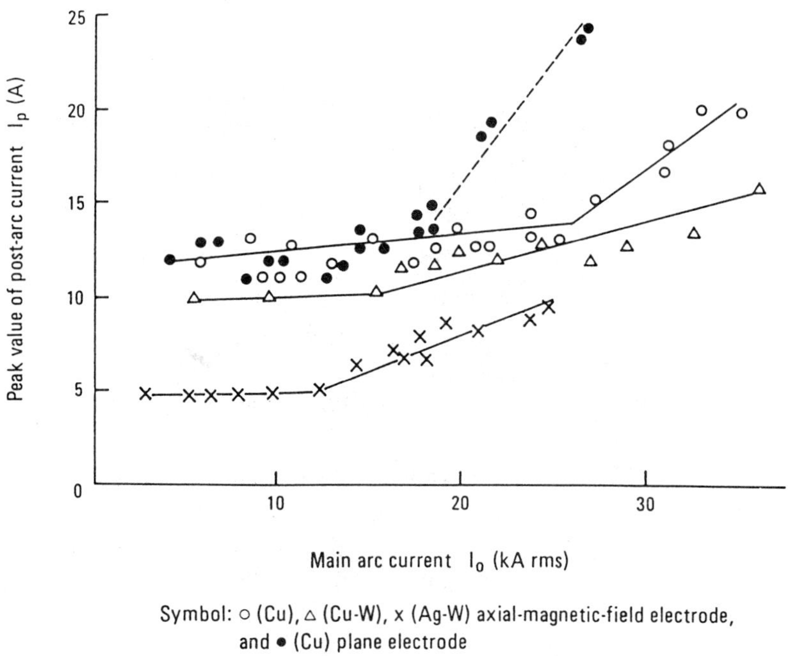

Fig. 3.39 Post-arc current after high-current interruption in vacuum. (Yanabu et al., 1985. © IEEE.)

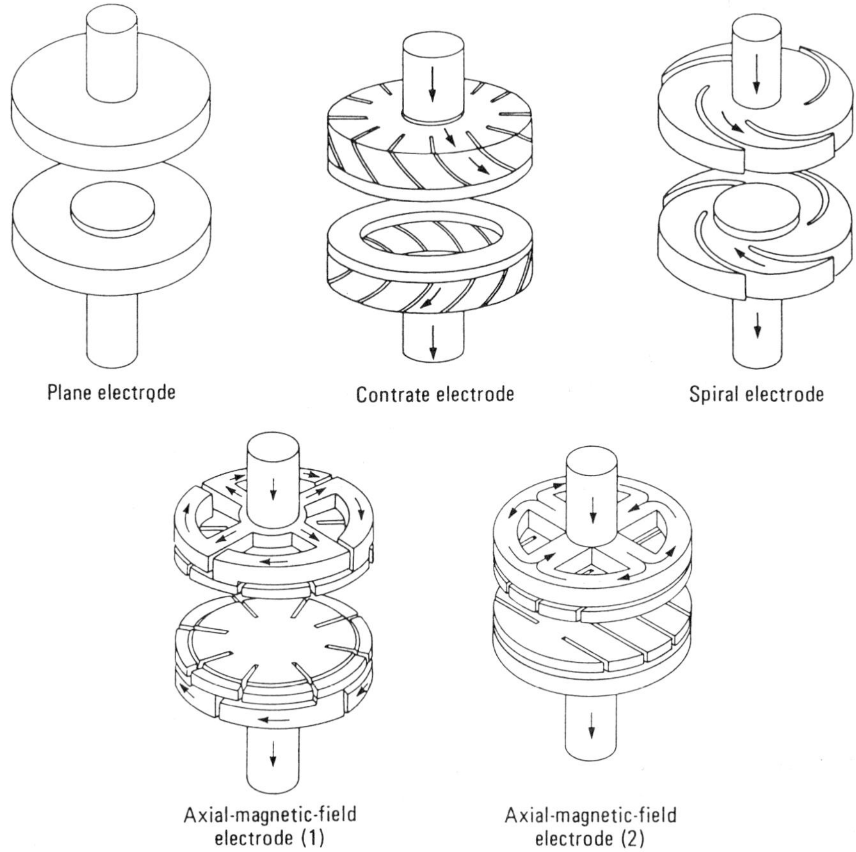

Fig. 3.40 Electrode configurations.

Figure 3.40 shows typical structures of these three types of electrodes. The vacuum interrupter for the first VCBs is composed simply of plane electrodes butted in a vacuum enclosure. Ordinarily, to increase interrupting currents, large electrode diameters are employed. With a current greater than 10 kA, the vacuum arcs collect and remain localized, causing the spot to be heavily melted as the electrode temperature rises. Thus it is difficult to interrupt current above 10 kA by using plane electrodes. For these reasons, plane electrodes are ordinarily used for vacuum contacts with interrupting currents below 8000 A and other similar equipment.

Driving magnetic field electrodes are employed to prevent local electrode melting by driving arcs through the use of a transverse magnetic field. One of these is a spiral electrode designed to cause a driving magnetic field through a current flow along grooves provided in it. The arcs are continuously moved over the electrode surface by interaction between the magnetic field and arcs between the electrodes. The development of spiral electrodes gave a considerable increase in interrupting capacity, permitting 40-kA currents to be interrupted. Spiral electrodes have little effect in interrupting currents above 40 kA. Recently, a contrate electrode designed to intensify the driving magnetic fields in a spiral has been developed, enabling interruption of currents greater than 40 kA.

A more effective arrangement is an axial magnetic field system that applies a magnetic field in the axial direction of the positive-vacuum arc column. With VCBs using electrodes in this system, interruption of currents above 200 kA has been achieved in the laboratory (Yanabu et al., 1981; Kurosawa et al., 1980), and it is considered possible to interrupt still larger currents. In the initial phases of development, a VCB having a coil for generating a magnetic field on the exterior of the vacuum interrupter enclosure was designed. Thereafter, another VCB designed to have the electrode itself form an axial magnetic field was invented in the rapid progress of VCBs toward large capacities.

Figure 3.41 shows the relation between electrode diameters and interrupting currents for magnetically driven and axial magnetic field electrodes.

The electrodes of VCBs must be able to conduct current and high withstand voltages in addition to interrupting current. Generally, the electrodes described above have their own merits and disadvantages. Selection from among these should be made in accordance with the intended use while considering other factors such as cost.

While electrode materials with high withstand voltages were being studied, studies on insulation in vacuum and techniques for treating and machining materials advanced, permitting the withstand voltage per vacuum interrupter to rise remarkably. VCBs that achieve interruption of 66–77 kV per break are already widely used, and it has been reported that a vacuum interrupter model for 169 kV, 31.5 kA has been produced in a laboratory (Yanabu et al., 1986).

Figure 3.42 shows an example of the analysis of electric field distributions in a vacuum interrupters for determining the structure of the interrupter.

3.4.4 Switching Overvoltages

Early in the development of VCBs, it was noted that they exhibit current-chopping phenomena due to their high arc extinction capacity, and they tend to cause current-chopping surges. This overvoltage, however, is effec-

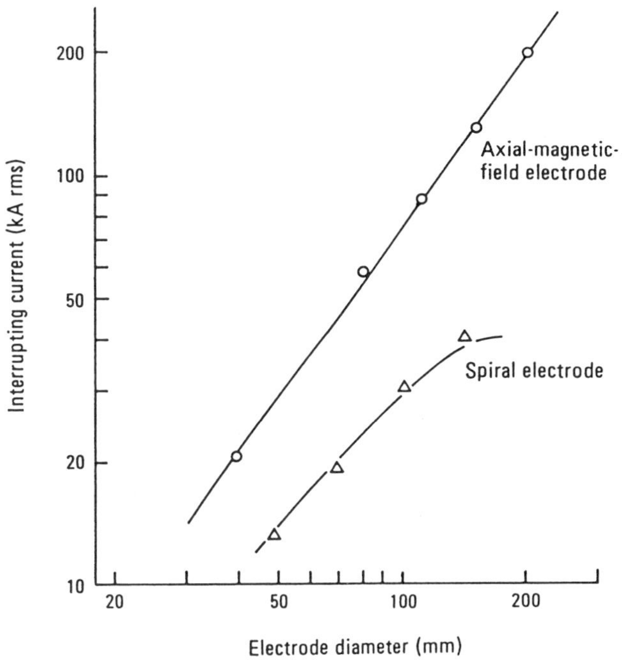

Fig. 3.41 Relation between diameters and interrupting currents for magnetically driven and axial-magnetic-field electrodes. (Yanabu et al. (1981). © IEEE.)

tively no longer problematic owing to the recent development of contact materials with low chopping current.

Multiple reignition, voltage escalation, and virtual current-chopping phenomena have since been observed and studied by many investigators (Nishiwaki and Yanabu, 1977; Murano et al., 1974; Matui et al., 1988; Greenwood and Glinkowski, 1988; Colombo et al., 1988). Even if the current is interrupted after the electrodes of a circuit breaker open near a current-zero point, reignition occurs if the transient recovery voltage directly after current interruption exceeds the dielectric strength of the circuit breaker. If high-frequency current flowing at reignition cannot be interrupted, it leads directly to an arc and is interrupted at the next ac current-zero point. However, if the high-frequency current interrupting capacity of the circuit breaker is great and the current level and interrupting phase are suitable, this high-frequency current may be interrupted directly after reignition. If the endurance capability is not sufficient for the transient recovery voltage occurring subsequently, a repetitive reignition phenome-

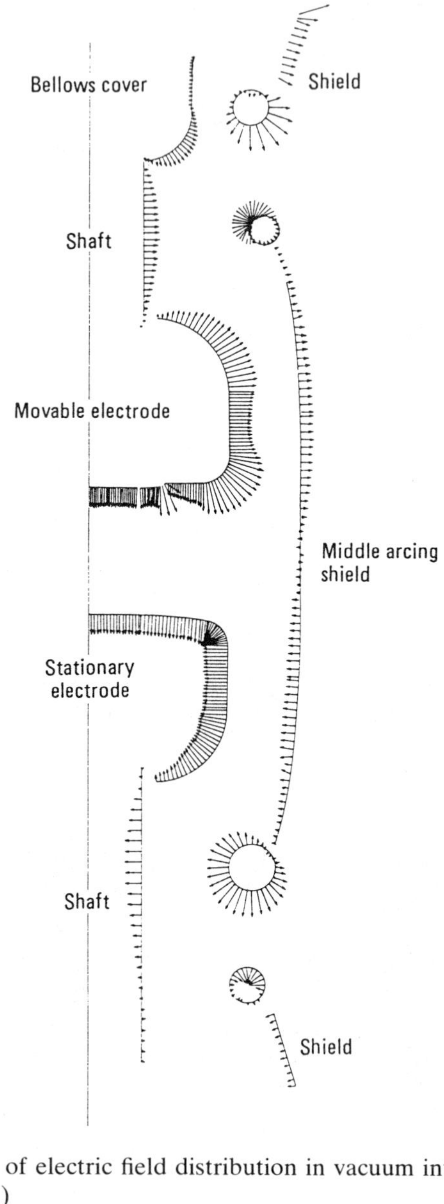

Fig. 3.42 Analysis of electric field distribution in vacuum interrupter. (Shioiri et al. (1982). © IEEE.)

non follows. The surge generated here, termed *multiple reignition surge,* presents problems: its amplitude gradually increases, and it is repeatedly applied to the load equipment. Figure 3.43 illustrates repeated reignition. As reignition is repeated, the level of the load inductance current rises, causing the surge voltage amplitude to increase. The surge voltage amplitude will not increase without limit, however. Ultimately, when the interelectrode withstand voltage exceeds the surge voltage, reignition ceases. Also, the level of the load inductance current will not increase without limit. It will finally decrease, and the surge voltage amplitude will also decrease, causing the repetition of reignition to stop. Detailed and precise descriptions are presented in the book by Broune (1989).

As repetitive reignition occurs and the surge voltage rises in a three-phase circuit, the high-frequency current at reignition can, under certain circuit conditions, transfer into other phases, causing current zero to be formed forcibly in these phases and ultimately resulting in current interruption. This is termed *three-phase simultaneous interruption* (virtual chopping) and is shown in Fig.3.44. In such cases, surge voltages are likely to be larger than with repetitive reignition.

To protect power equipment from the surge voltages described above, various methods have been developed, including the improvement of electrode materials and circuit conditions by providing certain protective means. For example, the development of electrode materials with very small chopping currents nearly eliminated the problem of chopping surge. With VCBs in particular, electrode materials are so effective that their development has been eagerly pursued while considering withstand voltage and interrupting performance, successfully enabling surges caused only by current chopping to be controlled to a safe level.

When the interrupting current reaches a certain level, the number of charged particles that are emitted from the cathode spots becomes less than the number diffused into vacuum. The arc becomes unstable because of the lack of charged particles, and current chopping occurs. Therefore, increasing the number of neutral vapor atoms emitted from the cathode spots may be effective in reducing the chopping current. Figure 3.45 shows the relation between amount of materials added and level of chopping current. It is evident that suitable additives included in electrode materials greatly reduce the level of chopping currents. Such well-designed electrode materials were first used in circuit breakers with small interrupting current, and their development has gradually been extended to applications in circuit breakers with large interrupting currents. Surges causes by repetitive reignition and three-phase simultaneous interruptions have also been controlled by improving electrode materials.

Generally, low-surge VCBs whose interrupting ability is not high are

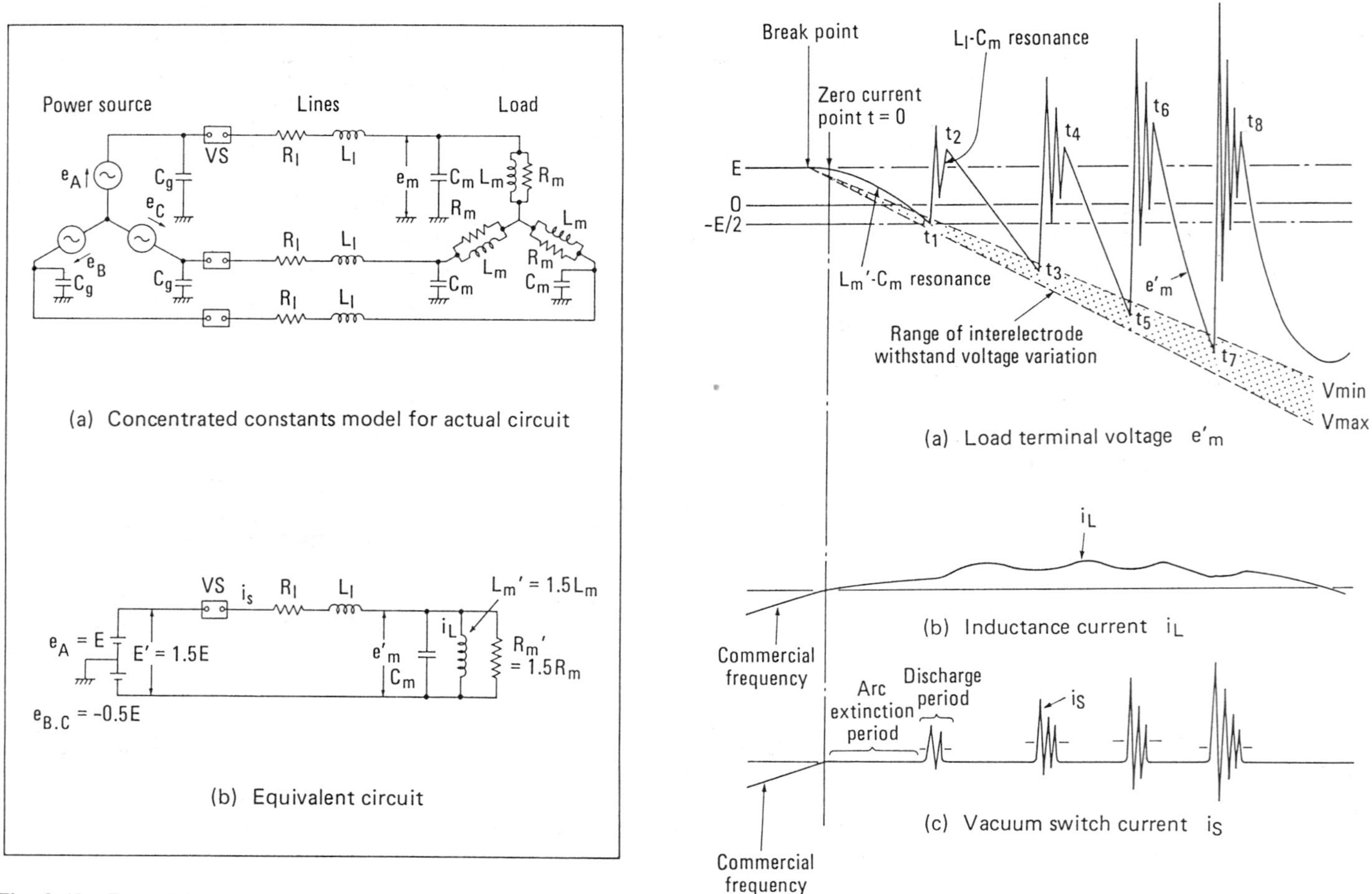

Fig. 3.43 Repetitive reignition.

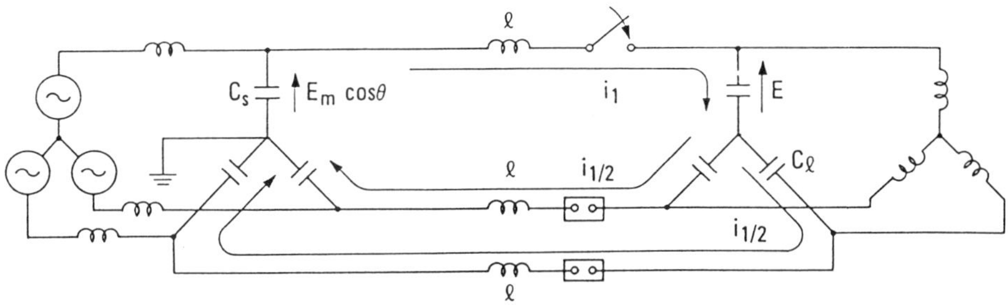

Fig. 3.44 Three-phase simultaneous interruption.

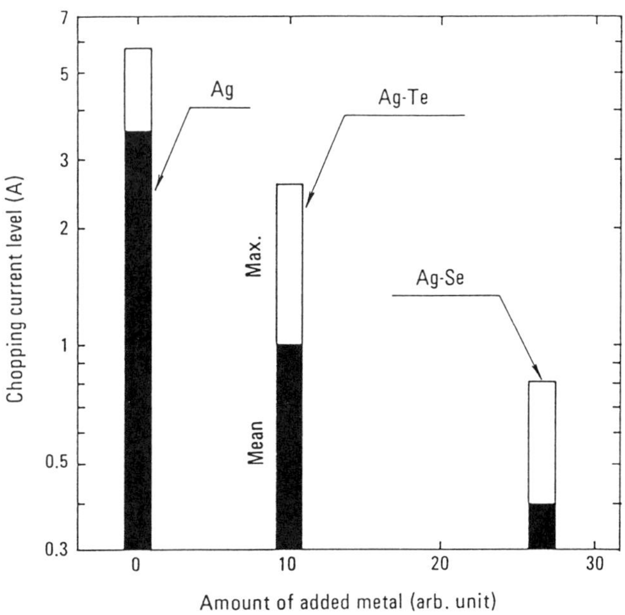

Fig. 3.45 Relation between materials and chopping current levels. (Kurosawa et al. (1985). © IEEE.)

used without protection. But VCBs that must interrupt high current should be provided with some type of protection, such as

1. Capacitor surge suppressor
2. C-R surge suppressor
3. Lightning arrester
4. ZnO surge suppressor

The capacitor surge suppressor consists of a capacitor mounted on the load side of a VCB. It reduces the surge impedance on the load side, thus decreasing the peak of overvoltage and decreasing the rate of rise of recovery voltage (Murano et al., 1974). However, it has such disadvantages as an increase in chopping current level. Because the effective capacitance across the breaker increases, the arc becomes unstable.

In contrast, the C-R suppressor (consisting of a capacitor connected in series with a resistor) can reduce the reigniting current even if reignition occurs, thus suppressing a voltage buildup or three-phase simultaneous interruption. Figure 3.46 shows cumulative distributions of surges generated when a conventional VCB is used to switch a 3.3 kV, 45 kW induction motor. This shows the significant effects of these protective methods (Yanabu et al., 1986).

A lightning arrester is useful in limiting the absolute value of overvoltage because of its discharge at the sparkover voltage. However, a conven-

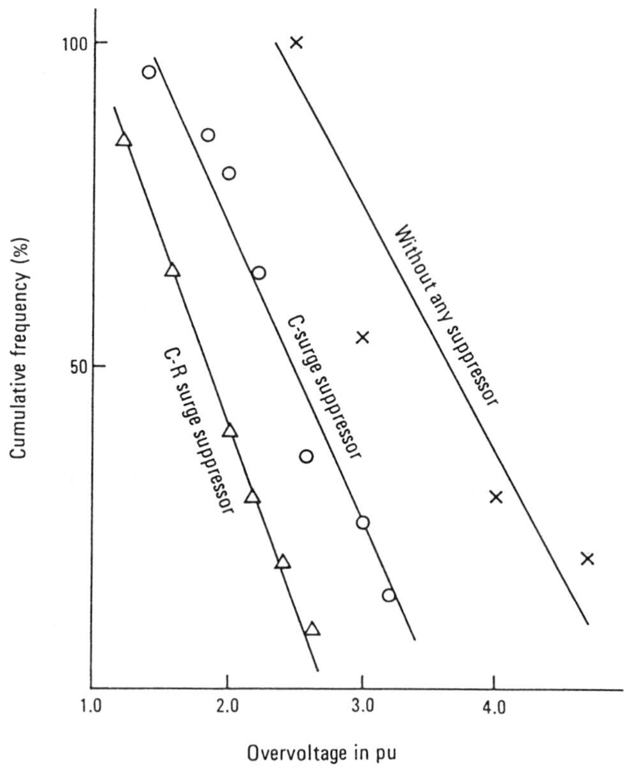

Fig. 3.46 Effects of protective methods. (Kaneko et al. (1982). © IEEE.)

tional arrester has a high protective level and is not always effective as protection for a motor or other equipment having little margin in insulation because of the nature of the insulation design.

The recently developed ZnO surge suppressor is useful because it has a high-frequency response against steep surge voltage waves.

Recent studies on the occurrence of surges have been so advanced that it can be considered that VCBs can be operated at full performance level by taking adequate care in the selection of suitable contact materials for load circuits and by installing protectors.

3.5 HIGH-VOLTAGE DIRECT-CURRENT CIRCUIT BREAKERS (Y. Yoshioka, K. Hirasawa)

3.5.1 Brief History of Development

In 1954 the first high-voltage direct-current (HVDC) transmission system was put into operation between Gotland and the mainland of Sweden. Its system voltage and capacity were 100 kV and 20 MW, respectively (Rusck et al., 1950). Since then many HVDC transmission systems have been planned, constructed, or commissioned in more than 30 places worldwide, and their total capacity is close to 40 GW.

Most systems commissioned to date are two-terminal schemes, and HVDC breakers are not yet used in the high-potential main circuit of those systems, because the system is expected to perform well using only converter/inverter control even at a fault stage of the transmission line. However, even in a two-terminal scheme there are not a few merits in using an HVDC breaker when the system has two parallel transmission lines, that is, when it is a double-circuit system (Tokuyama et al., 1987).

A system connecting three or more converter stations with HVDC lines is referred to as a *multiterminal transmission system*. In multiterminal HVDC system, circuit breakers are needed to maintain reliability and a degree of freedom in operation, and various types of HVDC circuit breakers have been studied and developed.

The first HVDC circuit breaker was developed in the USSR, where the *inverse current-injecting method* (discussed later) was employed using air-blast breakers for ac interruption. The breaker ultimately succeeded in interrupting 100 kV, 250 A direct current after numerous improvements (Kukekov et al., 1959).

Up to now, various concepts have been developed and reported for HVDC interruption, and the highest ability reaches 500 kV in rated voltage and 8 kA in interrupting current. In this section, typical developments of

HVDC circuit breakers are described. Low-voltage dc breakers such as those used in railway networks will not be mentioned here, and descriptions are limited to breakers to be used in HVDC systems of 100 kV and above.

3.5.2 Technical Subjects in HVDC Interruption

Alternating current circuit breakers interrupt current at its natural zero point, but there is no zero point in direct current. The most important concern in dc interruption, therefore, is to provide the breaker itself with an active means to reduce the current reaching zero. Reducing the current requires the breaker to dissipate the energy stored in the circuit. This energy amounts to $(\frac{1}{2})LI^2$, where L is the total inductance of the transmission lines and dc smoothing reactors and I is the fault current to be interrupted. After the interruption, overvoltage caused by sudden depression of current in the inductive circuit should be low enough that it cannot harm the system or apparatus.

Thus, technical subjects in dc interruption can be summarized as follows: (1) creation of current zero in dc current, (2) dissipation of the energy stored in the inductive circuit, and (3) suppression of overvoltage caused by interrupting current. Both dissipation of energy and suppression of overvoltage are realized by using zinc oxide resistors.

3.5.3 Classification of Interrupting Methods

Figure 3.47 shows the classification of dc interrupting methods based on the way current zero is created. Figure 3.48 shows typical waveforms of current and voltage across the breaker during interruption.

The *inverse-voltage-generating method* reduces the current to zero by making the arc voltage higher than the source voltage, which is con-

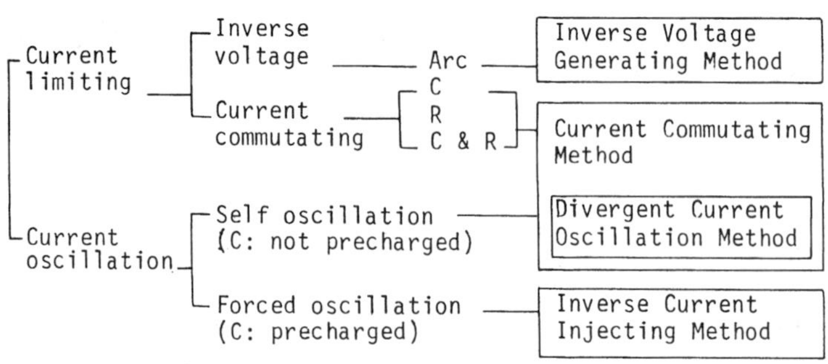

Fig. 3.47 Classification of DC interrupting method.

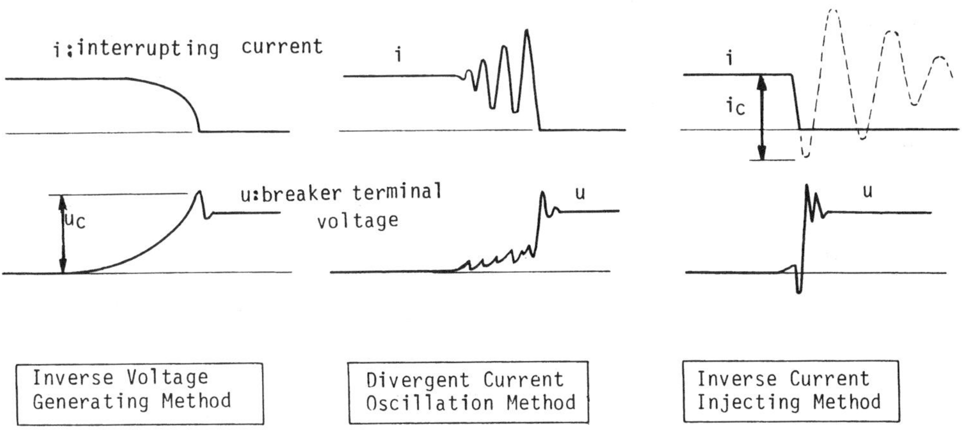

Fig. 3.48 Typical waveforms of current and voltage during dc interruption.

ventional in low-voltage dc breakers of a few kilovolts. Although the method is ideal in the sense that both the reduction of current and the dissipation of energy are done within an interrupter, its application to HVDC systems is difficult because the arc voltage and dissipated energy are very large.

In general applications, the energy is dissipated by commutating the current to a resistor and/or a capacitor partly with the aid of the current-limiting effect of arcing. This is the *current-commutating method.*

The *divergent current oscilation method* creates current zero by magnifying the amplitude of the high-frequency oscillating current in the commutating loop composed of an *L-C* circuit and the arc itself. This oscillation is due primarily to the volt-ampere characteristics of the arc, where the arc voltage decreases with increasing current, which are called the *negative characteristics* of the arc. Oil, air, SF_6, or vacuum can be selected as the interrupting medium, and the ac interrupting technique is applicable to this method, but the interrupter may have to be modified to minimize the parallel capacitance. This method is also classified as a current-commutating method.

The *inverse-current-injecting method* creates current zero by superimposing a high-frequency inverse current on dc current to discharge a precharged capacitor through an inductor. There is little need for the interrupter to dissipate the stored energy, and thus a conventional ac circuit breaker can be selectively utilized in this method, but an auxiliary power source is necessary to precharge the capacitor, and the circuit configuration tends to be more complicated than those of other methods.

3.5.4 Examples of Current-Commutating Methods

Typical examples of current-commutating methods are shown in Table 3.2. A breaker of maximum capacity can interrupt 250 kV, 8000 A when this method is used.

(A) Penning Discharge Tube Method

In low-pressure gas-discharge tubes, it is intrinsically impossible to maintain the discharge at a pressure less than the Paschen minimum. When a magnetic cross-field is applied, however, electrons are trapped and plasma

Table 3.2 HVDC Breakers Employing the Current-Commutating Method

No.	Circuit Configuration	Interrupting Operation	Test Results
(1)	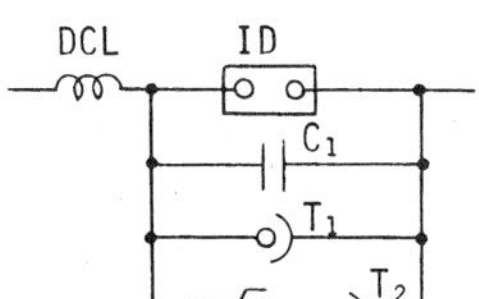	① Open the in-line switch(ID) and raise its arc voltage, ② ignite the cross-field discharge tube (T_1) and transfer the current into T_1, ③ distinct T_1 after recovery of withstand voltage of ID, ④ ignite another tube (T_2) for the non-linear resistor(NR) to absorb the energy.	100kVp 1kA
(2)	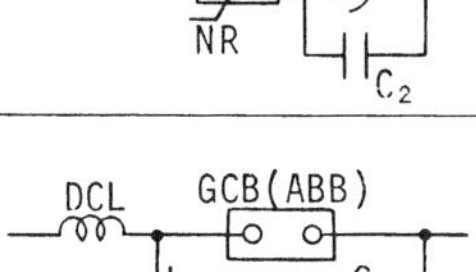	① Open the commutating breaker (GCB), ② magnify the amplitude of high frequency current oscillation in the loop of the capacitor(C) and the inductance (Ls), ③ residual current is reduced in the zinc-oxide resistor(R).	400kV 2kA 250kV 8kA
(3i)	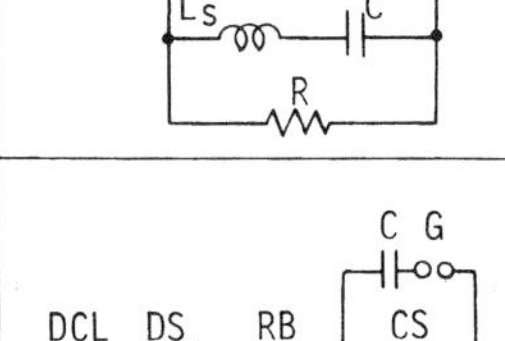	① Open the minimum-oil breaker(CS) and ignite the gap(G) by its arc voltage, transferring the current in CS to the capacitor(C) and the resistor(R), where the current is reduced, ② open another breaker(RB) and transfer the current to the non-linear resistor (NR), thus the energy is dissipated in R and NR, ③ interrupt the small residual current by the switch(DS).	85kVp 2.5kA
(3ii)	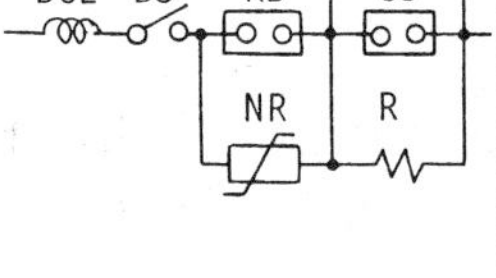	① Open the liquid SF_6 breaker(CB_1) and transfer the current to the parallel circuit(CB_2-R_1), ② open another breaker CB_2 and interrupt the current in R_1, ③ dissipate the energy by R_1.	109kVp 1kA

Source: (1) Reed (1972); (2) Vithayathil et al. (1985) and Tokuyama et al. (1982); (3i) Ekstrom et al. (1976); (3ii) Gabin et al. (1974).

is formed between the electrodes. The plasma has a glow discharge mode, and the application of a magnetic field has an effect similar to an increase in pressure. This discharge is known as a *Penning discharge.* By removing the magnetic field, the plasma is rapidly quenched, and the tube can recover its dielectric withstand voltage across the electrodes according to Paschen's law.

Table 3.2(1) shows an HVDC breaker employing *Penning discharge tubes* (Hofmann et al., 1976). Normally, current flows through an inline switch (ID) and the tubes (T_1 and T_2) are not discharging. To start the interrupting operation, ID is opened and its arc voltage is increased. When the arc voltage reaches several hundred volts, T_1 discharges by the application of the magnetic field and the current through ID is commutated to T_1. After the recovery of dielectric withstand in ID, the magnetic field in T_1 is removed and the discharge is quenched (i.e., the current is interrupted). Just after the interruption of T_1, T_2 is ignited, and a nonlinear resistor (NR) dissipates the energy stored in the circuit including the system. The principle of this method was confirmed in the 1960s, and the first proposal of its use in a breaker of the 100-kV class was published in 1972 (Reed, 1972). Field tests in Pacific Intertie showed that the breaker interrupted 93 kVp, 840 A and the dissipated energy was 805 kJ (Hofmann et al., 1976). The interruption of 100 kVp, 2 kA was also confirmed in their laboratory tests, and the energy dissipated appeared to be about 1 MJ. Since magnetic control becomes ineffective after the discharge mode transfers from glow to arc, limits on the performance and life of breakers of this type are determined by the mode transfer.

(B) Divergent Current Oscillation Method

This is one of the current-commutating methods using the negative characteristic of the arc. An example of this method is shown in Table 3.2(2). The dc interruption is made possible by creating a zero point in the current flowing through the commutating breaker GCB (ABB) to magnify the amplitude of the high-frequency current oscillation between a parallel capacitance C and an inductance L_s of the commutating circuit by means of the negative characteristic of arcing. The irregular fluctuation of arc length plays an important role in the initiation and growth of oscillation.

A patent application for this method appeared in 1912 for low-voltage use (Burstyn, 1912), but much progress was made in high-voltage and large-capacity technologies in the 1980s, and test results interrupting 400 kV, 2000 A (Vithayathil et al., 1985) and 250 kV, 8000 A (Tokuyama et al., 1982) were reported. The advantages of this method are the simple configuration of the circuit and the applicability of an ac interruption technique in the commutating breaker. The size of the system is determined mainly by the capacity of the parallel capacitor, which has to be increased as the dc current increases.

(C) Interruption with the Aid of Arc Voltage

Although it is difficult to interrupt a large HVDC current only by the current-limiting effect of arc voltage, step-by-step commutation of current is possible, provided a few appropriate commutating circuits are connected in parallel.

Two typical examples of this method are shown in Table 3.2(3i) and (3ii). One uses a minimum oil circuit breaker (Ekstrom et al., 1976), and the other uses a liquid SF_6-blast circuit breaker (Gabin et al., 1974). In the first example (3i), the minimum oil circuit breaker (CS) is opened and its arc voltage is increased. When the gap (G) is ignited due to the increase in arc voltage, the current in CS is transferred to the capacitor (C). A fraction of the current flows into the resistor (R) and is gradually reduced. After that, another circuit breaker (RB) is opened, and the current is transferred to a nonlinear resistor (NR), where it is reduced further. The energy stored is absorbed by R and NR. Finally, the resultant small residual current is interrupted by a switch(DS), and the interruption is completed.

In the second example (3ii), the liquid-SF_6-blast circuit breaker (CB_1) is opened and the current is transferred to a branch circuit composed of another liquid-SF_6 circuit breaker (CB_2) and a resistor (R_1) in series. Final interruption is accomplished by CB_2. Transient recovery voltage is controlled by capacitors C_1 and C_2.

It has been reported that breakers (3i) and (3ii) have interrupted 40 kV, 2400 A and 120 kV, 1000 A, respectively. Since this method relies, at least partially, on the current-limiting effect of arc voltage, there is a limit to the higher rated voltage, and its circuit configuration is a little complicated because two or more sets of breakers must be employed.

3.5.5 Inverse-Current-Injecting Method

Table 3.3 shows typical examples of developments using the inverse-current-injecting method. This method has the longest history of use and provides the highest capacity dc interruption among the many HVDC breakers developed. There are some variations in circuit configuration with this method, but they follow the same principle of interrupting the dc current by injecting a discharge current from a precharged condenser to create zero points.

Item 1 in Table 3.3 uses a vacuum circuit breaker and three gaps, which

Table 3.3 HVDC Breakers Employing the Inverse-Current-Injecting Method

No.	Circuit Configuration	Interrupting Operation	Test Results
(1)	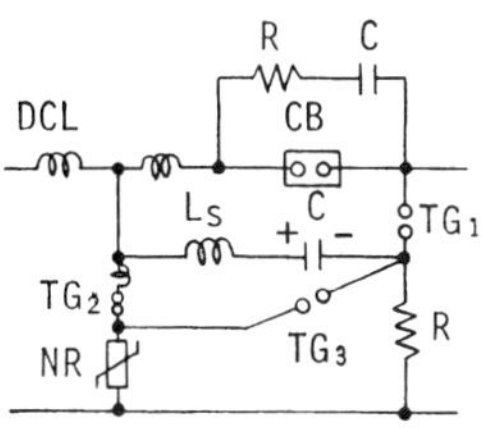	① Open the vacuum breaker(CB) and then ignite the gap(TG_1), discharging the capacitor(C) to create current zero-points, ② interrupt the current in CB at its zero-point, ③ ignite the gap(TG_2) to dissipate the energy in the non-linear resistor(NR), ④ ignite the gap(TG_3) to interrupt the current through NR.	100kVp 10kA
(2)	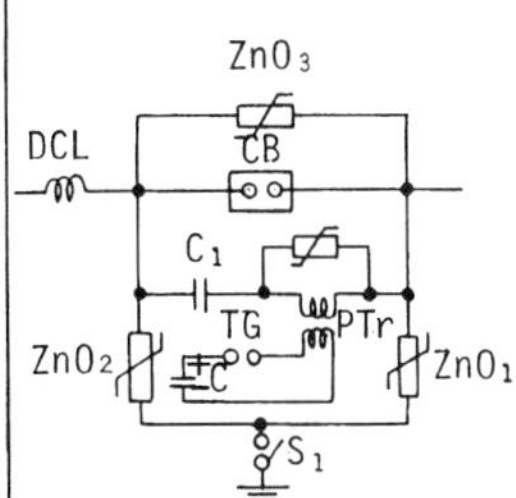	① Open the hybrid circuit breaker (CB) composed of a vacuum interrupter and an SF_6 blast interrupter in series, ② ignite the gap(TG) to inject a high frequency current into CB by discharging the capacitor(C) through the pulse transformer(PTr) and the blocking capacitor(C_1), ③ interrupt the current in CB at its zero-point, ④ close switch (S_1) and the energy is absorbed by the non-linear resistors (ZnO_1, ZnO_2)	380kVp 8kA 250kA 1.4A (GIS type)
(3)	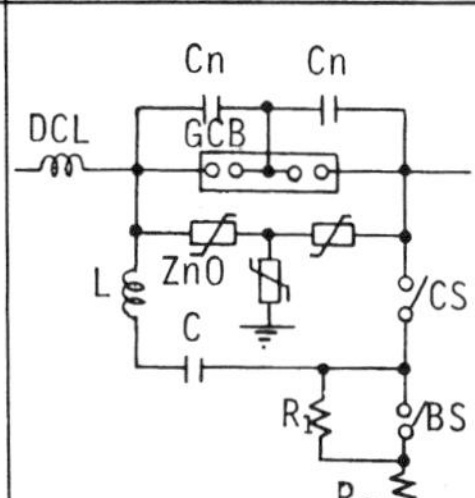	① The capacitor(C) is normally charged by the system voltage through the resistors(R_1 and R_2), ② open the breaker(GCB) and close switch(CS) to discharge C through the reactor(L) super-imposing a high frequency current on the DC current, ③ interrupt the current in the GCB at its zero-point, ④ open CS and close the switch(BS) to charge C quickly up only through R_2.	250kV 8kA (GIS type)

Source: (1) Premerlani (1982); (2) Funahashi et al. (1982); (3) Tokuyama et al. (1985).

are ignited in turn by inserting the inverse current and a (nonlinear) resistor. This breaker has interrupted 100 kVp, 10 kA (Premerlani, 1982).

Item 2 in Table 3.3 uses a vacuum breaker and a gas-blast breaker in series. A special feature of this example is that it uses a pulse transformer (PTr) and a precharged capacitor is set up on the low-voltage side of PTr. A breaker of this type has interrupted 380 kVp, 8 kA (Funahashi et al., 1982) and has reportedly been used as a breaker for HVDC gas-insulated switchgear (GIS) rated 250 kV, 1.4 kA. This was the first time that SF_6 gas insulation was used for an HVDC application.

Item 3 in Table 3.3 uses a gas-blast breaker and a capacitor precharged by the system itself. The breaker of this scheme interrupted 250 kV (400 kVp), 8.8 kA (Tokuyama et al., 1985) and employed SF_6 gas insulation like item 2 in application to GIS.

3.5.6 An Example of HVDC Breaker Design

Figure 3.49 shows the physical assembly of breaker No. 3 of Table 3.3 (Tokuyama et al., 1985). The target of its ratings is an interrupting current of 8000 A, commutation voltage of 400 kV, recovery voltage of 250 kV, and energy dissipation of 75 MJ. The interrupting current is determined to be 1.6 times the rated current of 5000 A of an HVDC transmission system expected in the future. The commutation voltage is determined by taking account of the balance between the required insulation level for the system and the voltage generated by the breaker itself. It is generally 1.6–1.7 times

Fig. 3.49 High-voltage dc breaker prototype, 250 kV 8 kA (Tokuyama et al., 1985. © IEEE)

the system voltage. The recovery voltage is equal to the system voltage in normal operation. The energy dissipation is calculated from the energy stored in the dc circuit and the number of consecutive opening operations. The stored energy is $(\frac{1}{2})LI^2$, assuming that $L = 0.8$ H and $I = 8000$ A. Three consecutive operations are considered, because the HVDC breaker is required to perform a high-speed reclosing operation taking multiple line faults into account.

The interrupter of the HVDC breaker is a puffer-type SF_6 gas interrupter for an ac circuit breaker rated at 362 kV and 63 kA. Each break of the interrupter is provided with a capacitor (Cn in Table 3.3) for grading or suppressing transient recovery voltage. This interrupter has the *di/dt–dv/dt* characteristics shown in Fig. 3.50, and is therefore suitable for dc current interruption by means of the inverse-current-injecting method in the sense that the larger ratio of *dv/dt* to *di/dt* requires the smaller commutating capacitor. On the basis of the characteristics in Fig. 3.50, the commutating capacitance (C) and inductance (L) are chosen as 4 μF and 2 mH, respectively, and therefore the frequency of the inverse current is about 1.8 kHz. In order to charge the commutating capacitor up to HVDC line voltage,

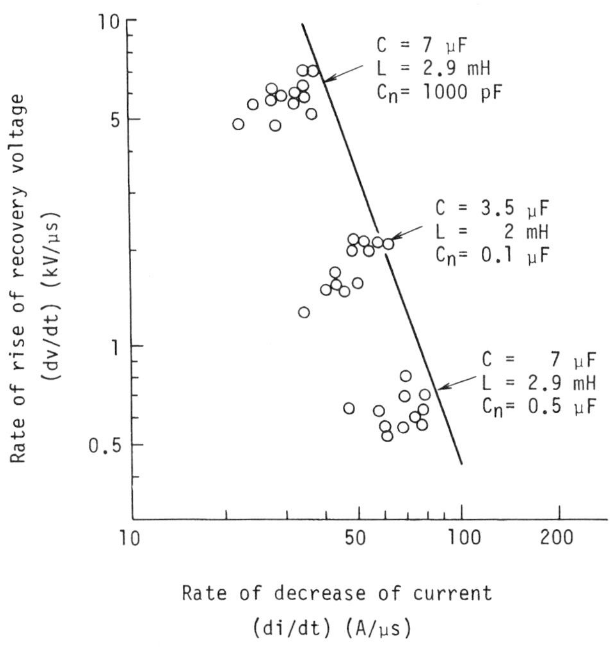

Fig. 3.50 *di/dt* vs. *dv/dt* characteristics of GCB interrupters. (Tokuyama et al., 1985. © IEEE)

two linear resistors are connected in series with the capacitor. One (R_1) is of high resistance with a bypass switch (BS), and the other (R_2) is of low resistance. The bypass switch is normally opened to charge the commutating capacitor through the high- and low-resistance resistors. When high-speed reclosing is required, the bypass switch is closed to charge the commutating capacitor quickly through just the low-resistance resistor. The time constant of the low-resistance charging circuit is determined so that recharging can be completed within 10 msec. The closing switch (CS) is similar in physical construction to the interrupter except for the omission of the puffer chamber and some modifications to the operating mechanism. The energy-absorbing resistor (ZnO) is composed of zinc oxide elements with nonlinear characteristics stacked in series and parallel.

The interrupting performance of the HVDC breaker is confirmed by various equivalent test methods. Commutating performance is verified by a direct test method using an ac short-circuit generator at reduced rotating speed as a current source so that quasi-dc (ac 10 Hz) current is applied to the tested breaker. The test results shown in Fig. 3.50 may also reveal the optimum circuit parameters for minimizing the commutating capacitance. To obtain high dc recovery voltage, an inductive element is used as a voltage source, and this method is regarded as an inductive test circuit (Hofmann et al., 1973). The inductance of the element is around 10 mH, which is much smaller than that of an actual HVDC system (around 0.8 H), but the equivalency of this test circuit is clarified through computer simulations. In this test, the HVDC breaker successfully interrupted current up to 8.8 kA at a commutating voltage of 400 kV. The maximum arcing time was 16 msec, which was incidentally equal to the opening time, and the total commutating time was 32 msec.

Various types of behavior of the HVDC breaker in actual systems is simulated by means of the Electro-Magnetic Transient Program (EMTP), and the effectiveness of application of the breaker is confirmed for the system with two parallel monopolar transmission lines. Figure 3.51a shows a simplified circuit diagram of an HVDC system to be simulated. The system is continuously rated at 2500 MW, +500 kV, and 5000 A, and each line has a continuous current-carrying capability of 5000 A. The rectifier station is under constant current control, and the inverter station is under constant voltage control supplemented with constant margin angle control. Figure 3.51b shows a typical result of simulation under the condition of a faulted line disconnection. In this case a permanent ground fault is set at the middle point M of the line at the moment of 150 msec on the time scale. The direct current at the rectifier station immediately increases and is restored to the normal level by the function of the constant current control. In contrast, dc current at the inverter station immediately decreases and is

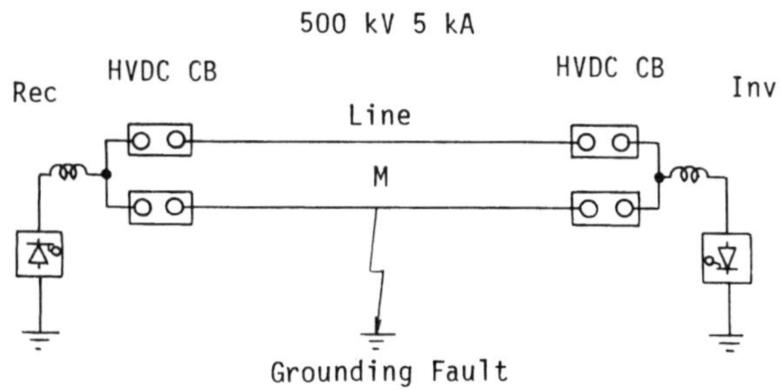

(a) HVDC system with two monopolar transmission lines for simulation analysis

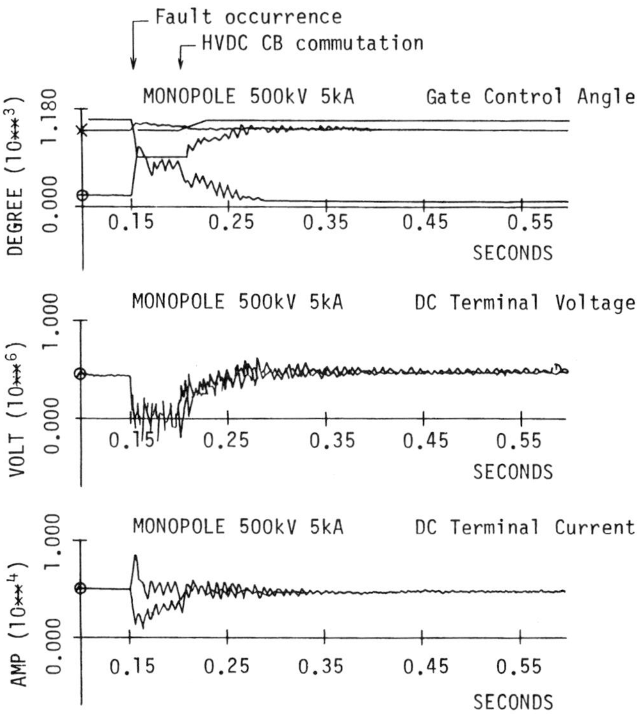

(b) Results of simulation on fault line disconnection

Fig. 3.51 Simulation of HVDC breaker operation. (Tokuyama et al., 1987. © IEEE)

restored to the normal level. The fault is detected by a protective relay system, and the breakers at both terminals of the faulted line are simultaneously opened to transfer the current to the unfaulted or sound line at 50 msec after the fault occurs. It is recognized from the results in Fig. 3.51 that the HVDC breaker is able to disconnect the faulted line without interrupting power transmission and to restore the HVDC system completely within 150 msec after the fault occurs. Another merit of this application is a big reduction of energy dissipation in the breaker during switching because the existence of a parallel line provides an extra current-commutating path. More detailed simulation results under various conditions are presented in paper by Tokuyama et al. (1987).

3.5.7 Other Switching Equipment for HVDC Systems

There are other types of dc switching equipment in operation besides the HVDC breakers for main circuits. These are the (1) high-speed bypass switch (HSBS); (2) low-voltage high-speed switch (LVHS); (3) metallic return protecting breaker (MRPB); and (4) metallic return transfer breaker (MRTB). These switches use any of the interrupting methods already mentioned and work as dc breakers, but they do not interrupt as high a voltage or as large a current. Table 3.4 shows some applications of these switches.

(A) High-Speed Bypass Switch

The high-speed bypass switch (HSBS) is connected in parallel with a converter and is used to bypass a dc current, which is closed at the *operation stop* of part of the converter. It is also used to interrupt and transfer the current at the *operation start* of part of the converter (Hausler et al., 1980). Since the transfer is easily achieved by an arc voltage of the valve drop higher than a few kilovolts, ac breakers such as minimum oil breakers, which have relatively high arc voltage, are employed for HSBS. Thus, a simple method like an inverse-voltage-generating method is sufficient as the interrupting principle for this application.

(B) Low-Voltage High-Speed Switch

A low-voltage high-speed switch (LVHS) is set up on the earth side of a converter and is used to clear the ground fault between a converter transformer and valves, or it is used on the low-voltage side of the converter. The LVHS transfers the fault current to the earth return circuit, withstanding a transient voltage of a few tens of kilovolts across the contacts. This voltage is higher and more difficult to interrupt than that on the HSBS. Nevertheless, an inverse-voltage-generating method is employed using ac circuit breakers (Hausler et al., 1980).

Table 3.4 Examples of the Practical Use of Various Types of DC Switching Equipment

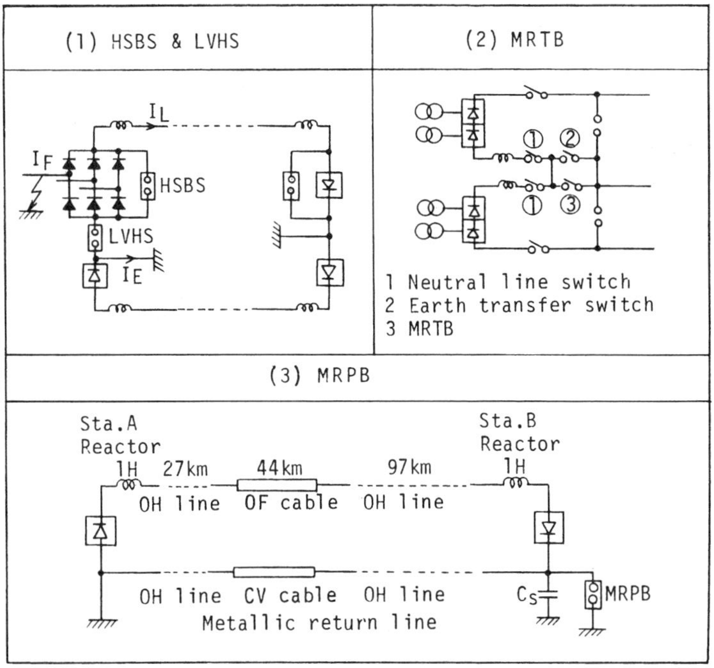

Source: (1) Hausler et al. (1980); (2) Courts et al. (1982), © IEEE; (3) Sakai et al. (1981), © IEEE.

(C) Metallic Return Protecting Breaker and Metallic Return Transfer Breaker

In steady-state operation, little current flows in the earth return circuit of bipolar two-terminal systems. When one of the poles stops, however, all the current in the other pole flows into the earth return circuit. It is undesirable to allow dc current to flow into the earth for a long time because of the electrolysis effect. Thus an overhead (OH) line of the stopped pole is used as a metallic return, and a metallic return transfer breaker (MRTB) is used to interrupt the earth current and transfer that current to the metallic return.

Another application has been proposed in Japan. In the system employing the metallic return scheme, when the ground fault occurs in the metallic return, a metallic return protecting breaker (MRPB) is used to interrupt the fault arc and then transfer the current from the earth into the metallic return, like the MRTB, without stopping power transmission.

As an example of an MRTB, a switch that uses a vacuum breaker under a high-frequency magnetic cross-field has been developed (Courts et al., 1982). It generates a divergent oscillation in the arc voltage as well as the current to be interrupted. In field tests, the MRTB of this method succeeded in transferring a current of 1585 A to generate an inverse voltage of 81 kVp.

Another practical example of an MRPB, also a switch, uses a divergent current oscillation method with an air-blast breaker. It has been used in operation after current-transferring field tests of 42 kVp, 1160 A (Sakai et al., 1981).

REFERENCES

Sections 3.1 and 3.2

Baker, B.P., and Wilcox, H.W. (1930). The Use of Oil in Arc Rupture, *Trans. Am. Inst. Electr. Eng., 49:* 431–446.

Biermanns, J. (1929). Hochleistungschalter ohne Oel, *Elektrotech. Z., 1929-2:* 1073–1079, 1114–1119.

Biermanns, J. (1938). Fortschritte im Bau von Druckgasschaltern, *Elektrotech Z., 1938-1:* 165–168.

Biersack, J., Ebert, H., and Lehmann, W. (1966). 220 kV Greifertrenn Schalter, *Siemens Z., 278:* 228.

Cooper, F.S.(1940). U.S. Patent 2,221,671.

Cromer, C.F., and Friedrich, R.E. (1956). A New 115-kV, 1000MVA Gas Filled Circuit Breaker, *Trans. AIEE Pt. III (Power Appar. Syst., PAS-75:* 1352–1357.

Denkigakkai (1989). Denkigakkai Gijutsu Houkoku (Tech. Report of IEE of Japan), Category-2, No. 290. In Japanese.

Easley, C.J., and Telford, J.M. (1964). A New Design 34.5 kV to 69 kV Intermediate Capacity SF_6 Circuit-Breaker, *IEEE Trans. Power Appar. Syst., PAS-83:* 1172–1177.

Einsele, A. (1964). The New Siemens F Circuit Breaker for 220 kV, 15 GVA, *Siemens Rev., 31,* No. 7.

Friedrich, R.E., and Yeckley, R.N.(1959). SF_6 Circuit-Breaker—A New Design Concept, *Westinghouse Eng.,* pp. 51–55.

Kleman, J.N. (1907). U.S. Patent 874,601.

Leeds, W.M. (1941). A High Power Oilless Circuit Interrupter Using Water, *Electr. Eng., 60:* 85–88.

Lingal, H.J., and Owens, J.B. (1953). A New High Voltage Outdoor Load Interrupter Switch, *Trans. AIEE, Pt. III, 72:* 293–297.

Lingal, H.J., Strom, A.P., and Browne, T.E., Jr. (1953). An Investigation of the Arc Quenching Behaviour of Sulphur Hexafluoride, *Trans. AIEE, Pt. III, 72:* 242–246.

Lingal, H.J., Hills, F., and Browne, T.E., Jr. (1956). U.S. Patent 2,757,261.

Natsui, K., Kurosawa, Y., Hakamata, Y., Hirasawa, K., and Yoshioka, Y.(1988). Voltage Distribution Characteristics of Series Connected SF_6 Gas and Vacuum

Interrupters Immediately After a Large AC Current Interruption, *IEEE Trans. Power Delivery, 3:*(1).

Noda, Y., Kashiwagi, Y., Yoshioka, N., Kitakizaki, K., and Sakuma, S.(1987). 145 kV High Current Interruption by a Vacuum Interrupter, IEE of Japan, Conf. Publication. In Japanese.

Ozaki, Y., Ikeda, H., and Uenishi, T.(1988). Recent Trends and Prospects for SF_6 Gas Insulated Switchgear Technique, *Toshiba Rev., 43*(6).

Prince, D.C., and Skeats, W.F. (1931). The Oil Blast Circuit-Breaker, *Trans. AIEE, Pt. III, 81:* 629–639.

Shioiri, T., Oshima, I., and Okumura, H.(1983). Insulation of 168 kV Vacuum Interrupter, IEE of Japan Conf. Publication. In Japanese.

Slepian, J.(1929). Theory of the De-ion Circuit-Breaker, *Trans. AIEE, 48:* 523–553.

Sorensen, R.W., and Mendenhall, H.E.(1926). Vacuum Switching Experiments at California Institute of Technology, Trans. AII, 45: 1102–1105.

Walty, W.(1935). The Brown Boveri Air Circuit-Breaker, *BBC Eng., 22:* 199–207.

Whitney, W.B., and Wedmore, E.B.(1926). Br. Patent 278,764.

Whitney, W.B., and Wedmore, E.B.(1930). Br. Patent 366,998.

Yeckley, R.N., and Cunningham, R.H. (1958). A New 46 kV Low Capacity Circuit Breaker for Multiple-Reclosing Duty, *Trans. AIEE Pt. III (Power Appar. Syst.), PAS-76:* 402–406.

Yoshizumi, T., Ibuki, K., Hosomi, M., Yonezawa, T., and Kawasaki, H. (1989). Hot Gas Flow in SF_6 Gas Circuit Breaker During the Short Circuit Interruption, *IEEE Trans. Power Delivery, 4,*(3): 1757–1764.

Section 3.3

CIGRE Working Group 33/13-09 (1988). Very Fast Transient Phenomena Associated with Gas Insulated Substations, CIGRE 33-13.

Ikeda, H., Ueda, T., Kobayashi, A., Yamamoto, M., and Yanabu, S. (1984). Development of Large-Capacity, SF_6 Gas Interruption Chamber and Its Application to GIS, *IEEE Trans. Power Appar. Syst., PAS-103*(10): 3038–3043.

Kobayashi, S., Yamagata, Y., Nishiwaki, S., Okubo, H., Kawaguchi, Y., Murakami, Y., and Yanabu, S. (1987). Particle Initiated Flashover Caused by Disconnector Restriking Surge in GIS, 5th Int. Symp. on High Voltage Engineering, 12.03 Braunschweig, pp. 24–28.

Nishiwaki, S., Kanno, Y., Sato, S., Haginomori, E., Yamashita, S., and Yanabu, S. (1983). Ground Fault by Restriking Surge of SF_6 Gas Insulated Disconnecting Switch and Its Synthetic Tests, *IEEE Trans. Power Appar. Syst., PAS-102*(1): 219–227.

Ogawa, S., Haginomori, E., Nishiwaki, S., Yoshida, T., and Terasaka, K. (1983). Estimation of Restriking Transient Overvoltage on Disconnecting Switch for GIS, *IEEE Trans. PWRD-1*(2): 95–101.

Suzuki, K., Mizoguchi, H., Shimokawa, N., Murayama, Y., and Yanabu, S. (1984). Current Interruption by Disconnecting Switch and Earthing Switch in GIS, *Proc. IEE, Pt. C,* 131 No. *(2).*

Suzuki, M., Ikeda, H., Mizoguchi, H., Ozaki, Y., and Yanabu, S. (1989). Investigation of Interrupting Performance of Newly Developed 300 kV 3-Phase-in-One-Tank-Type GCB, *IEEE Trans. Power Delivery, 4*(1): 362–67.

Yanabu, S., Mizoguchi, H., Suzuki, K., Kamata, I., and Murayama, Y. (1984). Development of Loop Current Interrupting Disconnecting Switch gor Gas Insulated Switchgear, *IEEE Trans. Power Appar. Syst., PAS-103*(12): 3561–3568.

Yanabu, S., Murase, M., Aoyagi, H., Okubo, H., and Kawaguchi, Y. (1988). Estimation of Fast Transient Overvoltage in Gas Insulated Substation, IEEE PES Summer Meeting, 88 SM 628-0.

Yanabu, S., Ohishi, M., Ozaki, Y., and Murayama, Y. (1986). Development of UHV Gas Circuit-Breaker and Disconnecting Switch, and Their Performance Tests, Cigre Symposium 06-85, 320-02.

Section 3.4

Colombo, E., Costa, G., and Piccarreta, L. (1988). Results of an Investigation on the Overvoltages Due to a Vacuum Circuit-Breaker When Switching an H.V. Motor, *IEEE Trans. Power Delivery, PWRD-3:* 205–213.

Greenwood, A., and Glinkowski, M. (1988). Voltage Escalation in Vacuum Switching Operations, *IEEE Trans. Power Delivery, PWRD-3:* 1698–1705.

Hakamata, Y., Kurosawa, Y., Matui, K., Hirasawa, K., and Kurita, K. (1988). Post Arc Current of Vacuum Interrupter After Large Current Interruption, *IEEE Trans. Power Delivery, PWRD-3:* 1692–1697.

Kurosawa, Y., Sugawara, H., Kawakubo, Y., Abe, N., and Tsuda, H. (1980). Vacuum Circuit Breaker Electrode Generating Multi-pole Axial Magnetic Field and Its Interruption Ability, *IEEE Trans. Power Appar. Syst., PAS-99:* 2079–2085.

Kurosawa, Y., Iwasita, K., Watanabe, R., Andoh, H., Takasuna, T., and Watanabe, H. (1985). Low Surge Vacuum Circuit Breakers, *IEEE Trans. Power Appar. Syst., PAS 104:* 3634–3642.

Matui, Y., Yokoyama, T., and Umeya, E. (1988). Reignition Current Interruption Characteristics of the Vacuum Interrupter, *IEEE Trans. Power Delivery, PWRD-3:* 1672–1677.

Murano, M., Fujii, T., Nishikawa, H., Nishiwaki, S., and Okawa, M. (1974a). Voltage Escalation in Interrupting Inductive Current by Vacuum Switches, *IEEE Trans. Power Appar. Syst., PAS-93:* 264–271.

Murano, M., Fujii, T., Nishikawa, H., Nishiwaki, S., and Okawa, M. (1974b). Three-Phase Simultaneous Interruption in Interrupting Inductive Current Using Vacuum Switches, *IEEE Trans. Power Appar. Syst., PAS-93:* 272–280.

Nishiwaki, S., and Yanabu, S. (1977). Overvoltage Due to Current Chopping in Nongrounded Neutral Three-Phase System, Third International Symposium on Switching Arc Phenomena.

Okumura, H., and Kaneko, E. (1982). Recently Developed Vacuum Interrupter Construction and Performance Improvement, *Toshiba Rev. 141:* 5–10.

Reece, M.P. (1963). The Vacuum Switch, *Proc. IEE, 110:* 793–811.

Sugawara, H., and Takasuna, T. (1972). Some Investigation of High Current Vac-

uum Arcs and the Development of 24 kV–36 kV High Power Vacuum Circuit Breakers, IEEE Winter Meeting, Paper No. C72 528-8.
Sorensen, R.W., and Mendenhall, H.E. (1926). Vacuum Switching Experiment at California Institute of Technology, *Trans. AIEE, 45:* 1102–1105.
Yanabu, S., Okumura, H., Aiyoshi, T., and Kaneko, E. (1981). Novel Electrode Structure of Vacuum Interrupter and Its Practical Application, *IEEE Trans. Power Appar. Syst., PAS-100:* 1966–1973.
Yanabu, S., Homma, M., Tamagawa, T., and Kaneko, E. (1985). Post Arc Current of Vacuum Interrupter, *IEEE Trans. Power Appar. Syst., PAS-104:* 166–172.
Yanabu, S., Satoh, Y., Tamagawa, T., Sohma, S., and Kaneko, E. (1986). Ten Years' Experience in Axial Magnetic Field-Type Vacuum Interrupters, *IEEE Trans. Power Delivery, PWRD-1:* 209–214.

Section 3.5

Burstyn, W. (1912). Verfahren zur Loschung des Schaltern auftretenden Unterbrechungslichtbogens nach Patent 260903, Ger. Patent 268 889.
Courts, A.L., Vithayathil, J.J., Hingorani, N.G., Poter, J.W., Gorman, J.G., and Kimblin, C.W. (1982). A New DC Breaker Used as Metallic Return Transfer Breaker, *IEEE Trans., PAS-101*(8): 4112–4121.
Ekstrom, A., Hartel, H., Lips, H.P., Schultz, W., Joss, P., Holfeld, H., and Kind, K. (1976). Design and Testing of an HVDC Circuit Breaker, CIGRE paper 13-06.
Funahashi, T., Tamagawa, T., Kaneko, E., Takahashi, N., Horiuchi, T., and Yanabu, Y. (1982). Circuit Breaker for DC Transmission—Interrupting Test Results, National Conf. IEE of Japan, paper 873. In Japanese.
Gabin, J., Gratzmuller, J., Gailly, M., Lacoste, A., and Lorcet, P. (1974). Description et essais d'un appareil de coupure destine au reseau a courant continu a haute tension, *Rev. Gen. Electr. 83*(9): 561–569. In French.
Hausler, M.P., Kolsch, H., and Pauli, M. (1980). Application and Switching Duties of High Speed Switches in the DC Switchyard of HVDC Stations, CIGRE paper 14-02.
Hofmann, G.A., Long, W.F., and Knauer, W. (1973). Inductive Test Circuit for a Fast Acting HVDC Interrupter, *IEEE Trans., PAS-92:* 1605–1614.
Hofmann, G.A., LaBarbera, G.L., Reed, N.E., and Shillong, L.A. (1976). A High Speed HVDC Circuit Breaker with Cross-Field Interrupters, *IEEE Trans., PAS-95*(4): 1182–1193.
Kukekov, G.A., Sorokin, P.G., and Shipulina, N.A. (1959). Switchgear for H.V.D.C. Lines, *Direct Current, 4*(5): 123–126.
Premerlani, W.J. (1982). Forced Commutation Performance of Vacuum Switches for HVDC Breaker Application, *IEEE Trans. PAS-101*(8): 2721–2727.
Reed, N.E. (1972). A Novel Switching Device for Use in an HVDC Circuit Breaker, IEEE-PES Summer Meeting, C72-447-1.
Rusck, A., Rathsman, B.G., and Glimstedt, U. (1950). The High Voltage D.C. Power Transmission from the Swedish Island of Gotland, CIGRE paper 406.
Sakai, M., Kato, Y., Tokuyama, S., Sugawara, H., and Arimatsu, K. (1981). Devel-

opment and Field Application of Metallic Return Protecting Breaker for HVDC Transmission, *IEEE Trans., PAS-100*(12): 4860–4868.

Tokuyama, S., Abe, N., Suzuki, K., Sugawara, H., Yoshioka, Y., Arimatsu, K., and Saito, K. (1982). Interruption of High Direct Current—Self-Oscillation Method. Joint Conference of the IEEE of Japan, S5-2. In Japanese.

Tokuyama, S., Arimatsu, K., Yoshioka, Y., Kato, Y., and Hirata, K. (1985). Development and Interrupting Tests on 250kV 8kA HVDC Circuit Breaker, *IEEE Trans., PAS-104*(9): 2453–2459.

Tokuyama, S., Hirasawa, K., Yosioka, Y., and Kato, Y. (1987). Simulation of Interruption of Circuit Breaker in HVDC System with Two Parallel Transmission Lines, *IEEE Trans., PWRD-2*(3): 772–778.

Vithayathil, J.J., Courts, A.L., Peterson, W.G., Hingorani, N.G., Nilsson, S., and Poter, J.W. (1985). HVDC Circuit Breaker Development and Field Tests, *IEEE Trans., PAS-104*(10): 2693–2705.

4
TESTS AND DIAGNOSTICS

Eiichi Haginomori, Hisatoshi Ikeda, and Susumu Nishiwaki, Toshiba Corporation, Tokyo and Kawasaki, Japan

4.1 STANDARDS AND SPECIFICATIONS (E. Haginomori)

Developing a circuit breaker with the ability to cover a set of ratings entails an extremely large expense, and therefore standard circuit breakers covering standard ratings are selected and applied in most systems. Standard ratings generally cover the majority of system requirements, and thus, in the trade, standard rating values can generally be specified.

The typical world standards are those of the *IEC* (International Electrotechnical Commission) and *ANSI* (American National Standard Institute), and their main sections concerned with current switchings are explained in this section. For ease of understanding, typical rated voltages of 145 kV, 245/242 kV, and 525/550 kV are mainly taken up for IEC/ANSI standards, respectively.

4.1.1 Making and Breaking Short-Circuit Currents

(A) Rated Breaking Currents (Rated Interrupting Currents in ANSI)

Circuit breakers must safely break the *rated short-circuit breaking currents* under specified conditions. As both ac and dc components of breaking currents have time-varying characteristics, rated values are specified with respect to the instant of contact separation of the circuit breaker by both the IEC and ANSI. In short, the term "rated breaking currents" often includes only ac components.

In the IEC standard, *ac components* of rated breaking currents are selected from the following rms values: 6.3, 8, 10, 12.5, 16, 20, 25, 31.5, 40, 50, 63, 80, 100 kA, and higher values in the R10 series. *Direct-current components* are prescribed in percentages of ac peak values as shown in Fig. 4.1. The figure corresponds to the time constant of a dc component decrement of 45 msec. Unless otherwise specified, the value of the horizontal axis is equal to the sum of one-half cycle (minimum relay time) and the opening time of the circuit breaker.

In the ANSI standard, combined with other rated values, rated breaking (interrupting) currents are listed in the rating tables as shown in Table 4.1 for typical rated voltages of 145, 242, and 550 kV. DC components are also shown in Fig. 4.1, and they are very close to the IEC curve. In systems

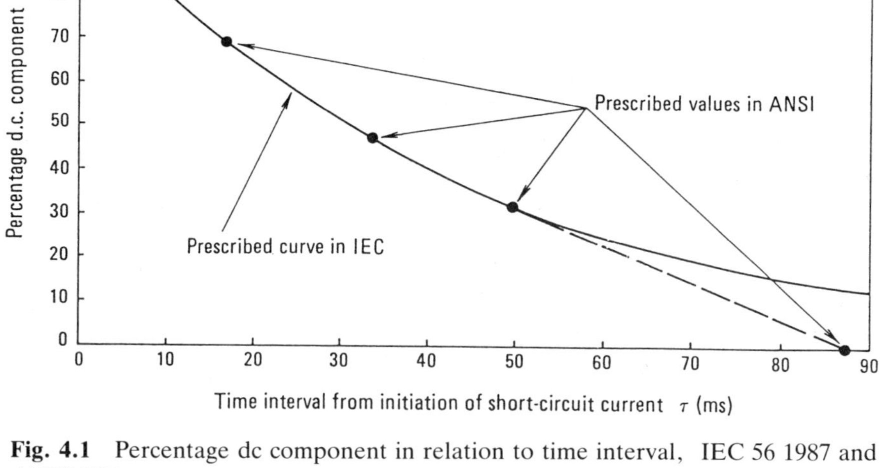

Fig. 4.1 Percentage dc component in relation to time interval, IEC 56 1987 and ANSI C37.

Table 4.1 Ratings for Rated Voltages of 145, 242, and 550 kV

		Transient Recovery Voltage					General-Purpose	Definite-Purpose Circuit Breakers			
									Shunt Capacitor Bank or Cable		
										*Back-to-Back Inrush Current	
Rated Max Voltage kV, rms	Rated Continuous Current at 60 Hz Amperes, rms	Rated Time to Point P T2 μs	Rated Rate R kV/μs	Rated Delay Time T1 μs	Rated Interrupting Time Cycles	Max Symmetrical Interrupting Capability kA, rms	Overhead Line Current & *Isolated Current Amperes, rms	Overhead Line Current Amperes, rms	*Isolated Current Amperes, rms	Peak Current kA	Frequency Hz
145	1200	331	1.7	3.2	3	20	63	160	315	16	4250
145	1600	310	1.8	3.2	3	40	80	160	315	16	4250
145	2000	310	1.8	3.2	3	40	80	160	315	16	4250
145	2000	302	1.8	3.2	3	63	80	160	315	16	4250
145	2000	298	1.8	3.2	3	80	80	160	315	16	4250
145	3000	310	1.8	3.2	3	40	80	160	315	16	4250
145	3000	302	1.8	3.2	3	63	80	160	315	16	4250
145	3000	298	1.8	3.2	3	80	80	160	315	16	4250
242	1600	529	1.8	4.1	3	31.5	160	200	400	20	4250
242	2000	529	1.8	4.1	3	31.5	160	200	400	20	4250
242	3000	529	1.8	4.1	3	31.5	160	200	400	20	4250
242	2000	517	1.8	4.1	3	40	160	200	400	20	4250
242	3000	517	1.8	4.1	3	40	160	200	400	20	4250
242	3000	503	1.8	4.1	3	63	160	200	400	20	4250
550	2000	1325	1.6	5.4	2	40	400	500	500	-	-
550	3000	1325	1.6	5.4	2	40	400	500	500	-	-

* Applicable to shunt capacitor bank and cable current

Source: ANSI C37. © 1987, IEEE.

with longer time constants of the dc decrement, the performance of circuit breakers may be affected. Typical ANSI guides that correspond to reduced breaking capabilities for such cases are shown in Fig. 4.2.

For rated voltages up to 72.5 kV, *k factors* are introduced. Within the voltage range between the rated highest voltages and the rated highest voltages divided by *k*, breaking currents can be inversely proportional to the (recovery) voltages. This means that for lower voltages, circuit breakers can cover higher breaking currents. This characteristic generally corresponds to the performance of oil circuit breakers and increases the flexibility of applications. Actual values of *k* are

1.19–1.36, *2.07 for rated max. voltage of 4.76 kV
1.25, *3.59 8.25 kV
1.30, *1.29–3.75 15.0/15.5 kV
*1.0, *2.15 25.8 kV
1.0, 1.65 38.0 kV
*1.21 48.3 kV
1.10, *1.21 72.5 kV

where the starred values are applicable only to oil circuit breakers.

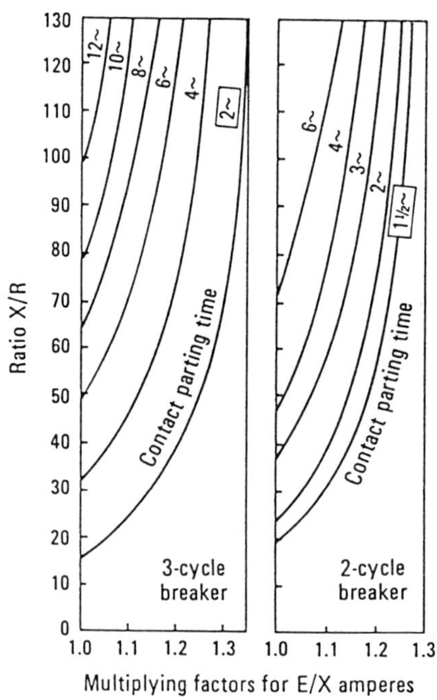

Fig. 4.2 Breaking severity multiplying factors. (ANSI C37. © 1979, IEEE.)

Table 4.1 shows the coordinations of various rating values of ANSI, and preferred IEC coordinations are listed in Table 4.2.

The actual maximum breaking current of each circuit breaker in a certain station may differ from the others. For safety and standardization, generally, the bus-short-circuit current—that is the maximum short-circuit current of the station—is considered for every rated breaking current of the circuit breakers.

(B) Transient Recovery Voltage

Transient recovery voltage (TRV) is voltage that appears across a circuit-breaker's terminals immediately after current interruption. It is as important for current-breaking severity as the magnitude of the breaking current. The specifications in the IEC standards are based on CIGRE Report 13-10, 1968, which is a report on the investigation of European HV and EHV power systems. For rated voltages equal to and above 100 kV, TRVs are represented by three segments, that is, by the *four-parameter method,* which is based on the principle that the envelopes of the various TRV waveforms investigated, drawn superimposed on one diagram, are well represented by three segments (see Fig. 4.3).

The initial rate-of-rise-of-recovery voltages (RRRV) for breaking of 100, 60, and 30% of rated breaking currents are 2, 3, and 5 kV/μsec, respectively. For 10% of the rated breaking current, mainly considering the breaking current through a single transformer, RRRVs of 5.5–12.6 kV/μsec are specified. As for the initial parts of TRVs, time delays of 2–5 μsec

Table 4.2 Coordination Table of Rated Values for Circuit Breakers

Rated voltage U (kV)	Rated short-circuit breaking current I_{sc} (kV)	Rated normal current I_n (A)				
145	12.5	800	1250			
	20		1250	1600	2000	
	25		1250	1600	2000	
	31.5		1250	1600	2000	3150
	40			1600	2000	3150
	50				2000	3150
245	20		1250	1600	2000	
	31.5		1250	1600	2000	
	40			1600	2000	3150
	50				2000	3150
525	40				2000	3150

Source: From IEC 56, 1987. © 1987, IEC.

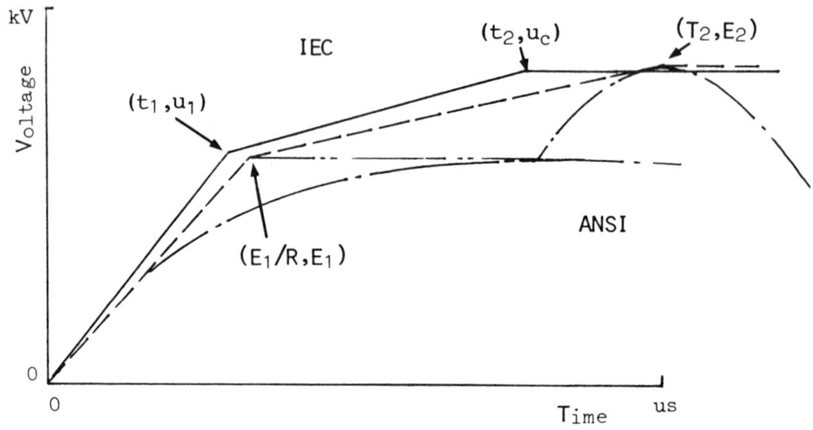

a) Representation in IEC and ANSI

Fig. 4.3 Transient recovery voltage waveforms.

are specified due to the lumped capacitances connected to the busbar sides of the circuit breakers. For outdoor substations, corresponding to oscillations caused by reflections at major discontinuities along the line of the bus sides of the circuit breakers, initial transient recovery voltages (ITRVs), which have small amplitudes, and relatively high RRRVs are specified. For example, for 525 kV and 40 kA, the peak voltage is 5.56 kV, and the time to the voltage peak is 1 μsec. Actual applications of ITRVs in breaking tests are mentioned in the subsection on short-line-fault breaking tests.

The peak factors of TRVs are the products of first-pole-to-clear factors and amplitude factors. The former are 1.3 and 1.5 for neutral effectively grounded systems and neutral noneffectively grounded systems, respectively. The amplitude factors are 1.4–1.5.

For rated voltages up to 72.5 kV, TRVs are more simply specified by applying the *two-parameter method.*

When breaking currents have dc components, lower TRVs appear. In synthetic breaking tests, the voltage source must be adjusted, obtaining these lower TRVs beforehand. Publication 427 of the IEC (Synthetic Tests of High-Voltage Alternating Current Circuit Breakers) lists mathematically calculated values (see Table 4.3).

Typical values of ANSI TRVs are listed in Table 4.1, where the first reference voltage is

$$E_1 = 1.5 \times \sqrt{2/3} \times \text{rated max. voltage*}$$

(*phase-to-phase rms value) and the peak voltage is

$$E_2 = 1.76 \times \text{rated max. voltage}$$

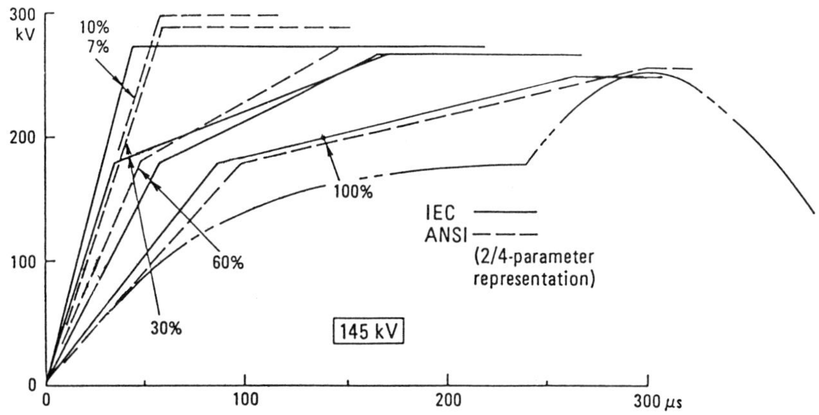

b) Rated voltage of 145 kV

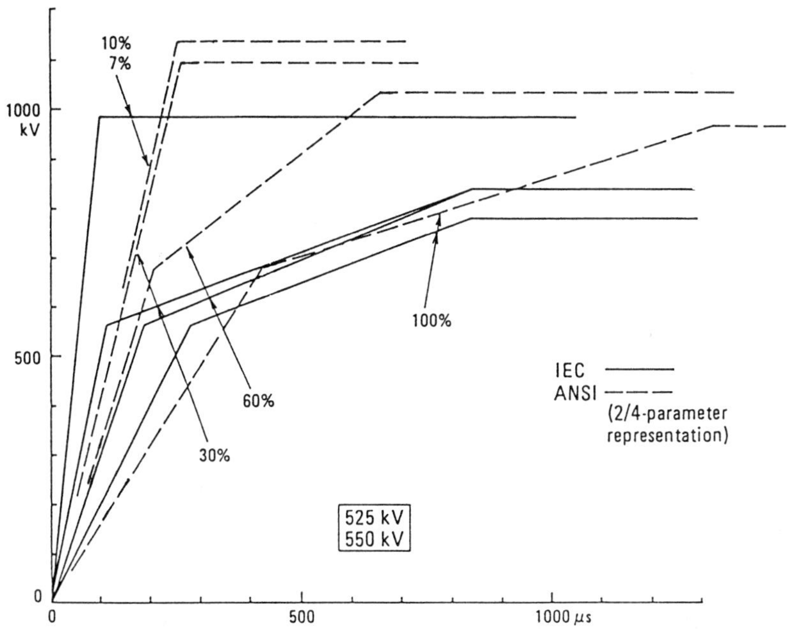

c) Rated voltage of 525kV/550kV

Table 4.3 Transient Recovery Voltage After Breaking Asymmetrical Current (60 Hz)

Voltage (kV)		245				525			
		①		②		①		②	
Time co-ordinate (μs)		t_1 130	t_2 390	t_1 130	t_2 390	t_1 279	t_2 837	t_1 279	t_2 837
		u_1	u_c	u_1	u_c	u_1	u_c	u_1	u_c
Asymmetry at current zero %	0	260	364	260	364	557	780	557	780
	5	260	363	259	364	557	773	556	785
	10	260	361	258	364	555	764	554	789
	15	258	358	256	362	551	753	550	790
	20	257	354	253	360	547	740	545	789
	25	254	349	250	356	540	724	539	786
	30	251	343	245	352	533	707	531	781
	35	247	336	240	346	523	688	521	774
	40	242	327	235	340	512	666	509	764
	45	236	318	228	332	499	641	496	752
	50	230	308	221	323	484	614	481	737
	55	222	296	212	313	467	584	463	720
	60	213	282	203	301	448	550	444	698
	65	203	267	192	287	426	512	421	673
	70	192	249	179	271	400	471	395	644
	75	179	229	165	252	371	423	366	609
	80	163	206	149	231	337	369	332	567

① After major loop ② After minor loop

Source: IEC427, 1989.

For rated voltages of 121 kV and above, and for 100% and 60% of rated interrupting (breaking) currents, ANSI's TRV waveforms are referred to as *exponential cosine envelopes.* This is because the actual TRV waveforms are considered to be (1−exp) forms for the initial parts due to the circuit parameters close to the circuit breakers and (1−cos) forms for the overall shapes due to oscillations of the circuit's short-circuit inductances and whole capacitances to ground.

Although representations of TRVs differ between IEC and ANSI standards, the actual values are quite similar to each other. The largest difference between the two standards is that in ANSI a *first-pole-to-clear factor* of 1.5 is introduced for all rated voltages by considering also three-phase ungrounded fault breaking, whereas the IEC standard also introduces 1.3 for rated voltages of 100 kV and above, mainly considering three-phase grounded fault breaking. The waveforms for the two standards for rated voltages of 145 kV and 525/550 kV are compared in Fig. 4.3. The difference between the two standards is particularly clear for 525/550 kV in Fig. 4.3c. The TRV parameter conversion coefficients specified by ANSI for breaking currents of 60, 30, and 7% of the rated breaking currents are listed in Table 4.4, where, because (1−cos) waveforms represent the TRVs for 30% and 7%, the initial rates of rise of the TRVs are zero.

As previously shown, TRV ratings in IEC and ANSI standards cover the majority of system requirements. Nevertheless, according to IEC, in the special cases listed below, care should be taken, and circuit breakers with higher breaking current ratings or with the ability to cope with higher TRV requirements should be applied.:

Circuit breakers in generator circuits
Circuit breakers directly connected to transformers without appreciable capacitance between the circuit breakers and transformers that provide more than 50% of the short-circuit currents
Circuit breakers adjacent to series reactors

In ANSI C37.011-1979, a detailed explanation of how to calculate TRVs according to system parameters is given according to the *current-injection principle.*

(C) Operating Sequences

Various kinds of operations are assigned to circuit breakers in power systems. The *rated operating sequences* represent various modes of operation of the circuit breakers in a system and circuit-breaker characteristics and performance must be considered. Operating sequences for short-circuit current-breaking tests take the easier test procedures into account and are therefore slightly different from the rated operating sequences.

Table 4.4 Transient Recovery Voltage Multipliers for Rated Parameters

Percent of Interrupting Rating	121 kV and Above		
	R	E_2	T_2
100	1	1.00	1
60	2	1.07	0.5
30	0	1.13	0.2
7	0	1.17	0.2

(For R, E_2, T_2, See Fig. 4.3)

Source: ANSI C37. © 1987, IEEE.

Rated operating sequences in IEC are

O–3m–**CO**–3m–**CO**	for circuit breakers not intended for rapid autoreclosing
O–0.3s–**CO**–3m–**CO**	for circuit breakers intended for rapid autoreclosing
CO–15s–**CO**	for circuit breakers not intended for rapid autoreclosing

where **O** means opening/breaking and **CO** means closing/making immediately followed by opening/breaking.

For short-circuit tests, operating sequences are specified as follows:

For test duties Nos. 1, 2, 3, and 5, which correspond to 10% (symmetrical), 30% (symmetrical), 60% (symmetrical), and 100% (asymmetrical) of rated breaking currents, respectively, test duties consisting of rated operating sequences confined solely to opening operations are specified. The **O** operation in the short-circuit test is the breaking of short-circuit current initiated by making of the making switch of the test plant, following a closing operation of the test circuit breaker. The **CO** operation is allowed instead of each **O** operation. Test duty No. 5 may be changed to be three times **O.**

For test duty No. 4, corresponding to 100% of rated breaking current, **CO** is specified as making a fully asymmetrical short-circuit current, followed by symmetrical current breaking after damping of the asymmetry; that is, the test sequence consists of three symmetrical breakings and two asymmetrical makings.

In short-circuit making and breaking tests, the gas pressure for quenching and the pressure/voltage of the operating device are kept at minimum conditions in order to represent the most severe conditions that could be expected during service. For circuit breakers with short arcing times during short-circuit current breakings, detailed point-on-wave settings of the contact separations are prescribed in order to represent system conditions within each test sequence with three breakings.

In ANSI, standard operating duty is prescribed simply as

CO–15s–**CO**

In actual applications to power systems, various kinds of operations and operating sequences are required. For such applications as multiple reclosing and/or reclosing within 15 sec time intervals, application of a circuit breaker with lower breaking capacity is prescribed in ANSI. Some examples are shown in Fig. 4.4. For instance, when a circuit breaker with a rated breaking ability of 63 kA is used for the operating duty of **O**–0s–**CO**–

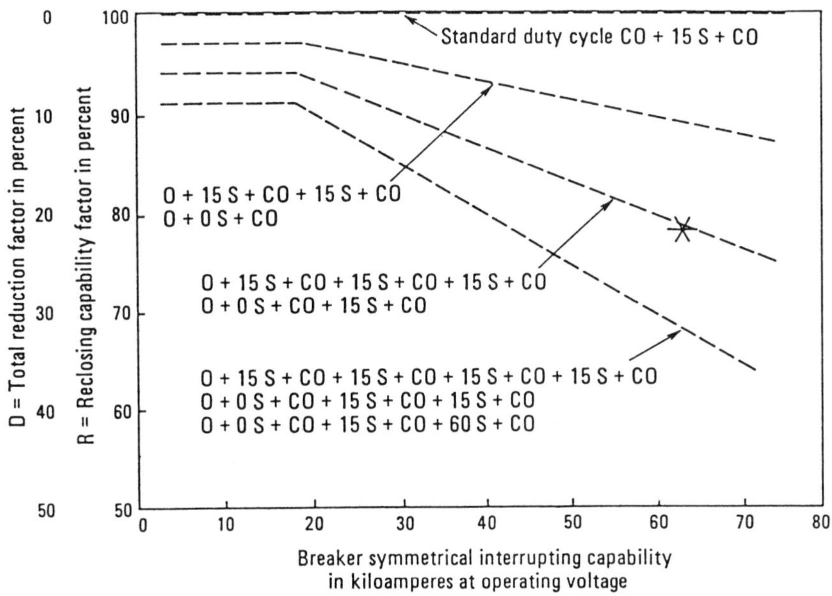

Fig. 4.4 Reduction factor in various operation sequences. (ANSI C37. © 1979, IEEE.)

15s–**CO**, it can be applied up to 78% of the rated breaking current (49 kA) as shown by the asterisk (*) in Fig. 4.4.

In short-circuit making and breaking tests, respective test duties are specified for circuit breakers of various kinds and ratings. The example shown below is for single-phase short-circuit tests of circuit breakers intended for rapid automatic-reclosing and a rated voltage of 121 kV or above.

O and **CO** 0.07**I**–0.13**I**(asym.), 0.2**I**–0.3**I**(sym.)
0.4**I**–0.6**I** (asym.)
CO–15s–**CO** **I** (asym.)
CO–15s–**CO**–15m–**CO**–15s–**CO**–1h–**CO** **I** (sym.)
CO–0s–**CO** *R***I** (asym.)
C–*T*s–**O** **I**
O–15s–**CO** 0.7**I**–0.8**I** (SLF test)

where

I = rated breaking (interrupting) current
R = reduction rate as shown in Fig. 4.4

Ts = duration of short-time current

asym. = asymmetrical breaking current

sym. = symmetrical breaking current

When testing, the pressure and voltage of the operating device should be the rated values at each start, and during each test, minimum conditions should also be included. Gas pressure for quenching should be minimum. Relay time during CO operation should be 0.5 cycle unless otherwise specified.

4.1.2 Short-Line-Fault Breaking Tests

Short-line-fault (SLF) breaking tests are specified for circuit breakers directly connected to overhead transmission lines. When breaking a line fault located a few kilometers away from the circuit breaker on the overhead line, the arc-quenching condition of the circuit breaker is often quite severe due to the high rate-of-rise-of-recovery voltage (RRRV) on the line-side terminal, owing to reflections on the short line. Breaking severity is dominated by the magnitude of the fault and breaking currents, the surge impedance of the line, and the amplitude factor of the line-side TRV corresponding to the damping. IEC and ANSI specify the rated values shown in Table 4.5.

In IEC, introducing the multiconductor's attraction performance under heavy fault current conditions, values for surge impedance and amplitude factor are prescribed for both single- and multiconductor lines in EHV systems. In ANSI, this phenomenon has not yet been introduced; therefore two sets of values are prescribed, one each for single- and multiconductor

Table 4.5 Short-Line Fault Breaking Test Conditions in IEC/ANSI Standards

	IEC	ANSI
Rated voltage range	$\leq$170 kV $\geq$245 kV	$\leq$242 kV $\geq$362 kV
Surge impedance	450 ohm	450 ohm 360 ohm
Amplitude factor	1.6	1.8 1.6
Line side delay time	0.2 μs 0.5 μs	-----
Breaking current	0.75*I* & 0.9*I*	0.7*I* --- 0.8*I*
Test duty	*O* -t'- *O* -t- *O*	*O* - Os - *CO*
Operating pressure/voltage	min.	from rated
Pressure of quenching gas	min.	min.

I: Rated breaking current

lines. Lumped capacitances and/or lower surge impedance portions in the vicinity of the line-side terminal of the circuit breaker (i.e., voltage transformer, voltage divider, current transformer, disconnector, post/hanging insulators, etc.) introduce delay time on the very first part of the line-side TRV. For example, with a surge impedance of 450 ohms and a lumped capacitance of 1000 pF, the delay time is 0.5 μsec. In IEC, this delay time is prescribed. In ANSI, no value is specified.

For short-line-fault breaking tests, IEC considers primarily the performance of SF_6 gas and air-blast circuit breakers, and therefore for test current, line lengths corresponding to 90% and 75% of the rated short-circuit breaking current are prescribed. From experience, these conditions are very severe for these circuit breakers. ANSI mainly considers air-blast circuit-breaker performance and specifies a test current between 70 and 80% of the rated breaking (interrupting) current; from experience, this corresponds to the most severe breaking condition for air-blast circuit breakers. In both standards, breaking test conditions are based on one line-to-ground fault in effectively earthed systems. There may be slightly more severe conditions, such as the first-pole-to-clear in a three-phase grounded fault, but considering the following actual conditions, the one-line-to-ground fault condition is thought to cover well the practical system requirements:

The frequency of occurrence of one-line-grounding faults is the highest.

There is a low probability of a short-line fault at a very critical length on the line.

The actual short-line-fault breaking current corresponds to the total bus short-circuit current minus the contribution of the concerned line, but in application of the circuit breaker, the total current, which is generally higher than the actual current, is considered.

The short-circuit current of the bus is always less than the circuit-breaker's rated short-circuit breaking current.

In the IEC standards, the source- or bus-side TRV values, especially in short-line-fault tests, are specified. These values are based on recovery voltages equal to the phase voltages, whereas the recovery voltages for breaker terminal fault conditions are based on phase voltages multiplied by 1.3–1.5, the first-pole-to-clear factor under the three-phase fault condition.

The prospective TRVs on the line-side and bus-side terminals and across terminals of the circuit breaker during breaking of a short-line fault are calculated from the previously mentioned surge impedance, amplitude factor, delay time of the line side, breaking current, TRV parameters of the source side, etc. One example is shown in Fig. 4.5. From this figure it is apparent that the source-side ITRV (initial transient recovery voltage) and the effect of the line-side T_{dl} (time delay) amply compensate each other.

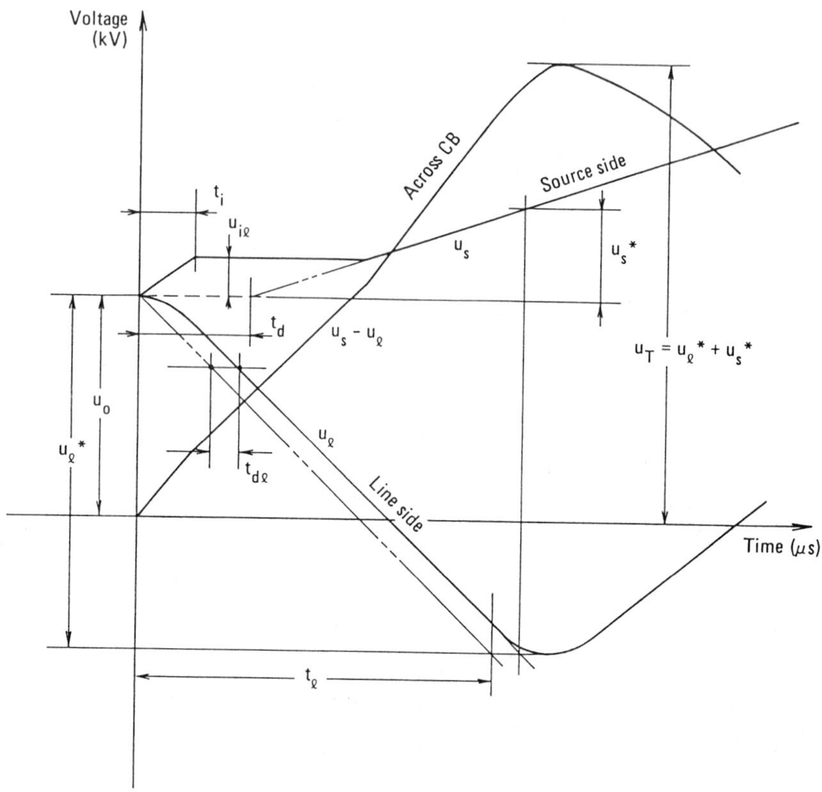

Fig. 4.5 Construction of TRV for short-line fault. (IEC 56 1987.)

Therefore, if the short-line-fault test is performed with a line without time delay and with a source-side TRV circuit without ITRV, the effect on the circuit breaker is thought to be identically as specified. In practical test procedures, a difficulty in applying the source-side ITRV circuit and line with time delay is foreseen. Therefore IEC approves short-line-fault tests without ITRV in the source side and without time delay in the line side. If such a short-line test is conducted, ITRV conditions in short-circuit breaking test duties No. 4 and No. 5 can also be omitted; that is, test duties No. 4 and No. 5 can be performed without ITRV.

4.1.3 Out-of-Phase Breaking Tests

Theoretically, the most severe *out-of-phase* breaking condition in a system is considered to be the following:

Breaking current	50% of total short-circuit current
Recovery voltage	2 × first-pole-to-clear factor × phase voltage = 2 × (1.3–1.5)**E** = (2.6–3.0)**E**
TRV peak value	Amplitude factor × recovery voltage = (1.4–1.5) × (2.6–3.0)**E** = (3.64–4.5)**E**

where **E** is the peak phase-to-earth voltage corresponding to rated voltage.

Nevertheless, such an extreme condition appears only when (1) system circuit parameters on both sides of the circuit breaker are almost symmetrical; (2) each side circuit has a high amplitude factor of TRV; and (3) the voltage sources on both sides are perfectly out of phase; that is, all voltage-source phase angles of one side differ by 180 electrical degrees from all voltage sources on the other side.

In practice, since "out-of-phase" rarely occurs just at the center of the system and an early stage of out-of-phase conditions (less voltage phase-angle differences) can be detected by a modern relay system, such severe breaking conditions have not been considered in IEC/ANSI. From actual experience, out-of-phase is further likely to occur at a point between a major part of the system and the rest of the system. In such a case, the breaking current is smaller and the amplitude of the TRV across the circuit breaker is smaller owing to the difference in the waveforms of the TRVs on the two sides mainly with respect to the time to peak of the TRVs.

The rated values and test conditions specified by IEC and ANSI are listed in Table 4.6. In producing these standards, parallel discussions on this subject were held simultaneously, so both standards have almost the same content:

Breaking current of 25% of the rated breaking current
TRV peak value significantly lower than the theoretical highest value.

For some circuit breakers, out-of-phase breaking ability is not required, but for the purpose of demonstrating restrike-free breaking under higher recovery voltage conditions, out-of-phase ratings are often specified for these circuit breakers also.

4.1.4 Capacitive Current Switchings

As mentioned in the preceding section, *capacitive current switching* corresponds to energizing/deenergizing no-load overhead lines, no-load cables, capacitor banks, and so on, where, after interruption of the current, the load-side terminal of the circuit breaker is energized by quasi-direct voltage, the value of which may be a certain amount higher than the peak

Table 4.6 Out-of-Phase Current Breaking Prescriptions in IEC/ANSI Standards

	IEC	ANSI
Rated breaking current	$0.25I$	$0.25I$
Recovery voltage	$2.5E$ for k=1.5 $2.0E$ for k=1.3	$2.5E$ for ungrounded syst. $2.0E$ for grounded syst.
TRV wave shape	2/4 parameters	(1 - cos) curve
peak value	$3.13E$ for k=1.5 $2.5E$ for k=1.3	($3.13E$ for ungrounded syst.) $2.5E$ for grounded syst.
time to peak	$2T_2$	$2T_2$
Breaking test duties	*O* and *O* at $0.25I$ x (0.2--0.4)	Two times *O* at 0.05 --- $0.10I$
	O and *CO* at 0.25 -- $0.275I$	*O* and *CO* at $0.25I$
Operation and quenching	minimum conditions	minimum conditions

I: Rated short-circuit breaking current
E: $\sqrt{2/3}$ times U (peak value of phase voltage)
k: First-pole-to-clear factor
T_2: Time to peak of TRV at breaking 100% of I

phase voltage caused by coupling with other phases. At the same time the other terminal of the circuit breaker is energized by the source-side power-frequency voltage. Therefore, voltage across the terminals of the open circuit breaker after current interruption is a certain amount higher than double the phase voltage with a waveform of the (1 − cos) form and lasts for a long time interval.

On the other hand, while making such a capacitive current, especially with residual voltage on the load side, or when restriking immediately after the current interruption, the highest switching overvoltage in the system is often created. Concurrently, a high inrush current may flow.

Capacitive currents are often switched almost daily, for example in switchings of shunt capacitors, no-load transmission lines, and no-load cables. Thus, especially for higher voltage circuit breakers, the safe switching of capacitive current, that is, restrike-free switching, is highly important.

(A) Rated Capacitive Switching Currents

The rated capacitive switching currents specified in IEC for rated voltages of 145, 245, and 525 kV are as follows:

	145 kV	245 kV	525 kV
Overhead line charging currents	50 A	125 A	500 A
Cable charging currents	160 A	250 A	500 A

As for single-capacitor-bank breaking current, IEC standards have no provisions for back-to-back capacitor bank breaking currents nor for capacitor bank inrush making currents for circuit breakers applicable to back-to-back conditions. Appropriate values should be selected from among R-10 series numbers for rated switching currents. Especially for back-to-back requirements, reactors connected in series to the capacitor banks introduce a significant reduction not only of the harmonic components of charging currents but also of inrush currents during back-to-back making.

Specifications in ANSI regarding capacitive current switchings are listed in Table 4.1 together with other ratings. For general-purpose circuit breakers, common values are specified for overhead line charging, cable charging, and shunt capacitor bank currents. For special-purpose circuit breakers, higher values for overhead lines, cables, and shunt capacitor banks are prescribed, and back-to-back requirements are also shown regarding peak currents and frequencies of inrush currents.

Except for back-to-back inrush requirements, the performance of modern SF_6 gas circuit breakers in breaking capacitive currents is not so significantly affected by breaking currents themselves, though various values are specified for individual system conditions in IEC/ANSI.

(B) Switching Currents in Type Tests

In both IEC and ANSI standards, two kinds of test switching currents are specified:

1. Switching currents equal to or higher than 100% of rated capacitive currents
2. Switching currents approximately 30% of rated capacitive currents

(C) Test Voltages

Test voltages for three-phase tests are the rated voltages of the circuit breakers. The test circuits have the appropriate specified zero- and positive-phase sequence parameters, so first-pole-to-clear recovery voltages are the same as those required for the systems. Today's high-voltage circuit breakers are usually tested by single-phase procedures, and test voltages for a *single-phase test* must be specified carefully, according to the system circuit parameters and conditions such as the ratio of zero/positive-sequence impedance, the existence of a ground fault, and so on. Generally, recovery voltages of the first pole to clear are the highest, so test voltages correspond to these conditions. In both standards, the prescribed voltage factors, which are multiplied with the phase voltages corresponding to the rated voltages to give the test voltages, are shown in Table 4.7.

In IEC standards, two source-side circuits are specified:

Source circuit A: The short-circuit current in it must be less than 10% of the rated short-circuit breaking current of the circuit breaker, with a voltage variation of less than 10% caused by capacitive current switching

Source circuit B: The short-circuit impedance must be minimum but not less than the impedance corresponding to the rated short-circuit breaking current, with a voltage variation of less than 5%.

In the IEC specifications, TRVs of the source circuits are also specified as shown in Fig. 4.6 and Table 4.8. These state the upper limits of TRVs, which are derived from the fact that the current is likely to be interrupted with shorter arcing time by the lower TRV that constitutes the initial part of the recovery voltage, and shorter arcing time is likely to cause restriking: that is, lower TRVs correspond to more severe conditions.

In ANSI, voltage variation during switching of the capacitive current is introduced, and the recovery voltage is required to be not less than a specified value. In single-phase tests, as an example, with the short-circuit current of the test circuit $\mathbf{I}_{sc}$ and test current $\mathbf{I}_c$, the test supply voltage is

$$\mathbf{E} = \mathbf{BV}\frac{2}{1 + \mathbf{A}}$$

where

$$\mathbf{A} = \frac{\mathbf{I}_{sc}}{\mathbf{I}_{sc} - \mathbf{I}_c}$$

and

$\mathbf{E}$ = test supply voltage
$\mathbf{V}$ = rated voltage divided by $\sqrt{3}$
$\mathbf{B}$ = voltage factor (Table 4.7)

Table 4.7 Voltage Factors for Capacitive Load Current-Switching Tests

System neutral	Load circuits/Switching conditions	IEC	ANSI
Grounded	- Shunt capacitor bank - Overhead line - Screened cable - Belted cable	1.0 1.2 1.0 1.2	1.0 1.2 1.0 1.2
Un-grounded****	- 2nd, 3rd poles are interrupted at the next current zeros - 2nd, 3rd poles are interrupted delayed	***1.4 ***1.4	1.25 1.5
Grounded	- Sound phase under 1 or 2 line(s) grounded	**1.4	*
Un-grounded	- Same as above	**1.7	

* In ANSI 37.012-1979-5 (Application guide), analytical calculation results shows that the maximum recovery voltage is 2.84E corresponding to 1.42 of the voltage factor.

** In IEC, tests under these conditions are prescribed to perform with the rated capacitive currents multiplied by 1.25 and 1.7 for neutral grounded and un-grounded systems respectively.

*** During the first 90 electrical degrees after current interruption the recovery voltage corresponds to 1.5E, and after 90 electrical degrees corresponds to 1.0E, resulting in the peak recovery voltage corresponding to 1.25E. Considering also the value of 1.5E for the case of delayed 2nd/3rd poles' interruption, the compromised value of 1.4 is introduced.

****In neutral un-grounded systems with significantly larger capacitances on the source side compared with those on the load side, the recovery voltages are close to those in neutral grounded systems. But in the both standards, severer conditions are considered. Therefore, higher values are introduced.

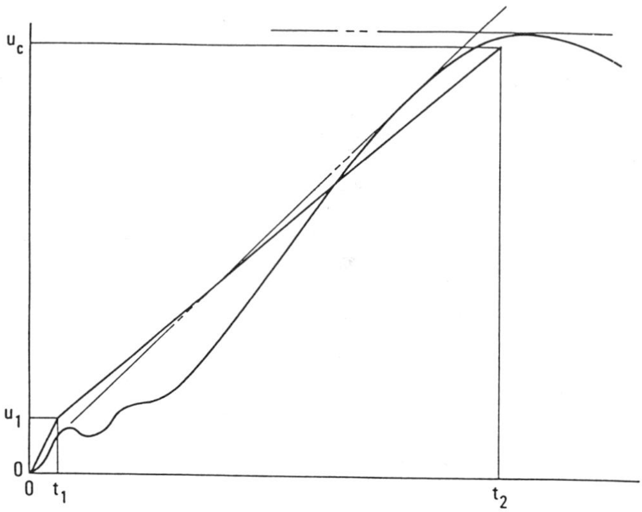

Fig. 4.6 Recovery voltage for breaking capacitive current. (IEC 56 1987.)

Table 4.8 Test-Circuit Parameters for Capacitive Current Switching

Source circuit	Recovery voltage in relation to the test voltage peak value		Time values		
			t_1	t_2	
	u_c	u_1		50Hz	60Hz
A	1.95	0.14	Time to peak in 30% of short-circuit current breaking	8.7ms	7.3ms
B	2.0	0.01	Time to peak in 100% of short-circuit ing current break-	8.7ms	7.3ms

Source: IEC 56, 1987.

(D) Test Duties

In the IEC standards, test duties are as follows:

Test source circuits	A and B
Test currents	100% and 20–40% of $\mathbf{I}_c$
Operation and quenching	Minimum conditions

It is recommended that point-on-wave control of contact separation be applied at each 30 electrical degrees, where the number of tests for each of the four duties is 12. If point-on-wave control is not applied, 30 tests are required for each duty. For a test current of 100%, two of the total 12 or 30 tests must be **CO** operations.

In ANSI, 24 tests are specified for each 100% and 30% of the current $\mathbf{I}_c$.:

Two for each 30 electrical degrees from 0 to 150 electrical degrees (a total of 12)
Six at the maximum restriking (reignition) arcing time ± 7.5 electrical degrees
Six at minimum restrike-free arcing time

Test duties for a single shunt capacitor bank and back-to-back capacitor bank for 30% and 100% of $\mathbf{I}_c$ must be **O** and **CO**, respectively. Test duties for an overhead line must be **O–C** and **CO** for 30% and 100% of $\mathbf{I}_c$, respectively. For circuit breakers intended for use in rapid auto-reclosing, **O–C** operations must be performed with maximum residual line-side voltages of opposite polarity to the source side. The operation conditions are rated, and as for gas pressure for quenching, three duties for 100% $\mathbf{I}_c$ are the minimum required to investigate the performance in the most severe conditions, and the others are with the rated conditions.

(E) Overvoltages

In IEC, as suggested values (not rated values), the allowable overvoltage factors listed in Table 4.9 are applicable to line-to-ground voltages based on the peak phase voltages.

In ANSI, as rated values, the following are specified:

Special-purpose circuit breakers with rated voltages ≥ 121 kV	2.0
General-purpose circuit breakers	3.0

4.1.5 Inductive Current Switching

Much trouble has been experienced in switching inductive load currents, such as currents of no-load or inductively loaded transformers and stalled or starting induction motors while inching, calling for the standardization of *inductive-current-switching tests* at the earliest possible time. Close and complicated relations exist between switching phenomena (namely overvoltages) and each of the circuit parameters; therefore, standardization work involves many difficulties. Nevertheless, through the energetic activity of WG-02 of CIGRE Study Committee No. 13 over a period of more than 10 years, the basic phenomena have been surveyed, and both the IEC and ANSI are using these results in formulating their standards. Therefore, both standards are expected to be almost the same.

The above-mentioned phenomena are greatly influenced by various parameters of the circuit, and when tests are being performed a good understanding of basic and detailed phenomena is most important. When inductive current switching is being done by modern circuit breakers, the possibility of current interruption is no longer of interest. Successful interruption itself is thought to be a matter of course. On the other hand, the magnitude of the overvoltages in the system that may be caused by switch-

Table 4.9 Suggested Permissible Overvoltage Factors in IEC

U(kV)	LIWV* (kV)	A	B
145	550 650	3 3.5	2.5 2.5
245	850 950 1050	2.7 3 3	2 2 2
525	1425 1550	2.1 2.3	2 2

A: Circuit-breakers for general use
B: Circuit-breakers for special use in power systems with means of special insulation co-ordination
*: Lightning impulse withstand voltage

Source: IEC 56, 1987.

ing are of great interest. Overvoltages are greatly influenced by current chopping, reignition, and multiple reignition, which are also greatly influenced by circuit parameters. Therefore, test results with standard test circuits must be adapted to actual circuit conditions. The main purpose of type tests is to give enough information for application to actual system circuits. For details, see Chapter 1, Sections 1.3 and 1.4.

The following concerns discussions (which are almost at the point of reaching conclusions) of the IEC standardization committee.

(A) High-Voltage Motor Switching Tests

As for the scope, the standard is intended for application to switching of starting or stalled motors with rated voltages of 1–17.5 kV, which are considered most likely to create high overvoltages.

Tests are mostly performed as laboratory tests, and motor-equivalent circuits consisting of lumped *LRC* as the load are generally used in the tests, thus covering relatively severe conditions with respect to overvoltage. Actual motors, of course, can be used if they meet test requirements. The results of field tests are applicable only to circuit breakers used in circuits almost identical to the test circuits. Figure 4.7 shows a typical test circuit and the main parameters. For representing three-phase coupling effects, "bus representation" is introduced.

Regarding test duties, random (not point-on-wave control) tests consist of 40 **CO**, or **C** and **O**, operations. When point-on-wave control is introduced, 20 operations are specified. For an actual motor in the load circuit, a time interval of 200 msec between **C** and **O** is specified.

For test reports, details of the test circuit, overvoltage characteristics, reignition characteristics, and so on are requested. These are to be used in converting test overvoltages to actual values in the system.

(B) Shunt Reactor Switching Tests

Similarly to motor switching tests, shunt reactor switching test standards are expected to be established. However, because of the higher voltage ratings of the objects, more detailed and accurate information is needed.

The standard is intended to be applicable to circuit breakers intended for switching shunt reactors with rated voltages ranging from 12 kV to ultrahigh voltage.

Especially for shunt reactor switching tests, due to the close relationships between circuit parameters and test results, field test results are applicable only to circuit breakers used in circuits almost identical to the test circuits. Consequently, laboratory tests are generally performed on test circuits that will allow the wide application of test results. Figure 4.8 shows a typical test circuit and the main current parameters. For circuit breakers

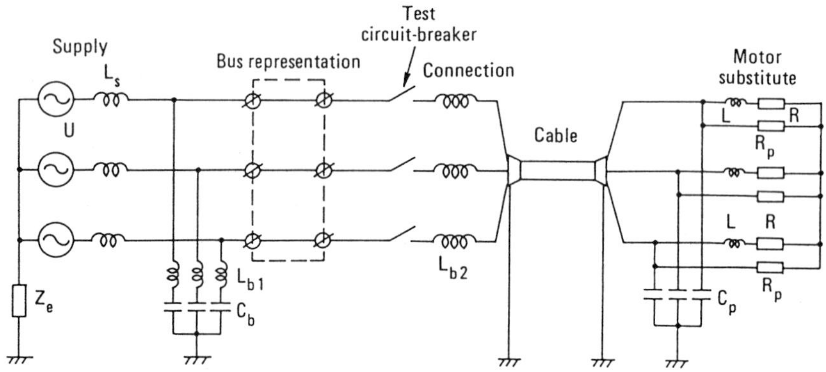

Z_e – Earthing impedance	– high enough or even infinite
L_s – Source impedance	– ⩽ 0.1 L, but not less than corresponding short-circuit impedance
C_b – Supply side capacitance	– 0.03 to 0.05 μF for supply circuit A 1.5 to 2.0 μF for supply circuit B
L_{b1} – Inductance of capacitors and connections	– ⩽ 2 μH
Bus representation	– length 5 to 7 meters
L_{b2} – Inductance of connection	– ⩽ 5 μH
Cable	– 100m, screened, Z_0 = 30 to 50 ohms
L – Motor substitute inductance	– 100A for load circuit A 300A for load circuit B
R – Motor substitute resistance	– cos φ ⩽ 0.2
C_p – Motor substitute capacitance	– frequency 10 to 15 kHz
R_p – Motor substitute parallel resistance	– amplitude factor 1.6 to 1.8

Fig. 4.7 Motor-switching test circuit and summary of parameters. (IEC 17A(S) document.)

with rated voltages of 100 kV and above for use in neutral grounded systems, single-phase tests are allowed. For rated voltages below 100 kV, only three-phase tests are allowed. The other major test parameters can be derived from the following table.

Rated voltage **U** (kV)	TRV peak (kV)	Time to TRV peak (μsec)	Frequency of $\mathbf{C}_s$-($\mathbf{Lb}_1$ + $\mathbf{Lb}_2$)-$\mathbf{C}_t$ loop (kHz)
12–170	$1.5 \times 1.9 \times \sqrt{2/3}\,\mathbf{U}$	$(5\text{–}7)\sqrt{\mathbf{U/I}}$	[a]
≥ 245	$1.9 \times \sqrt{2/3}\,\mathbf{U}$	$(5\text{–}7)\sqrt{\mathbf{U/I}}$	[a]

[a]Not specified but should be as high as possible.

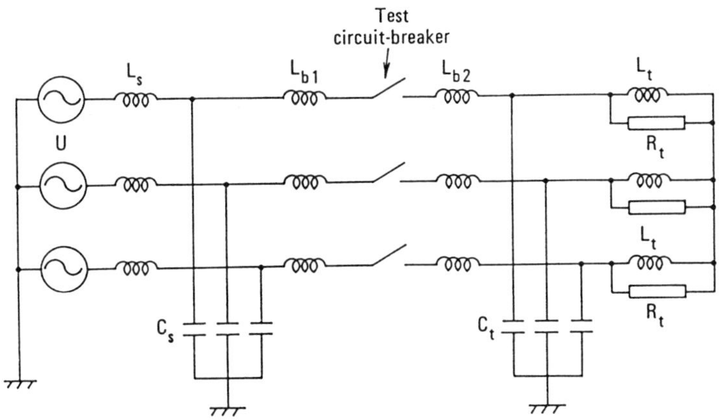

a) Three phase test circuit

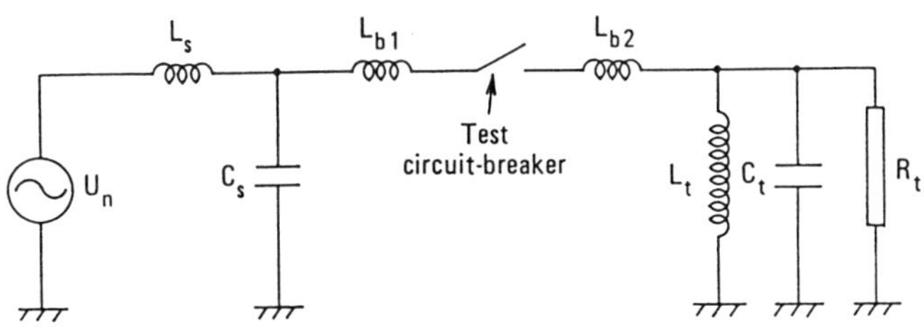

b) Single phase test circuit

L_s	– Source inductance	– less than $0.1 L_t$ but not less than corresponding short-circuit inductance
C_s	– Source capacitance	– not less than 10 times of C_t
L_{b1}, L_{b2}	– Inductance of connection	
L_t	– Load inductance	
C_t	– Load capacitance	
R_t	– Corresponding to damping	– amplitude factor = 1.9

Fig. 4.8 Shunt-reactor switching test circuit and summary of parameters. (IEC 17A(S) document.)

It has been proved that the frequency and the damping of oscillation of the **C_s-(Cb_1+C_2)-C_t** loop play significant roles in multiple reignition phenomena that can cause the most severe electrical stress to the shunt reactor windings. Here, **I** is the test current. Some examples are given in the following table.

Rated voltage (kV)	Load circuit 1 (kA)	Load circuit 2 (kA)
12–36	1.5	0.5
52–72.5	0.5	0.2
≥ 100	0.3	0.1

Regarding test duties, the statistical nature of the phenomena will require a large number of tests. For example, in a single-phase test, applying point-on-wave control of the contact separation timing, 20 tests with an interval of 18 electrical degrees, and a further 15–20 tests with the arcing time that caused the maximum overvoltage during the previous test are specified.

For the purpose of supplying the information needed for converting test results to actual system conditions, detailed information such as parameters of the test circuits, various overvoltages, chopping currents, insulation recovery across circuit breaker contacts after current interruption, details of the reignition phenomena, and so on are required in the test reports.

When the standard is published, the values listed here may have been revised to some degree.

4.2 DIRECT TEST METHODS (H. Ikeda)

4.2.1 Introduction

As described in Chapter 2, attempts to comprehend switching arc phenomena experimentally and analytically have been promoted energetically. As a result, it has become possible to comprehend switching arc phenomena fairly quantitatively. However, tests must still be conducted to prove that when a circuit breaker is installed in a power system its performance will be satisfactory.

Tests to verify the performance of circuit breakers and switching equipment preferably should be conducted on breakers directly connected to the power systems in which they will be installed. However, there are limits to the feasibility of such tests because of the need to protect the power systems, and the results obtained in one power system are not always applicable to another power system. For these reasons, it is customary to conduct tests in high-power testing laboratories equipped with such power sources as short-circuit generators and capable of safely simulating fault conditions in power systems.

Developing proof test methods capable of realizing the same conditions as those in power systems in accordance with standards plays an important role in the development of high-voltage circuit breakers, together with the development of other excellent equipment and the recognition of power system conditions where the equipment is to be used. As described in the preceding section, circuit breakers are required to perform a wide variety of switching duties. This section deals with direct test methods.

4.2.2 Outline of Direct Test Methods

A direct test is conducted to prove the performance of equipment by using a circuit similar to that of the power system in which the equipment will be installed. Under American national standards for circuit breakers (ANSI/IEEE C37.081-1981), this is defined as "a test in which the applied voltage, current, and *recovery voltage* are obtained from a single power source, which may be comprised of generators, transformers, networks, or combinations of these." For circuit breakers, the voltage–current characteristics of switching arcs are nonlinear, and there are complicated electrical interactions between a circuit breaker and the circuit to which it is connected. Thus, a direct test method should be used in which a test circuit can be composed of the same circuit constants as those of the power system.

To provide such conditions, a high-power testing laboratory was equipped with large-capacity power sources, current-controlling reactors, short-circuit transformers for voltage setting, and auxiliary equipment. Circuit conditions for testing are produced by suitably combining these units for the voltage, the current, and the test conditions.

Figure 4.9 shows system construction of a high-power testing laboratory (Harada and Yanabu, 1978). This laboratory is at the world's largest level institution, KEMA, which has four short-circuit generators as power sources. Four generators are connected through reactor N for synchronization, which is located after back-up circuit breaker B. This is the type of power sources most commonly used. Other testing laboratories use power systems directly drawn in or precharged capacitor banks consisting of groups of capacitors connected in series or in parallel. Where a power system is used, large load fluctuations can occur at a short circuit, produc-

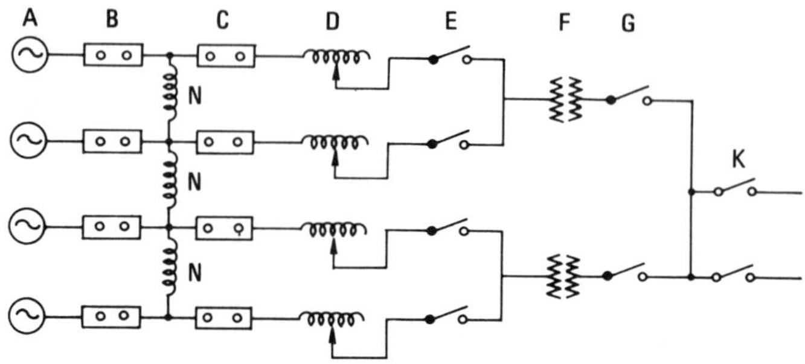

Fig. 4.9 An example of the construction of a high-power testing laboratory.

ing an adverse effect on system stability. However, current supplied by capacitor banks shows large attenuation, which results in doubt as to equivalence.

Tests of circuit breakers require power sources with much larger capacities than those for other switching equipment. In particular, *short-circuit tests* to prove a breaker's ability to make and break short-circuit currents at short-circuit faults require large-capacity power sources. With a direct test method, a short-circuit test of a circuit breaker at a rated voltage of 245 kV and rated interrupting current of 40 kA, for example, requires a power source with a 17-GVA three-phase short-circuit capacity. Table 4.10 shows power-source capacities of representative high-power testing laboratories throughout the world. All of them are lower than 17 GVA. Thus, direct test methods are used mainly for short-circuit tests of so-called medium-voltage circuit breakers with rated voltages below 72 kV. For high-voltage circuit breakers, synthetic test methods, described in Section 4.3, are used.

For proof under conditions equivalent to those in power systems, it is preferable that direct test methods be employed, and for some tests, direct test methods are essential. Various special schemes are followed to use direct test methods with limited power-source capacities.

4.2.3 Special Test Under Direct Test Method

Direct test methods are applied to all test duties involving medium- and low-voltage circuit breakers. Tests for switching surges in small-current switching and tests under reduced voltages are described below as special examples.

Table 4.10 Short-Circuit Testing Laboratory

From power system			
Nation	Laboratory name	System voltage (kV)	Short-circuit capacity (MVA)
France	EdF Renar-dieres	420	3000
U.S.S.R.	U.S.S.R. Laboratory	110 220 400	5000 4000 8000
Canada	IREQ	735	7200
Italy	CESI	225	3000
West Ger.	PEHLA	220	500

From short-circuit generator					
Nation	Laboratory name	Nominal capacity (MVA)	Short-circuit capacity (MVA)*	Max. generated current**	No. of units
Japan	CRIEPI	150	4550	–	1
Japan	Toshiba Corp.	125 200	3000 7400	200 kA - 12 kV	1 1
Nether-lands	KEMA	–	1000 1500 3400	330 kA - 14 kV	1 2 4
Repub. of Korea	KERI	180	6000	111 kA - 18 kV	1
U.S.A.	PSM High Power Laboratory	–	1750 2500 3000	110 kA - 22 kV	1 1 1
France	EdF Renar-dieres	–	6000	–	2

* Symmetrical at short-circuit instant
** 0.06 ms after single-phase short-circuiting

(A) Tests of Surges Involved in Switching Small Currents

Interrupting small inductive currents such as switching of no-load transformers and shunt reactors, and interrupting small capacitive currents such as switching of no-load lines and shunt capacitors, involve the hazards of overvoltages caused by current chopping and multiple reignition. It is known that in such interruptions, circuit construction is closely related to the occurrence of overvoltages, where direct test methods must be used.

High-voltage circuit breakers are normally expected to be *restriking-free* in interrupting small capacitive currents, and their tests are conducted not to check overvoltages caused by restriking but simply to check the restriking-free performance. For this purpose, equivalent tests under synthetic test methods (see Section 4.3) are used. On the other hand, for the interruption of small inductive currents, it is necessary, even with high-voltage circuit breakers, to check generated overvoltages under a direct test method. Shunt reactor switching tests are described here as an example.

(1) Shunt Reactor Switching Tests

Shunt reactor current-interruption tests are characterized by requiring large power-source capacities because of relatively high interrupting currents among inductive current switching, and requiring sophisticated measurement because of the high-frequency current interruption involved (Kawada et al., 1987).

Test method. Figure 4.10 shows the single-phase test circuit of a 275-kV shunt reactor switching. With the 11-kV output voltage of a short-circuit generator stepped up to 173 kV equivalent to phase-ground voltage of the power system, the current fed to the air-core reactor in simulation of a shunt reactor is switched by a circuit breaker. An actual shunt reactor is occasionally used instead of an air-core reactor.

Under standards, the load-side transient recovery voltage is determined by the frequency and the amplitude factor (see Fig. 4.8). The frequency is adjusted by inserting a capacitance (load-side capacitance) between the circuit breaker and the reactor. The amplitude factor is adjusted by adding a resistor in series to the load-side capacitance. The transient recovery voltage on the power-source side is adjusted by power-source-side capacitance so that transient recovery voltage in short-circuit current interruption satisfies the standards requirement.

Also, a high-frequency current caused by local oscillation around the circuit breaker upon reignition of the circuit breaker has an effect on the overvoltages. The frequency of oscillation is adjusted by the power-source-side and load-side capacitances and by the frequency-adjusting reactors inserted in both branches between the circuit breaker and the capacitance. When attenuation of local oscillation is taken into consideration, resistors are inserted in series with the frequency-adjusting reactors. As it is affected

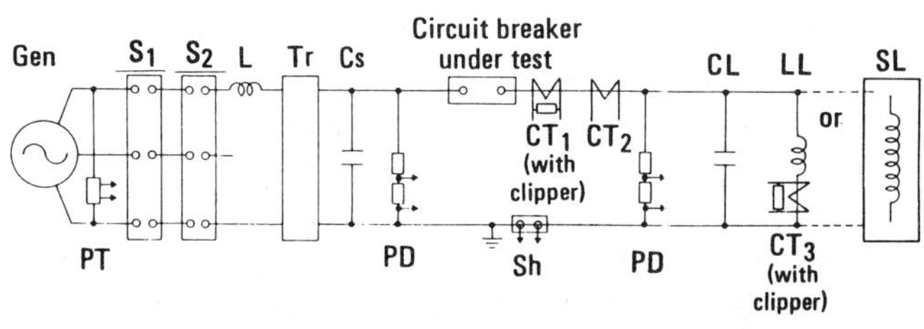

Gen : Short-circuit generator
S_1 : Back-up circuit breaker
S_2 : Making switch
Tr : Transformer
Cs : Supply-side capacitor
L : Current adjusting element
LL : Load-side reactor
CL : Load-side capacitor
CT : Current transformer
PT : Potential transformer
PD : Potential divider
Sh : Shunt
SL : 275kV-150MVA shunt reactor

Test item: Shunt reactor current switching test

Test method: Single-phase direct test

Fig. 4.10 Diagram of test circuit. (Kawada et al., 1987. © IEEE.)

by stray capacitance and inductance, the frequency must be checked by measurement. Since measurement is possible only at reignition during testing, circuit breaker reignition is simulated by designing an ideal switch such as a mercury switch and charging the power-source and load sides with low dc voltages. Figure 4.11 shows measured high-frequency voltages obtained with the circuit in Fig. 4.10.

Test results. Figure 4.12 is an oscillogram for reactor current interruption. Tests are conducted 12 times with the breaking instant varied by 30° steps in the phase angle, controlling the phase with ordinary tripping signals. Overvoltages due to chopping currents can be measured well from relatively low speed oscillograms.

For overvoltages caused by single and multiple reignitions, high-speed measurements are necessary. Figure 4.13a shows the potential to ground on the circuit-breaker terminal measured with a high-speed measuring instrument. Figure 4.13b shows the current waveform upon reignition. In high-speed measurement, difficulty has been encountered in synchronizing the triggering operation of the measuring instruments. Measurement of current waveforms is hindered by a noise problem because sensitivity is raised for measurement at a low level near the current-zero point. When current waveform measurement is difficult, overvoltages are calculated from reignition voltages obtained from low-speed oscillograms like those in Fig. 4.12.

It is known that in interrupting shunt-reactor currents, the stray inductance and capacitance near the circuit breakers have a large effect on overvoltages. It is desirable to improve test techniques for high-frequency phenomena.

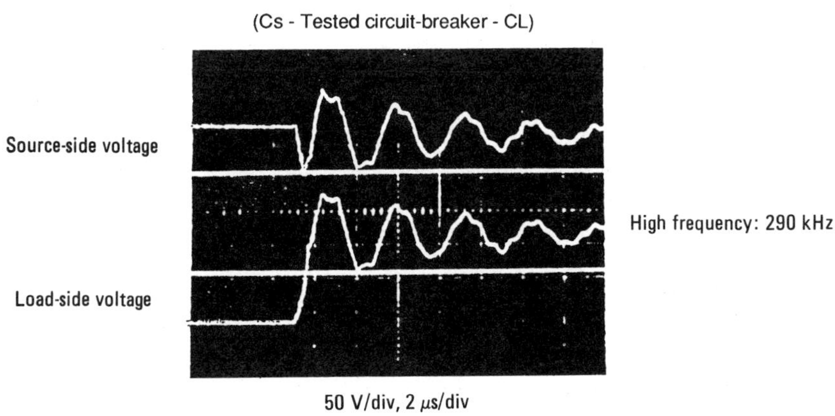

Fig. 4.11 High-frequency voltages measured at low dc voltage.

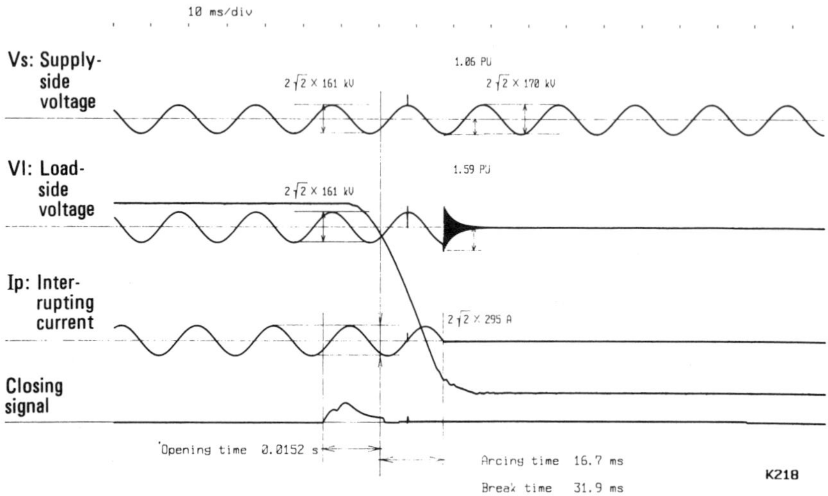

Fig. 4.12 Oscillogram for reactor current interruption.

(B) Tests Under Reduced Voltages

Typical examples of direct test schemes adopted for limited power-source capacities are unit tests and single-phase tests. The *unit test* checks circuit-breaker performance by using a unit in a *multibreak circuit breaker* composed of interrupting contacts connected in series. Test conditions depend on the voltage distribution ratio of the unit being tested. The *single-phase test* is used for checking breaker performance by using only one phase of a three-phase or two-phase circuit breaker. Since the transient recovery voltage and the recovery voltage of the first pole to clear are highest because of the existence of neutral grounding impedance in the power system, tests are conducted at these conditions. In any case, the equivalence of the single-phase test to tests with complete circuit breakers must be checked thoroughly.

Tests with reduced voltages and currents are another form of direct tests. Under certain conditions, proof tests are acceptable if either the current or the voltage requirement is fulfilled. In this chapter, making tests and *zero-missing passing current interrupting* tests are described as examples of reduced-voltage tests.

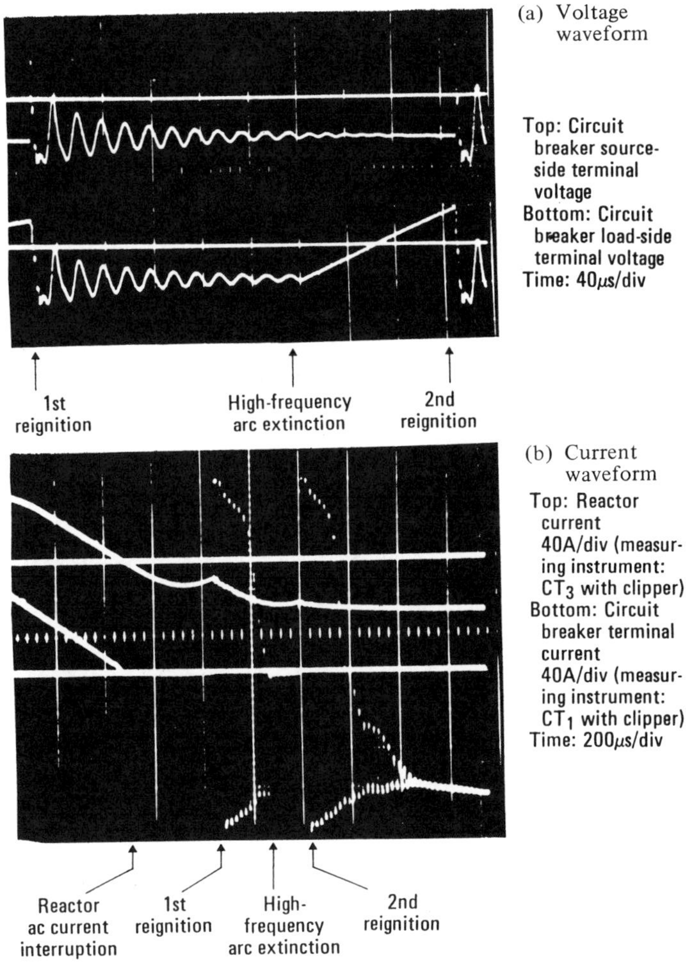

Fig. 4.13 High-frequency current and voltage waveform at reignition. (Kawada et al., 1987. © IEEE.)

(1) Making Tests

Making tests are conducted to check the contact performance against electromagnetic mechanical force. The most severe conditions are found in making tests at a *breaker terminal fault condition* with the largest currents in all duties. In the special case of having a high impedance in series, such as a resistor closing switch, making currents will be larger with higher

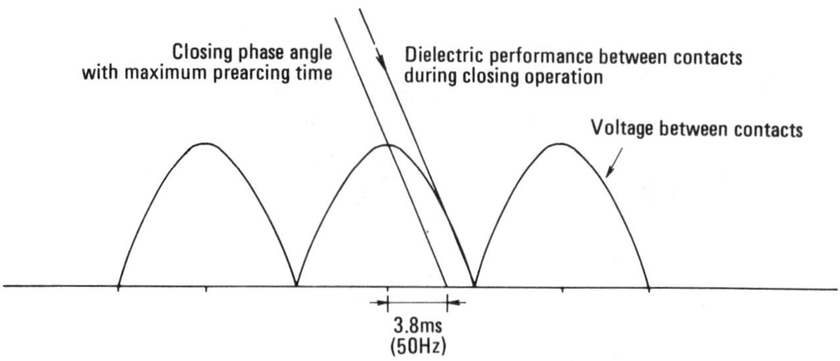

Fig. 4.14 Condition with contact making at voltage-zero point.

voltages, as in the out-of-phase condition, and a special test will be required. In general, making tests for ordinary contacts are duly conducted under breaker terminal fault conditions.

Making at the zero point of the system voltage causes an asymmetrical current with a dc component equivalent to the crest value of the ac component. Here, the current crest value is nearly doubled, and making conditions in this phase are considered the most severe. Figure 4.14 shows the relation between ac voltage and dielectric strength between contacts during the closing operation. As the contact closing speed increases, the gradient of the dielectric strength characteristics line also increases. Below a critical gradient, the line always crosses the ac voltage envelope, but it does not cross the voltage-zero point. Above this gradient, a condition for making at the voltage-zero point occurs.

With making operations with this critical gradient, crossing with the ac voltage occurs at various points of the voltage waveform. During the period from crossing of the voltage envelope to crossing of the voltage-zero line by the line, the current is carried through a discharge channel by a dielectric breakdown between contacts. This is termed *prearcing time.* It is longest when dielectric breakdown occurs at the crest value of the ac voltage, being 3.8 msec (50 Hz) and 3.2 msec (60 Hz) with the critical gradient.

As described earlier, a power source must have an impractically large capacity to supply a rated interrupting current at a rated voltage in a high-voltage circuit breaker. Therefore, making tests are conducted by using the synthetic test methods described in the following section. For tests of making at the voltage-zero point, reduced voltages are satisfactory. If the prearcing time is below 3.8 msec, a severe condition of making at the voltage-zero point actually occurs, and tests in this phase are satisfactory. The prearcing time is measured beforehand by a circuit that feeds the current limited by the testing apparatus with a rated voltage, for judging the applicability of the test with the reduced voltage. For prearcing times below 3.8 msec, tests are also conducted with a circuit that feeds a rated interrupting current with a reduced voltage; for times greater than 3.8 msec, a synthetic test may be applied.

(2) Interrupting Tests of Zero-Missing Passing Current

Switches with current-limiting functions, designed to interrupt short-circuit currents with the help of attenuation by utilizing arcing voltages, are widely used in distribution applications. They are tested under direct test methods with limited capacity. These switching philosophies are not followed with high-voltage circuit breakers because the ratio of arcing voltage to system voltage is so small that the current-limiting effect cannot be expected. In interrupting zero-missing passing currents, described in Chapter 1, a dc component attenuation effect is likely to occur with arcing voltages.

To test this effect, direct test methods are preferable, although the capacities of test apparatus are often insufficient. The current-interrupting performance at the current-zero point is checked by terminal-fault-interrupting tests. To check the current-limiting effect, direct test methods with reduced voltages are used.

Test methods. In testing, first it is necessary to generate a zero-missing passing current. It is known that zero-missing passing currents occur not only in the short-circuit phase with high current, but also in the sound phase with small current. Those under short-circuit conditions are simulated for tests (Ishikawa et al., 1984).

Figure 4.15 shows a test circuit. An *evolving fault* from a two-phase fault to a three-phase fault is simulated because it is most likely to cause zero-missing passing currents under short-circuit conditions. In three-phase circuit construction, a two-phase short circuit is first produced by switching on the making switch S_2 at the primary side of the transformer, and then this is extended to a three-phase short circuit by switching on the making switch DS at the secondary side of the transformer. An arc is generated by opening the test circuit breaker S_T connected to the phase in which a zero-missing passing current is generated, and the current and voltage are measured.

Test results. Figure 4.16 is the oscillogram of a measured zero-missing passing current. This current appears in phase *a*. In this oscillogram, the test circuit breaker is not opened, and the dc component exceeds the ac crest value over several cycles, producing a phenomenon without a current-zero point. Figure 4.17 shows a phase *a* current waveform with the test circuit breaker open. The process of rapid attenuation of the dc component upon opening of the circuit breaker can be seen from a comparison of Figs. 4.16 and 4.17.

This test is conducted with a reduced voltage, and not to prove the characteristics with actual system voltages. The analysis technique is evaluated on the basis of the test results and the characteristics are estimated at actual system voltages by using the analysis technique. The dashed line in Fig. 4.17 shows the results of calculation by the transient phenomenon

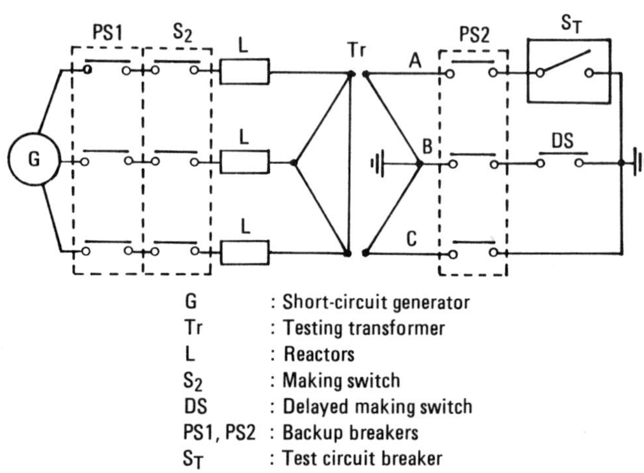

Fig. 4.15 Test circuit for zero-passing missing current interruption. (Ishikawa et al., 1984. © IEEE.)

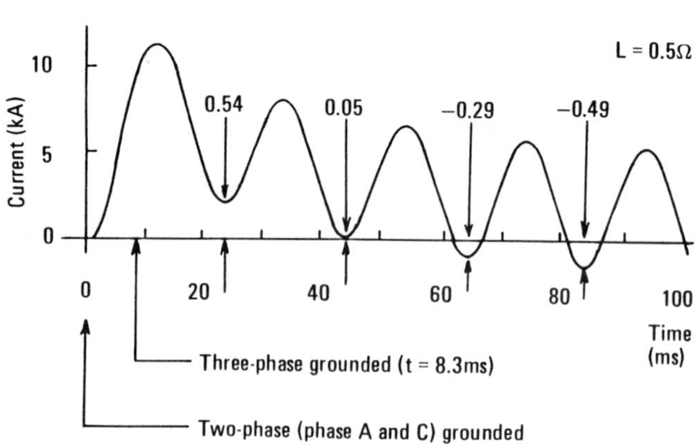

Fig. 4.16 Zero-miss current waveform (TCB closed). (Ishikawa et al., 1984. © IEEE.)

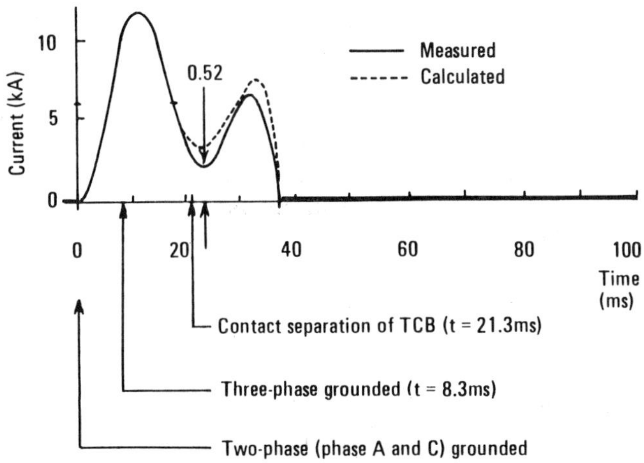

Fig. 4.17 Zero-miss current waveform (TCB open). (Ishikawa et al., 1984. © IEEE.)

analysis program (EMTP) with a simulated arcing voltage. These results agree well with the test results.

As explained above, the results of direct test methods using reduced voltages are used to establish techniques for accurately simulating interactions between arcs and circuits. Characteristics under system voltages are estimated by adopting this analysis technique.

4.2.4 Automation of Short-Circuit Tests

Direct test methods are generally well established, but automation is being promoted to conduct short-circuit tests efficiently. Automation also improves short-circuit tests by permitting complicated measurements and control, which are conventionally impossible by manual operations.

Automated functions related to short-circuit tests can be classified into three categories:

1. Preparation work for composing a circuit including generator setup and system construction, and clearing-away work.
2. Testing processes including sequence operations.
3. Measuring and processing test results.

The automatic operation of generators has already been put to practical use in power systems. In short-circuit tests, frequent starting and stopping impose such conditions that more protective functions are necessary to ensure safe operation. Such automatic operations have not been put into use in any high-power testing laboratories in the world. An advanced expert system and an excellent monitoring system will be required. Automation of system construction is under way in some test stations (Brasca et al., 1983). Remote control of disconnecting switches fed with a current above 100 kA has already been carried out, and automation of system construction can be achieved by a relatively simple system for operating and checking procedures. The full automation of processes from calculating a circuit structure on the basis of test conditions to designing a test circuit requires an expert system supported by knowledge of test apparatus.

Sequential control with a combination of a programmable controller and a digital counter is a technique already in wide use. In short-circuit tests, large high-voltage currents are switched, causing large overvoltages. It is important to take measures to eliminate malfunctioning of electronic devices.

The practical use of automation for measurement is most advanced, since equipment can be controlled by a shield at a low energy level. Details are described in Section 4.5. When measurement techniques are fully automated, the automatic preparation of test reports may become possible.

Short-circuit tests also use many units in complicated arrangements that are often confronted with equipment breakdown. Artificial intelligence for locating faulty parts and determining suitable repair time, as well as for indicating methods of correcting faults, is essential to automation (Horiuchi et al., 1989).

Short-circuit tests not only use high-energy units, but also are often destructive and cannot be repeated. A fault in one channel of a measuring instrument will nullify the entire test. For automation, it is essential to increase equipment reliability.

The most recent trends in direct testing have been described above. Most direct test methods are already established, but as systems become even more complicated and equipment becomes more advanced, new tests will be demanded. It is important to make the most efficient use of the limited equipment available. Automation is likely to be further promoted.

4.3 SYNTHETIC TEST METHODS (H. Ikeda)

4.3.1 Introduction

Because of increases in the interrupting capacity per break of circuit breakers, today it is impossible to conduct even unit tests under direct test methods. Thus, equivalent test methods are indispensable. *Synthetic test methods* are important equivalent methods that can be used for most tests of high-voltage circuit breakers. Because of their importance, a standard has been published on synthetic test methods by the International Electrotechnical Committee [IEC-427 (1987)].

Synthetic tests are those tests that have both voltage and current power supplies. In a broad sense, they include the *duplicate test scheme,* which uses current and voltage supplies separated from one power source by two transformers, proposed by Skeats in 1936.

A current is supplied with a reduced voltage, and voltage after interruption is supplied by a small-capacity voltage source. Large-capacity testing can be performed with this scheme.

To cope with the increasing capacities and ever more advanced functions of circuit breakers, new synthetic test methods are still being developed today. Also, improvements are being made to increase their equivalence to direct test methods. Synthetic test methods are outlined below, together with current trends.

4.3.2 Concept of Synthetic Test Methods

Synthetic test methods can be classified into *current-injection* and *voltage-injection schemes.* Typically, the difference between the two is that in the current injection scheme current is fed from a voltage supply for a short period before the current-zero point of interruption, whereas in the voltage injection scheme only voltage is fed from a voltage supply after the current is interrupted by the circuit breaker at the current-zero point. The current-injection scheme also has equivalence for the current and circuit impedances just before the current-zero point and is widely recognized as a test method capable of testing performance even in thermal reignitions. It is understood that the voltage-injection scheme is applicable to testing performance only in *dielectric reignitions.*

The voltage-injection scheme includes both types, using the capacitor bank and the secondary side of high-voltage transformers with the primary side connected to a current supply as a voltage supply. These are outlined below.

(A) Current-Injection Scheme

Figure 4.18a is a conceptual drawing of a test circuit under a current-injection scheme proposed by Weil and Dobke in 1953. It is, in fact, so widely used that today the term "synthetic test" implies this circuit. Under this scheme, while a power-frequency current fed from current supply G composed of a short-circuit generator, for example, is flowing, test breaker S_T and auxiliary breaker S_A are opened, producing a state of conduction

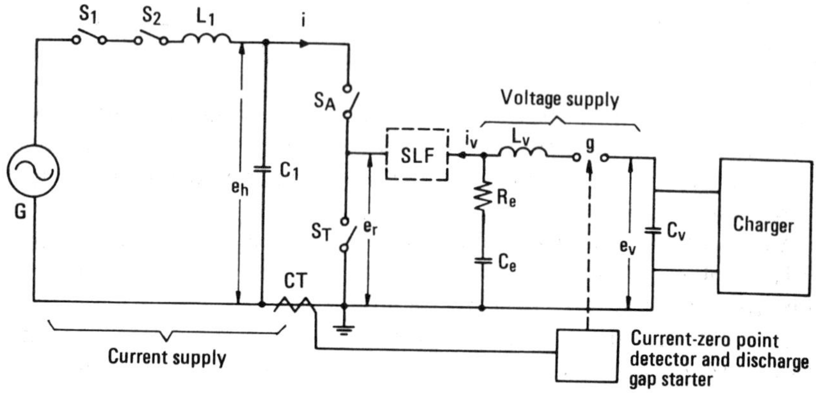

G : Current source
S_1 : Back-up circuit breaker
S_2 : Making switch
L_1 : Current adjusting reactor (current supply)
C_1 : Capacitor
S_A : Auxiliary breaker
S_T : Test breaker
C_v : Main capacitor
L_v : Current adjusting reactor (voltage supply)
C_e : TRV adjusting capacitor
R_e : TRV amplitude adjusting resistor
g : Triggering gap
i : Current (current supply)
e_h : Arcing voltage (S_A + S_T)
e_v : Voltage (voltage supply)
i_v : Current (voltage supply)
e_r : Arcing voltage and recovery voltage (S_T)
CT : Current transformer
SLF: Artificial short line

(a) Test circuit

Fig. 4.18 Weil synthetic testing method.

with arcing. The current-zero-point detector recognizes the point several hundred microseconds before the current-zero point at which interrupting performance is to be tested, causing the starter circuit of gap switch g to be triggered. This causes the charge initially stored in voltage-source capacitor Cv to be discharged via voltage-source current-controlling reactor Lv and test circuit breaker S_T. This point is shown as T_1 in the waveform in Fig. 4.18b. As a result, high-frequency current i_v determined by Cv and Lv is superimposed on the power-frequency current flowing through S_T.

As shown in Fig. 4.18b, the power-frequency current reaches a current-zero point at time T_0, and then auxiliary breaker S_A is interrupted. Test breaker S_T is disconnected from the current supply, allowing only i_v to flow. When i_v reaches a current-zero point at time T_2, S_T is interrupted. Compared with the power-frequency current, i_v has its frequency and crest value selected as about 10 and 1/10 times, respectively, and has its current gradient at the current-zero point adjusted to that of the power-frequency current.

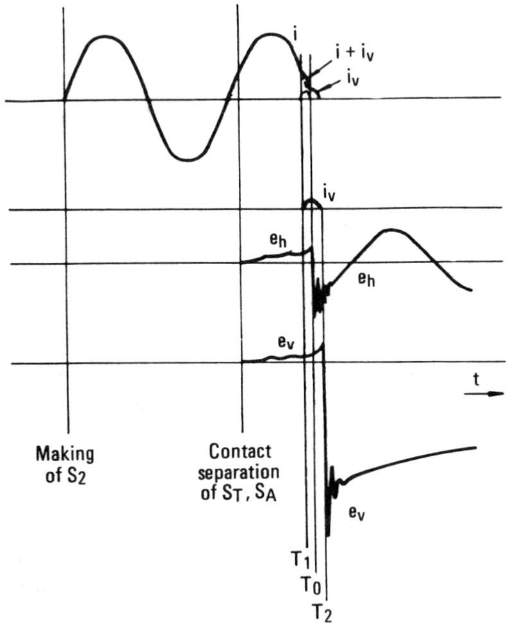

(b) Current and voltage waveforms

When current i_v is interrupted at the current-zero point, the polarity of Cv is reversed. After S_T interruption, the charge in Cv is transferred to transient recovery voltage (TRV) frequency-controlling capacitor Ce via Lv and TRV amplitude-factor-controlling resistor Re. At this point, the transient oscillation is applied as TRV on test breaker S_T. Also, directly after attenuation of oscillation, a dc voltage is applied on the test breaker. The charging voltage of Cv is preliminarily adjusted so that this voltage equals the crest value of the recovery voltage.

(B) Voltage-Injection Scheme Using a Capacitor

The basic circuit for voltage injection using a capacitor is the same as that of the current-injection scheme shown in Fig. 4.18a. Just after the zero point of the power-frequency current fed from current supply G, gap switch g is caused to be in a state of conduction by plasma by a starter circuit. Thus, without superposition of high-frequency current i_v, unlike the current-injection scheme, test breaker S_T and auxiliary breaker S_A are interrupted at the zero point of the power-frequency current. Simultaneously with or slightly after interruption, a transient recovery voltage and a recovery voltage determined by Cv, Lv, Re, and Ce are applied.

Since the high-frequency current is not superimposed as in current injection, there remains some doubt about equivalence near the current-zero point. To improve this condition, there can be certain suitable measures such as additionally applying the TRV near the current-zero point from the current supply with a low-impedance element, such as a large-capacity capacitor, connected in parallel with the auxiliary breaker (Fig. 4.19a) or superimposing the TRV of the current supply on the TRV of the voltage supply with a voltage-source circuit connected in parallel with the auxiliary breaker (Fig. 4.19b). However, it is widely recognized that it is impossible to test thermal reignitions. Also, the gap switch must be triggered with accurate synchronization achieved at the current-zero point. This makes test control very difficult.

Compared with the current-injection scheme, the voltage-injection scheme permits large-capacity circuit breakers to be tested with smaller-capacity capacitors. Studies have been made (Kurosawa et al., 1989) on a two-part method, which tests thermal reignition by applying only a low TRV peak value under a current-injection scheme while using a voltage-injection scheme for the full TRV peak value.

(C) Duplicate Test Scheme (Voltage-Injection Scheme Using a Transformer)

Figure 4.20 shows the basic circuit for the duplicate test scheme, which is also referred to as the *Skeats scheme.* As long as power-frequency current is fed from the current supply, the transformer of the voltage supply is short-circuited on the primary side. When the auxiliary and test breakers are interrupted at a current-zero point, transient recovery voltage and ac recov-

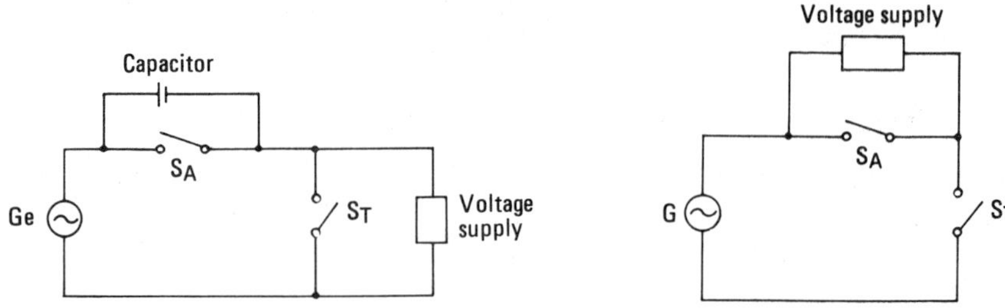

(a) Parallel capacitor with S_A (b) Parallel circuit

Fig. 4.19 Voltage-injection scheme.

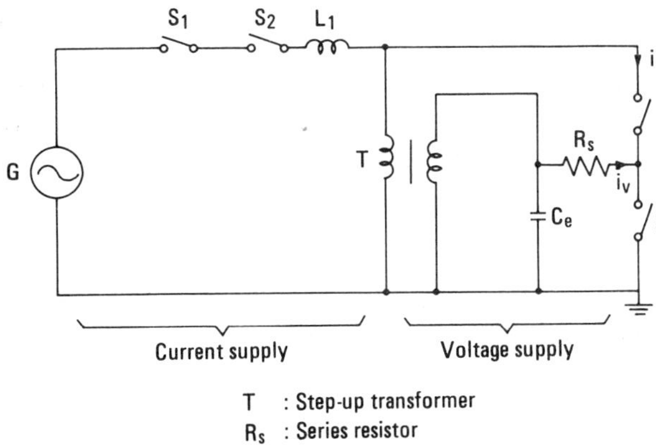

Fig. 4.20 Skeats synthetic testing circuit.

ery voltage are applied automatically. The equivalence near the current-zero point is as insufficient as, or more insufficient than, that of the voltage-injection scheme using a capacitor. If the transformer's stray capacity to earth is large, the rate of rise of the TRV is reduced and the initial delay time is so large that values specified by standards may not be attained.

The greatest advantage of this method is that there is no need for a gap switch, so tests can be conducted as stably as direct tests. Also, it has the advantage that an ac recovery voltage can be applied, as in a direct test.

When a direct test cannot be performed, synthetic test methods enable testing of circuit breakers with a very small capacity power supply. Development of the current-injection scheme is almost revolutionary in the history of short-circuit testing.

In addition to voltage sources and their starters, synthetic test methods require complicated equipment such as *arc-extending equipment.* Under a synthetic test method, the power-frequency current is fed from a current supply with reduced voltage. With a circuit breaker that has a long arcing time, an interruption test may be carried out at the current-zero point a few cycles after contact separation. Because of the reduced voltage, the circuit breaker is expected to be ready to interrupt at a current-zero point even before the intended one in the test, but measures must be taken to prevent interruption at current-zero points other than the intended one. The capacitor method is the most common. It produces a current with a steep rise by discharging a charged capacitor via a resistor. A current-zero point is passed at a large changing rate by superimposing the current just before the power-frequency current-zero point.

Another piece of equipment used in synthetic testing is an auxiliary breaker. This equipment is required to offer performance comparable to, or better than, that of test breakers. These breakers perform current interruption more frequently than any other switches in a high-power testing laboratory, so they must be reliable and have long contact lives.

Synthetic test methods need a wider variety of equipment than direct test methods and are more complicated. For smooth operation of synthetic tests, it is particularly important that the gap switches operate reliably. If arc-extension gap switches are included, the operation of several units must be synchronized. Various improvements will be necessary.

4.3.3 Special Tests Using Synthetic Test Methods

Since circuit breakers have come to be expected to have a variety of functions, they require unconventional new test methods for performance confirmation. A typical example is a new test scheme developed for testing the dielectric performance from pole to ground and between poles after interruption in circuit breakers used for gas-insulated switchgear.

New test schemes must also be developed because at differences between extinction and interruption methods. Tests for tank-type gas circuit breakers, vacuum circuit breakers, and dc circuit breakers are described below as an example. New test methods are required to cope with the special test duty. As an example, the development of a making test is outlined below.

(A) Tank-Type SF_6 Gas Circuit Breaker Tests

(1) Multibreak Circuit Breakers

With conventional multibreak circuit breakers where each break is housed in porcelain, only unit tests—that is, tests concentrating on the interruption performance of a single break—are conducted. With GIS circuit breakers, dielectric performance to ground under full voltage must also be proved. To carry out such proof tests, there are two methods as shown in Fig. 4.21: (a) *full-pole interruption test* and (b) *full-voltage interruption test* (Yamamoto et al., 1985).

The example shown in Fig. 4.22a is a full-pole test of a two-break, 550-kV circuit breaker. Since this method requires a huge capacitor bank, the voltage-injection scheme was used in this test, although it is necessary to conduct proof tests for thermal interruption performance between contacts, particularly near the current-zero point.

The example shown in Fig. 4.22b is a full-voltage test on an 800-kV circuit breaker. With an 800-kV TRV applied to the enclosure upon inter-

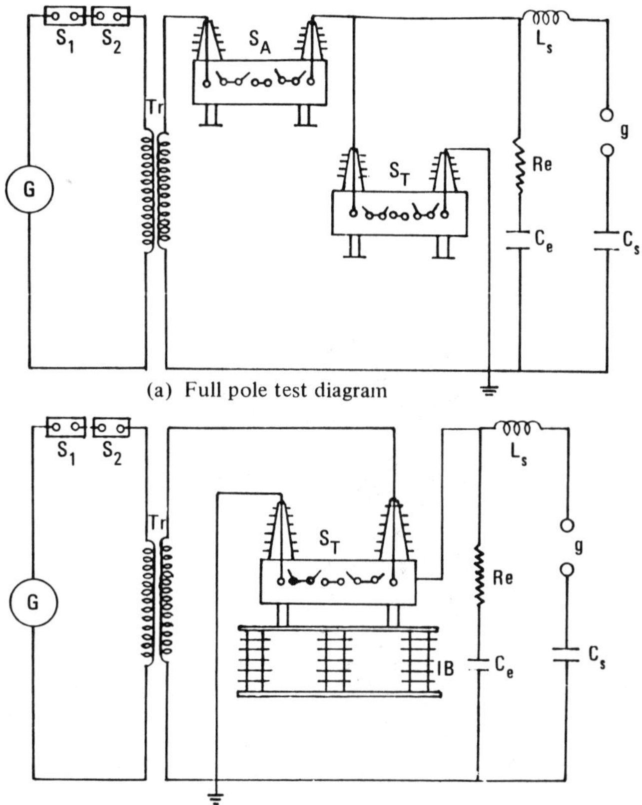

(a) Full pole test diagram

(b) Full voltage test diagram

C_e, C_s	: Capacitor	S_1	: Making switch
g	: Gap	S_2	: Backup switch
G	: Short-circuit generator	S_A	: Auxiliary breaker
IB	: Insulation base	S_T	: Test breaker
L_s	: Reactor	Tr	: Transformer
Re	: Resistor		

Fig. 4.21 Test circuit for a tank-type multibreak GCB. (Yamamoto et al., 1984. © IEEE.)

ruption of a 50-kA current fed from a low-voltage power source, dielectric performance to ground was checked. This system can check dielectric performance to ground relatively easily.

As a matter of course, thermal interruption performance between contacts must be checked separately. Besides the two typical methods just outlined, a combination of them has also been proposed (Schett, 1989).

(a)

Fig. 4.22 Test view of multibreak circuit breaker.

(2) Three-Phase-in-One-Tank-Type GCBs

In tests to prove a GCB's large-current-interruption performance, it is necessary to take into account not only interelectrode and interphase voltages but also interphase electromagnetic force and interactions with other phases such as hot-gas flows exhausted from the interrupting chamber, which can be prominent during high-current interruption.

To test short-circuit interruption performance of the 300-kV-class, extrahigh-voltage, large-capacity, three-phase-in-one-tank-type GCB, it is impossible to carry out three-phase direct tests because of the limitation on

(b)

testing facilities, making it essential to apply a synthetic test method. Several synthetic test circuits for three-phase-in-one-tank-type GCBs have been presented.

In CIGRE WG13-04, synthetic test methods for three-phase-in-one-tank-type GCB are summarized, and the following five requirements are given [CIGRE WG13.04 (1988)]:

1. Full three-phase current should be supplied to the three-phase circuit breaker under test.

2. Arcing duration in all three poles must correspond to the actual arcing conditions that would result in the equivalent full-power direct test. For this purpose, the technique of forced reignition may be applied at one or more phases.
3. A synthetic circuit capable of producing correct stresses must be applied to the first pole to clear.
4. The transient recovery voltage between the last poles to clear and the enclosure must have the proper amplitude and be in the correct phase relation. This can be attained by the use of synthetic circuits for which the requirements for equivalence during the interaction period may be relaxed.
5. The recovery voltage must be applied for a proper length of time and across at least two adjacent phases.

Any synthetic test circuit that meets these five requirements in one test must be provided with a three-phase current source and a set of voltage sources for three phases. Also, naturally, with the voltage source for three phases, at least one phase must have duties (*di*/*dt*, *dv*/*dt*) during the interaction period in conformity to standards, and at least one phase must be able to generate ac recovery voltage.

The circuit shown in Fig. 4.23 is an example. Testing facilities that can realize the 300-kV-rated, high-voltage GCB will be huge, and test control

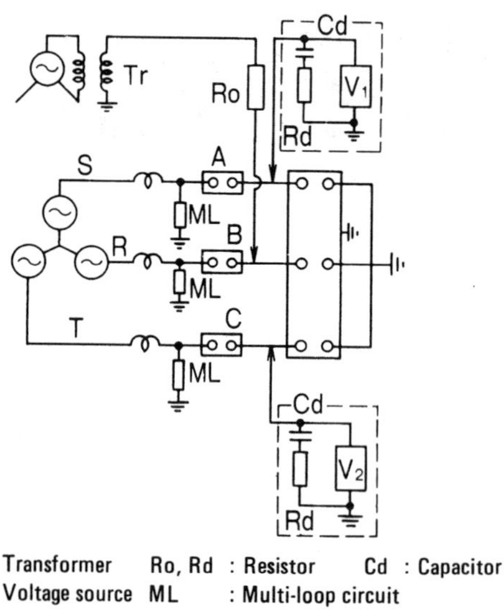

Fig. 4.23 Three-phase voltage source method.

will be very difficult. From the standpoint of test efficiency and cost, such test methods may be limited to final performance proofing or type approval tests.

On the other hand, some test methods that have been developed are not complete enough to meet all five requirements, but they can cover important interphase conditions such as interphase electromagnetic forces and exhausted hot-gas flows equivalent to or higher than those in actual systems.

Figure 4.24 shows a test system with two voltage sources combined for current and voltage superposition with single-phase current fed to phases coupled in series. The same current flows through all phases. If arcing time is adjusted to that of the last interruption phase in the case of a three-phase system, the electromagnetic force is twice as large as that of a three-phase current, and the arc energy is about 20% larger. The test can be controlled as easily as a single-phase synthetic test. However, this test method, using only single-phase current, is employed as a simplified method to check performance allowances in development stages or for other purposes.

Figure 4.25 shows a system that has a current source V-coupled to feed a two-phase current. It has two voltage sources for current and voltage injection, similar to the system in Fig. 4.24. This method is intended to efficiently check interruption capability and interphase dielectric performance. It verifies interphase dielectric performance for more than two phases in one test. Figure 4.26 is an oscillogram obtained during application of this test method (Ibuki et al., 1988).

Figure 4.27 shows a circuit developed particularly to prove second- and third-phase interruption in *effectively grounded systems*. The neutral points

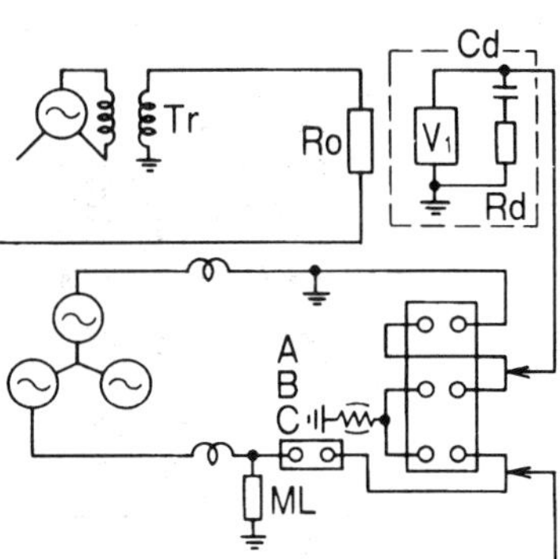

Tr : Transformer Ro, Rd: Resistor
Cd: Capacitor ML : Multi-loop circuit

Fig. 4.24 Three-phases-in-series test method.

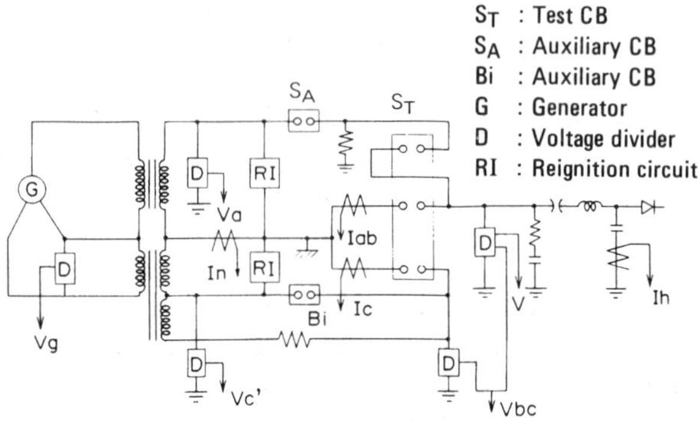

Fig. 4.25 Three-phase synthetic short-circuit test circuit.

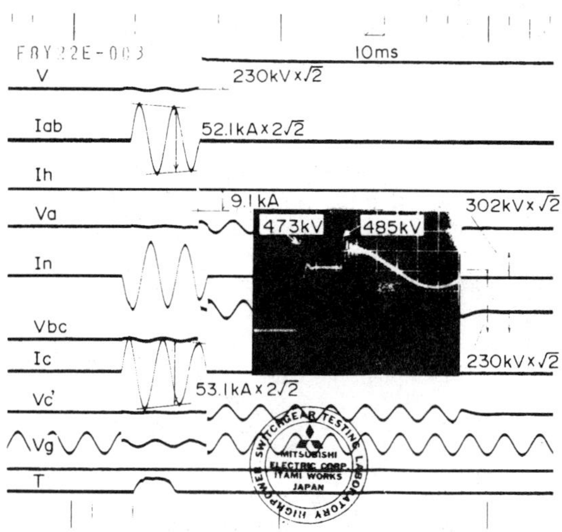

Fig. 4.26 Oscillogram from three-phase synthetic short-circuit test (300-kV three-phase common-enclosure-type GCB).

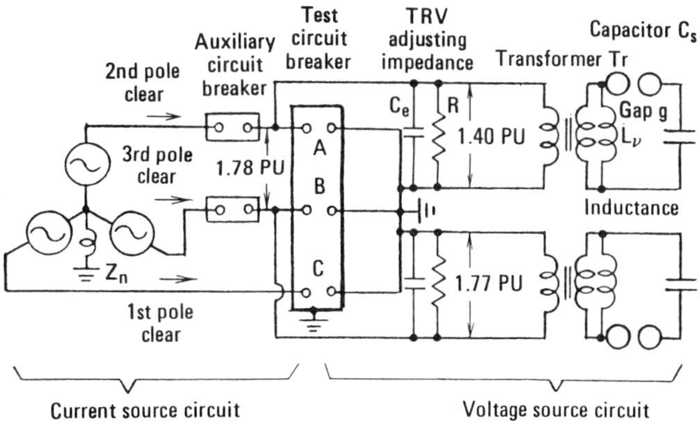

Fig. 4.27 Basic circuit diagram of second- and third-phase synthetic test. (Yamashita et al., 1989. © IEEE.)

of the current sources are set in accordance with the *earthing factor* of the effectively grounded system. This permits three-phase interruptions at different instants—second- and third-phase interruptions 4.3 msec and 6.7 msec, respectively, after first-phase interruption (in a 50-Hz system). The circuit in Fig. 4.27 uses two voltage sources via step-up transformers. Figure 4.28 is an oscillogram obtained during application of this test method (Yamashita et al., 1989).

These methods do not meet all five of the CIGRE requirements listed earlier. However, with such duties as not only interelectrode and interphase voltages, but also interphase electromagnetic forces and arc powers, set carefully, they do permit proof tests by applying stresses equivalent to or above those in actual systems. Further, with the ease of test control, test efficiency, and the scale of test apparatus taken into account, they can be effectively employed, contributing much to the development of EHV-class three-phase-in-one-tank-type GCBs.

(B) Vacuum Circuit Breaker Tests

A vacuum circuit breaker can interrupt the current even if the rate of current change at a current-zero point is very high. It is reported that interruption is possible at a current rate of change 10 times as large as those of SF_6 gas circuit breakers (Yanabu et al., 1981). Thus, with vacuum circuit breakers, unlike SF_6 gas circuit breakers, arc extension is so difficult that special measures must be taken.

Figure 4.29 shows an example. Here, after gap switch g_0 connected in series to the initially charged capacitor C_1 is made to be conductive, g_1 is

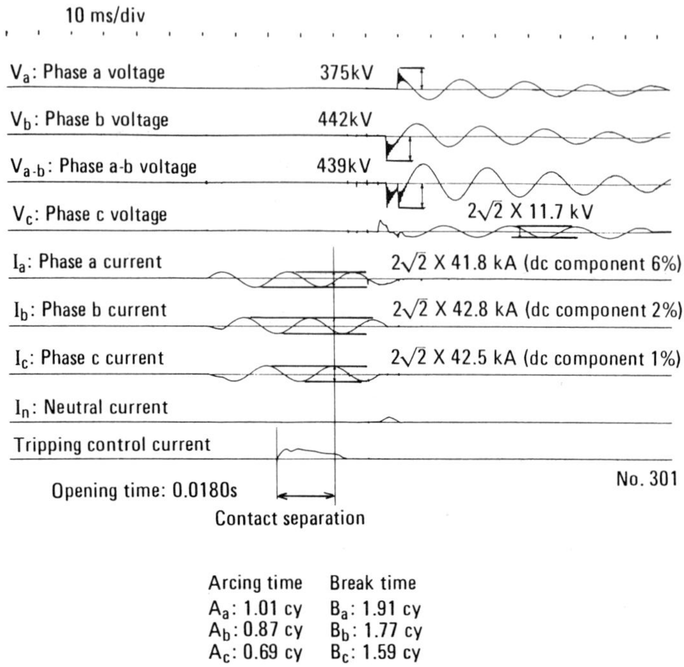

Fig. 4.28 Oscillogram of second- and third-phase synthetic test on 300 kV/40 kA three-phase common-enclosure-type GCB. (Yamashita et al., 1989. © IEEE.)

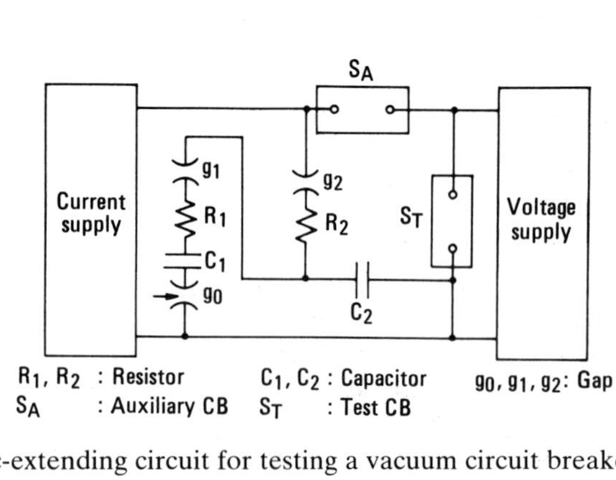

Fig. 4.29 Arc-extending circuit for testing a vacuum circuit breaker.

turned on, causing capacitor C_2 to be charged. When the voltage of C_2 rises, reaching the dielectric breakdown voltage of gap switch g_2 in series with C_2, g_2 is turned on, causing the charge of C_2 to be discharged via resistor R_2, auxiliary circuit breaker S_A, and test circuit breaker S_T.

Figure 4.30a shows the measured waveform of the current through the test circuit breaker; Figure 4.30b illustrates it schematically. Whereas only current from capacitor C_1 flows in ordinary arc-extension circuits, the quickly rising current from C_2 first flows in this circuit. The C_2 circuit can have low impedance, enabling steep currents to be obtained. In this example, pulses with current changing rates of several hundred amperes per microsecond can be realized (Kurosawa et al., 1987).

(C) Direct Current Circuit Breaker Tests

For dc circuit breakers utilizing current limited by arcing voltages or arc instability, direct test methods are necessary. For dc circuit breakers designed to make a current-zero point by high-frequency current injection, synthetic test methods are applicable.

Figure 4.31 shows a test circuit under a synthetic test method for dc circuit breakers using power-frequency ac current (Shimada et al., 1979). If no dc current is available as a current supply, a method that uses current around the crest value of an ac current as a dc current will be applicable. The synthetic test method here is basically a voltage-injection scheme. A large ac current is fed from low-voltage current supply G. Upon receiving an interruption command, test and auxiliary circuit breakers S_T and S_A open and are subjected to electric conduction by arcing. When the required gap length in S_T is obtained, the current-zero-point-making equipment with capacitor C_c operates to produce current-zero points artificially on S_T and S_A. S_T and S_A interrupt simultaneously, causing the current supply to be disconnected from S_T. At the same time, the voltage supply is connected by gap switch conduction with a signal from the current-zero-point-detecting circuit, causing a voltage to be applied to the test circuit breaker.

Under this test method, however, equivalence near the current-zero point is not obtained. This method should be used together with the equivalent test method shown in Fig. 4.32. Under this equivalent test method, no forced current-zero point is produced on S_A. This prevents S_A from being interrupted simultaneously with an S_T interruption, allowing the current supply to feed the current-zero-point-making equipment via S_A. Thus, near the zero point until S_A interruption this test is equivalent to a direct test method except that it lacks the proper recovery voltage.

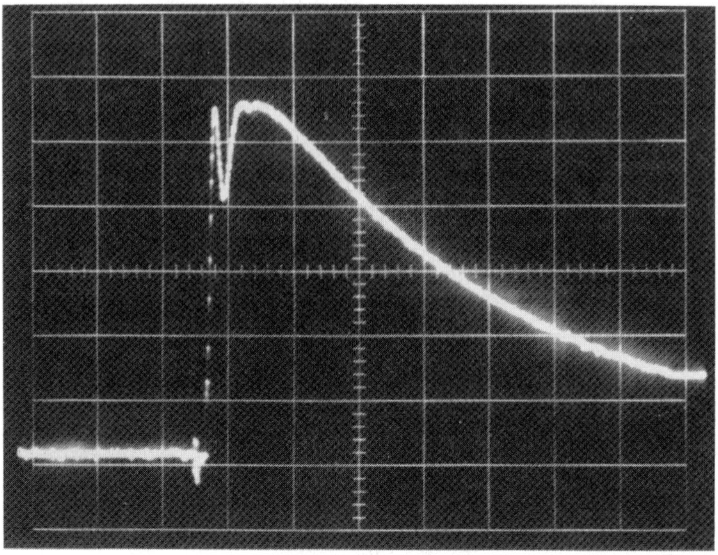

(a) 748 A, 100 μs/div

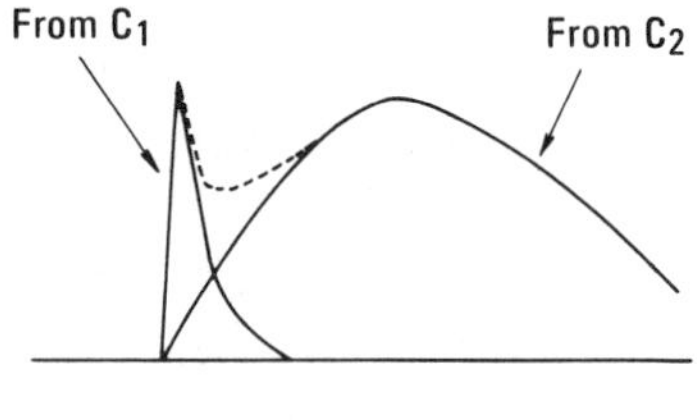

(b)

Fig. 4.30 Waveform for arc-extending pulse.

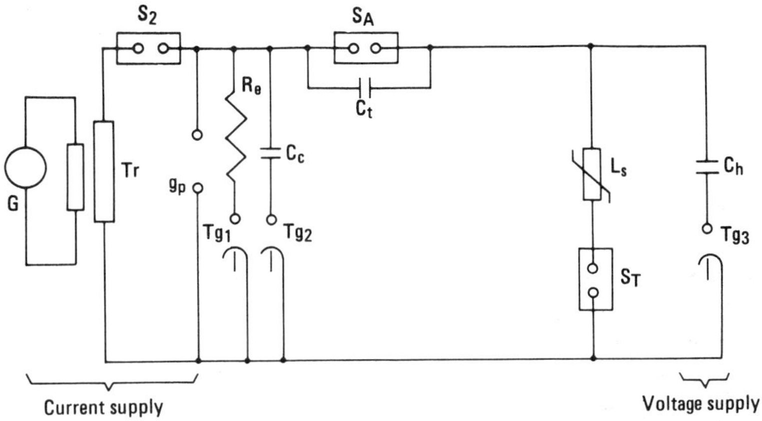

G : Generator
L_s : Saturable reactor
S_2 : Back-up circuit breaker
S_T : Test breaker
S_A : Auxiliary breaker
g_p : Protection gap
Tg_1, Tg_2, Tg_3 : Triggering gap
Tr : Transformer

C_t = 10 μF
C_c = 1100 μF
R_e = 0·2 Ω
L_s = 40 μH (saturated)
= 160 μH (unsaturated)
C_h = 150 μF

Fig. 4.31 Synthetic test circuit for dc circuit breaker. (Shimada et al., 1979. © IEEE.)

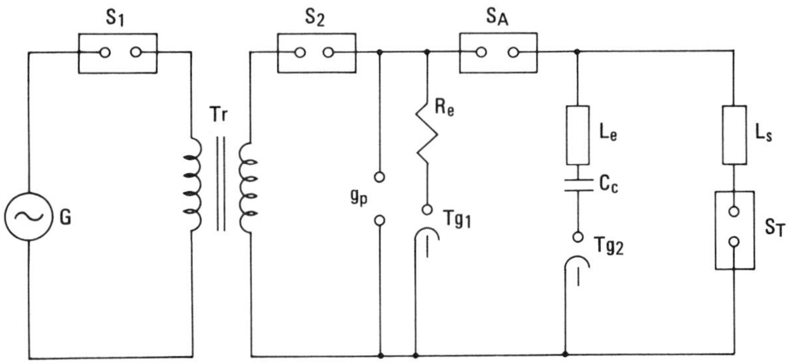

Tr : Transformer
G : Generator
Tg_1, Tg_2 : Triggering gap
S_1 : Making switch
S_2 : Back-up circuit breaker
S_T : Test breaker
S_A : Auxiliary breaker
g_p : Protection gap

L_c = 50 μH
C_c = 224 μF
L_s = 50 μH (saturated)
= 200 μH (unsaturated)
R_e = 0·1–0·6 Ω

Fig. 4.32 Simplified synthetic test methods. (Shimada et al., 1979. © IEEE.)

(D) Making Tests

If the contact-closing speed is not increased as the voltage per break of the circuit breaker is increased, the prearcing time is prolonged. The contact-closing speed, which is related to mechanical stress, cannot be increased in proportion to the voltage. This prevents making at a zero voltage in actual high-voltage circuit breakers; therefore making under a direct test method with a reduced voltage is inapplicable, because equivalency during the prearcing period is not ensured.

Figure 4.33 shows a *synthetic making test circuit* (Yamashita et al., 1976). Here, S_T making is performed with a voltage applied from a voltage supply. At the moment of making with prearcing, gap switches SM_1 and SM_2 are triggered for conduction, allowing a current to be fed from the current supply. Figure 4.34 is an oscillogram obtained in such a test. The gap switches are in conduction with arcs. To control contact wear, protective making switches are added in parallel, and gap switches are short-circuited by making the making switches several milliseconds after gap switch conduction.

Synthetic making tests are necessary not only for short-circuit making tests, but also for small-capacitive-current switching tests. Such tests necessitate simulating the steep discharge current of the capacitance on the power-source or load side of the circuit breaker. A complicated test circuit is required.

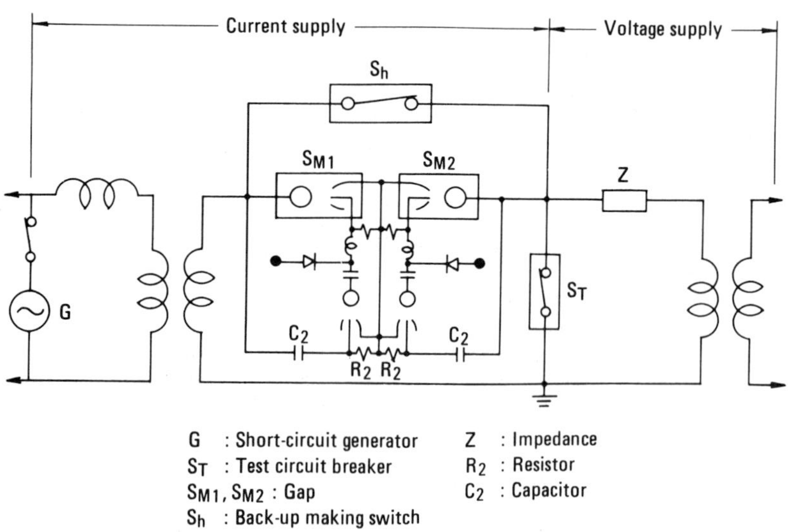

Fig. 4.33 Synthetic making test circuit.

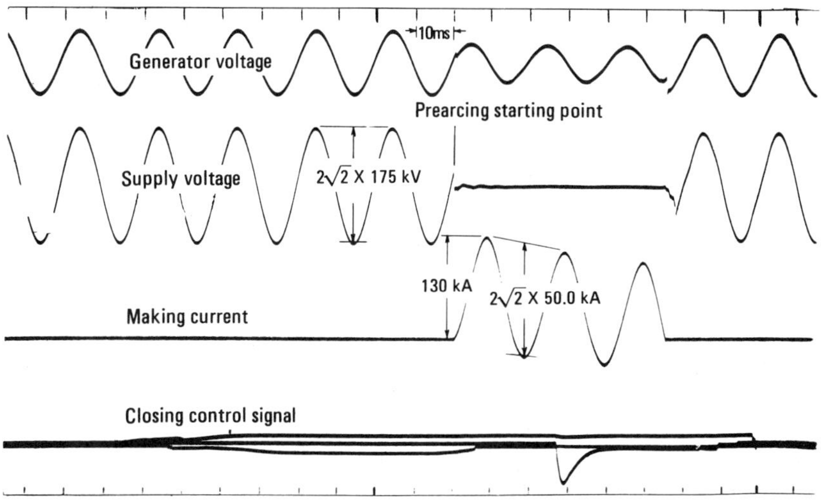

Fig. 4.34 Oscillogram at synthetic making test.

4.3.4 Equivalence of Synthetic Test Methods to Direct Test Methods

It is recognized that to obtain equivalence to direct test methods in synthetic test methods, the following requirements must be met (JIEE, 1982):

1. Current crest values must agree.
2. Current waveforms (gradient) before current-zero points must agree.
3. Voltage waveforms (rate of rise) after current-zero points must agree.
4. TRV waveforms (crest values) must agree.

As shown in Section 4.3.2, equivalence has been obtained by feeding a current with a reduced voltage from a short-circuit generator for requirement 1, above, and by injecting a high-frequency current under a current-injection scheme for requirement 2.

For requirements 3 and 4, the production of TRV is an important factor. As described previously, TRVs are divided into power-source-side voltage and line-side voltage. For a power-source-side TRV under synthetic test methods with a current-injection scheme, transient oscillation is used; this oscillation is caused when a part of the charge of voltage-source capacitor Cv is transferred to TRV frequency-controlling capacitor Ce via voltage-source current-controlling reactor Lv and TRV amplitude-factor-controlling resistor Re as shown in Fig. 4.18. This method realizes a TRV equivalent to that obtained in direct test methods.

However, TRVs in actual systems are complicated as described in Section 4.1. In the newer standards they are represented by envelopes.

Various measures have been taken to produce such envelopes of TRVs. An example is shown in Fig. 4.35. *Four-parameter TRVs* are produced by using several frequencies obtained from several oscillation circuits. There are also reports on other methods such as one that uses several voltage-source circuits in series (St-Jean, 1974) and one that uses a current-injection scheme and a duplicate test scheme in series (Maury and Vigreux, 1971).

For line-side TRVs, especially in SLF testing, as in direct test methods, an *artificial transmission line* composed of lumped-constant inductance and capacitance is inserted between a voltage-source circuit and the test circuit breaker. To compose simulated lines, an investigation has been conducted for obtaining a high equivalence with as small a number of elements as possible.

There is a report on experimental study on the equivalence between synthetic and direct test methods that meet the above requirements. The same circuit breaker was submitted to a short-line-fault (SLF) interruption test under synthetic and direct test methods, and a negligible difference was noted in the interruption limit between the two methods (Fujii et al., 1969).

Thus, it is recognized that synthetic test methods of the current-injection scheme have a high equivalence to direct test methods. However, the high equivalence is obtained on the assumption that the circuit breaker and power system have specific characteristics. Consequently, the equiva-

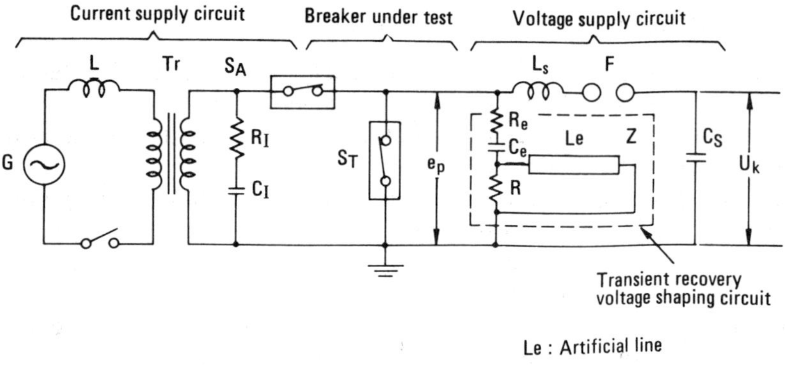

Fig. 4.35 Circuit for four-parameter transient recovery voltage.

lence varies as the circuit breakers and systems vary. The employment of synthetic test methods is unavoidable, and studies on them are likely to be continued.

Synthetic test methods for circuit breakers have only been outlined here. By applying basic schemes, various performance test methods have been developed for circuit breakers. It is expected that technological advances will accompany the development of circuit breakers in the future.

4.4 MEASURING METHODS (S. Nishiwaki)

4.4.1 Post-Arc Current

As is the case with SLF (short-line-fault) interruption, feeble *residual current* (also called post-arc current) sometimes flows between circuit-breaker electrodes after arc current passes the zero point during interruption when current is large and the rate of rise of TRV (transient recovery voltage) is also large. This phenomenon occurs because the space between circuit-breaker electrodes does not form the state of a perfect insulator directly after arc current passes the zero point. In many cases, residual current is observed near the interruption limit point. When the energy injected by residual current is greater than the energy loss caused by cooling between the circuit-breaker electrodes, thermal reignition occurs.

Residual current provides information that can be used in the evaluation of interrupting performance and at the same time offers an index for use with quantitative calculations of interruption results. Therefore, efforts are being continued to enable measurement of the residual current. While levels of such residual current vary depending on the type of circuit breaker, its intensity ranges from several hundred milliamperes to several tens of amperes. The continuation time of residual current spans from several microseconds to several tens of microseconds. The amount of this residual current on SF_6 gas circuit breakers is smaller than that on air circuit breakers. The difficulty associated with this type of measurement lies in the fact that such feeble current must be measured immediately after conduction of large short-circuit current of several tens of kiloamperes.

To overcome this difficulty, various means have been employed in attempting accurate measurement of residual currents. Figure 4.36 shows the principle of one method (Murano et al., 1975). In this figure, the resistance of shunt R_1 is selected to be a large value (50 mΩ, for instance) so that signals large enough for measurement against feeble residual current can be obtained. Further, a coaxial type of shunt is employed to improve rise-time response characteristics by reducing stray inductance. Also, for a synthetic test, one mechanical contact S_1 is used, whereas for a

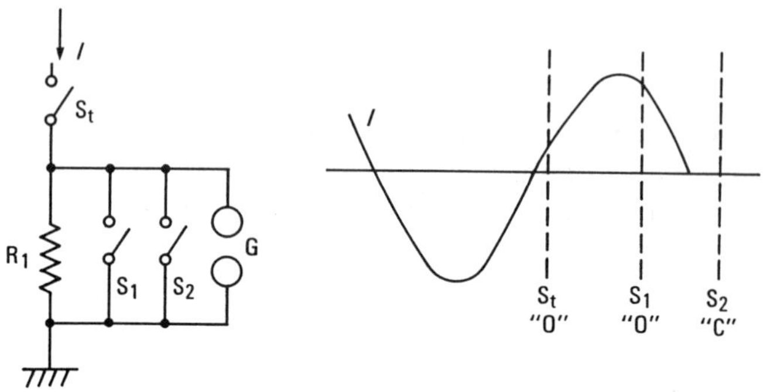

(a) Basic constitution of shunt circuit

(b) Operation of mechanical contacts

S_t: Testing breaker
S_1, S_2: Mechanical contact
R_1: Coaxial shunt
G: Protection gap

Fig. 4.36 Principle of measuring residual current by employing a shunt with mechanical contacts in parallel. (From Murano et al., 1975. © IEEE.)

direct test, two mechanical contacts—S_1 and S_2—as well as a protective gap G are employed. Contact S_1 is kept closed until directly before the measuring arc current reaches the zero point to protect the shunt circuit and prevent it from being broken down thermally or mechanically by a large short-circuit current of several tens of kiloamperes. The lower the arcing voltage at the time of opening of contact S_1, the smaller the current flowing through the shunt circuit, resulting in better protection of that circuit. As an ac arc current is lowered closer to the zero level, arcs of contact S_1 are completely extinguished, causing the entire current to flow through the shunt circuit. Thus, accurate current measurement becomes possible from at least 10 μsec before the current-zero point. In the case of a synthetic interruption test, only injection current from the voltage source flows into the shunt circuit even if reignition occurs on the circuit breaker. Therefore, no breakdown of the shunt circuit is caused even when second contact S_2 and gap G are not used. As for the direct interruption test, this second contact S_2 and gap G are essential for protecting the shunt circuit from large short-circuit current after the occurrence of reignition.

With contact S_1 connected in parallel with the shunt circuit, a vacuum circuit breaker can well satisfy the characteristics required on this type of contact. In other words, the following three requirements can be met satisfactorily:

1. Arcing voltage is low (several tens of volts or less) and free from extensive variations by current values.
2. Dielectric recovery between contacts is quick.
3. Conduction of a large current is possible.

If the sensitivity of a waveform measurement instrument is increased to enable the measurement of feeble residual current, it is possible that the instrument will malfunction; for example, that the zero level will shift, as a result of the high measuring signal voltage caused by the large current before the current-zero point or after reignition. To avoid this problem, a *clip circuit* is generally provided for measuring signals. Figure 4.37 shows an example of a clip circuit structured by using a 50-mΩ *coaxial shunt* circuit (Murano et al., 1975). On this circuit, the clipping function is made available by using a high-speed switching diode. Also, whenever necessary, the circuit is provided with a trigger device for triggering the waveform observation device directly before the current-zero point.

Shown in Fig. 4.39 is an example oscillogram of voltage and current near the current-zero point (Suzuki et al., 1980). The current and voltage were measured by using an SF_6 gas circuit breaker and an air circuit breaker on the synthetic interruption test circuit prepared by simulating the SLF shown in Fig. 4.38. R_1 in Fig. 4.38 is the coaxial-type shunt used for measurement of residual current. Other associated devices explained in Fig. 4.36 are not shown in this figure. The oscillograms in Fig. 4.39 indicate the following:

1-a, 2-a: Interruption achieved without accompanying residual current
1-b, 2-b: Interruption achieved by accompanying residual current
1-c, 2-c: Failure in interruption

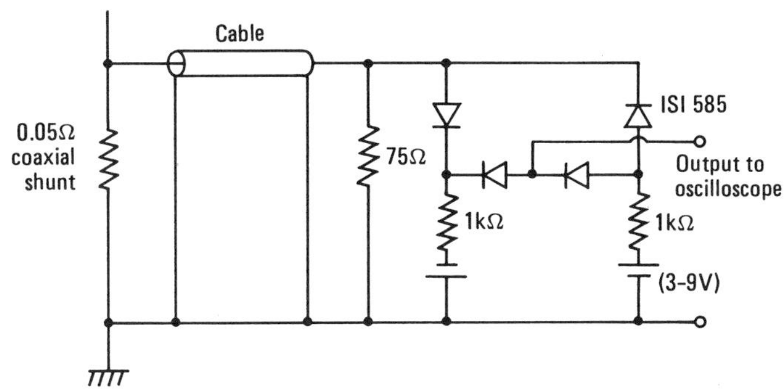

Fig. 4.37 Clip circuit for measuring signals. (From Murano et al., 1975. © IEEE.)

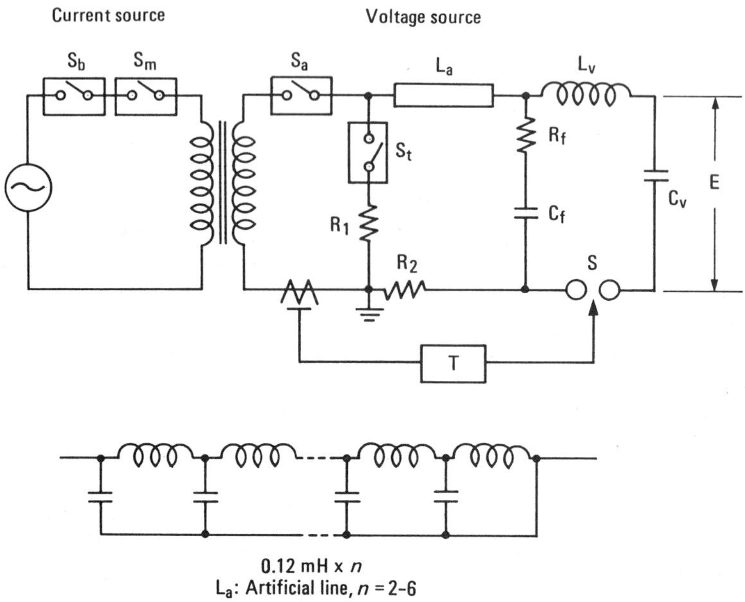

S_b: Back-up breaker
S_m: Making switch
S_a: Auxiliary breaker
S_t: Test breaker
L_a: Artificial line
L_v: Source-side circuit inductor
C_v: Main capacitor bank
R_f, C_f: Resistor and capacitor for source-side TRV
R_1, R_2: Shunt
T: Triggering device
S: Spark gap
E: Charging voltage (10–30 kV)

Fig. 4.38 Residual current measurement in synthetic interruption test circuit. (From Murano et al., 1975. © IEEE.)

From oscillograms 1-b and 2-b, it is evident that residual current of 12 A and 26 A, respectively, has flowed through the SF_6 gas circuit breaker and the air circuit breaker.

Illustrated in Fig. 4.40 are relations between the residual current crest values and the rate of rise of transient recovery voltage (RRRV) of SF_6 gas and air circuit breakers (Suzuki et al., 1980). This figure indicates classification of the case with interruption achieved by accompanying residual current and the case of failure in interruption. In the case of interruption achievement, the crest values of residual current are shown; and in the case of interruption failure, the first crest values of residual current are shown. As for RRRV, its values were obtained from the formula $Z\ dI/dt$, because distortion is created by the flow of residual current. Z is the surge imped-

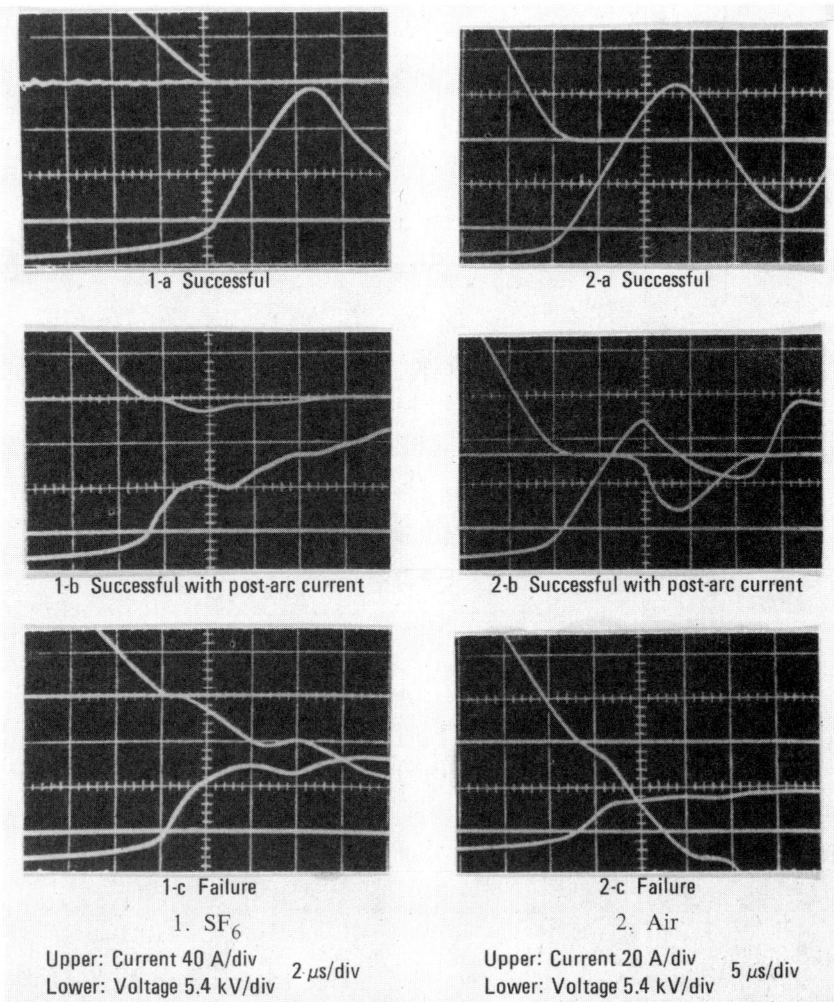

Fig. 4.39 Typical oscillograms around current-zero point. (From Suzuki et al., 1980. Copyright IEE.)

ance of the SLF artificial line. From Fig. 4.40, we are able to clarify the existence of the boundary on intensity of residual current that determines the success or failure of interruption.

Shown in Fig. 4.41 is the residual current of the vacuum circuit breaker obtained by the same measuring method as that shown in Fig. 4.36 (Yanabu et al., 1985). The RRRV differs between Figs. 4.41a and 4.41b.

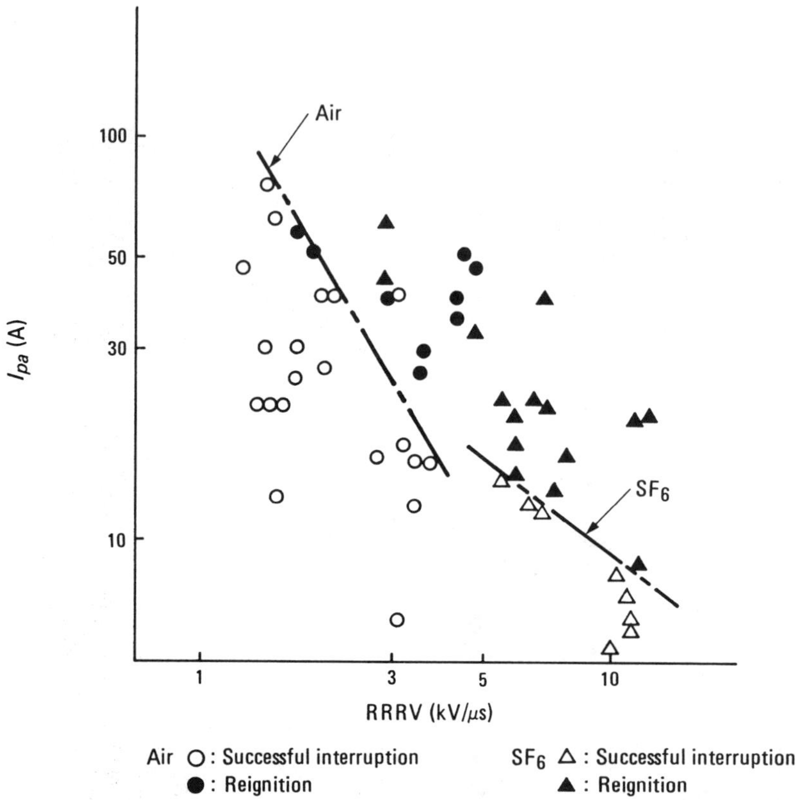

Fig. 4.40 Dependence of residual current crest values of air and SF_6 on RRRV. (From Suzuki et al., 1980. Copyright IEE.)

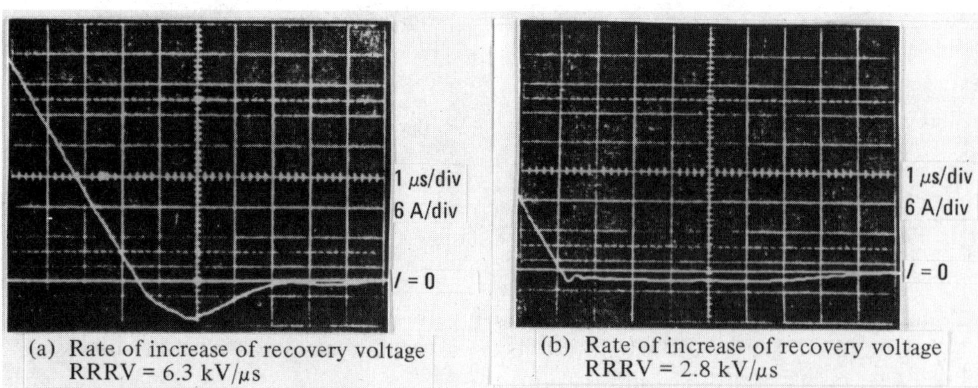

Fig. 4.41 Waveform of residual current of vacuum circuit breaker. (From Yanabu et al., 1985. © IEEE.)

Figure 4.42 is the synthetic test circuit used for measurement of residual current by connecting a *pyrobreaker* in parallel with the shunt circuit (Blez and Guilloux, 1989). During conduction of a large short-circuit current, the shunt circuit for enabling the measurement of residual current is kept shorted by a pyrotechnic device. The pyrobreaker is opened by electric control, and current is shifted to the shunt circuit. The circuit-opening command for the pyrobreaker is transferred by way of the optical link about 200 μsec before the current-zero point.

Figure 4.43 shows the case of providing shunt circuits for residual current measurement at the voltage source of a synthetic interruption test circuit and at the high-voltage section of a current-injection circuit (St-Jean et al., 1988). Since large short-circuit current from the current source does not flow into these shunt circuits, the thermal and mechanical duties on the shunt circuits can be reduced even without providing bypass circuits for them. In this case, however, there occurs a problem of observation; together with residual current, the extra current flows into stray capacitance C_0 of the high-voltage circuit. As a result, when the residual current is small it is hidden in the C_0 current. Therefore, the measuring current must be compensated to enable reading of only the residual current component. Since a shunt circuit is provided in the high-potential section, signals must be converted once to light so they can be measured with the measuring instrument remaining at ground potential.

Shown in Fig. 4.44 are the measurement results of residual current on the circuit illustrated in Fig. 4.43 when an SF_6 gas circuit breaker of 200 kV was used for the test (St-Jean et al., 1988). While the total current of 1.2 A

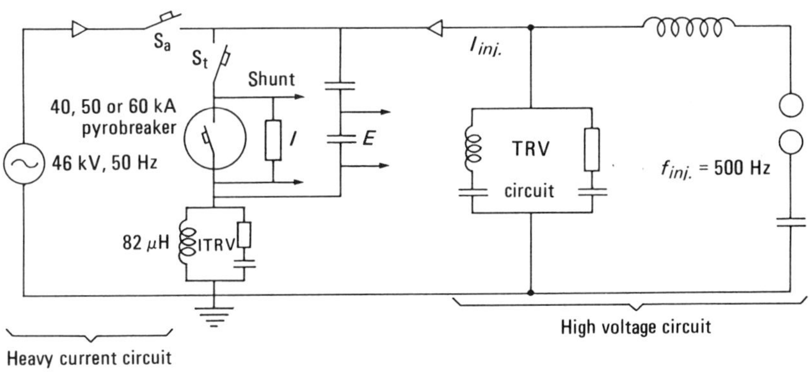

Fig. 4.42 Measuring residual current by employing a shunt with a pyrobreaker in parallel, in synthetic interruption test circuit. (From Blez et al., 1989. © IEEE.)

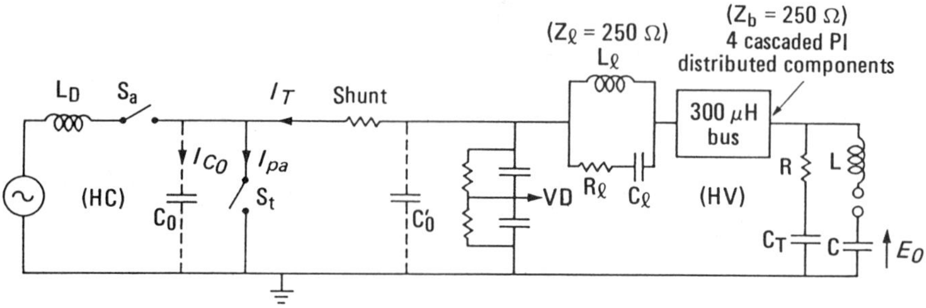

Fig. 4.43 Measuring residual current by employing shunts installed at voltage source of synthetic interruption test circuit and at high-voltage section of current-injection circuit. (From St-Jean et al., 1988. © IEEE.)

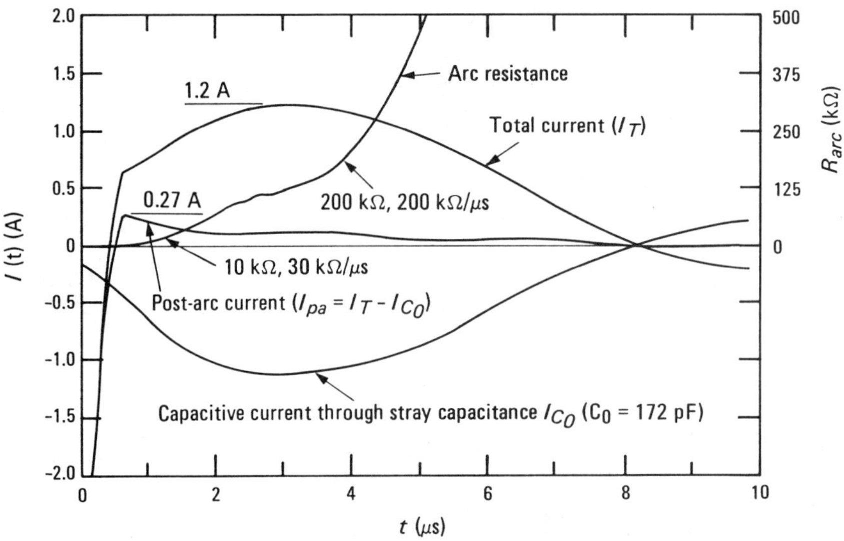

Fig. 4.44 Results of measuring residual current on circuit illustrated in Fig. 4.43 with 200-kV SF_6 gas circuit breaker. (From St-Jean et al., 1988. © IEEE.)

was observed, residual current of 0.27 A was calculated by subtracting the amount that flowed into stray capacitance $C_0 = 173$ pF.

Quantitative evaluation or estimation of the interrupting performance of circuit breakers is being conducted by utilizing the data that can be obtained by measuring residual current. For example, evaluation can be made by obtaining parameters A and α of arc time constant θ, and B and β of arc cooling loss N_o of Mayr's equation as shown below [Eq. (4.1)] and then solving a differential equation into which Eq. (4.1) and the circuits during interruption are combined.

$$\frac{1}{G}\frac{dG}{dt} = \frac{1}{\theta}\left(\frac{EI}{N_o} - 1\right) \tag{4.1}$$

where

θ = $A(1/G)^{\alpha}$ = arc time constant
N_0 = $B(1/G)^{\beta}$ = arc cooling loss
G = arc conductance
E = arc voltage
I = arc current

Figure 4.45 shows an example of evaluation achieved based on the measurement results shown in Fig. 4.44 (St-Jean et al., 1988). In this evaluation example, the values $\theta = 0.86(1/G)^{-0.09}$ (μsec) and $N_0 = 396(1/G)^{-0.49}$ (kW) were obtained at a stress level of 96%; by using these parameters, interruption at a stress level of 102% is estimated to be impossible. This estimation agrees with actual measurement results.

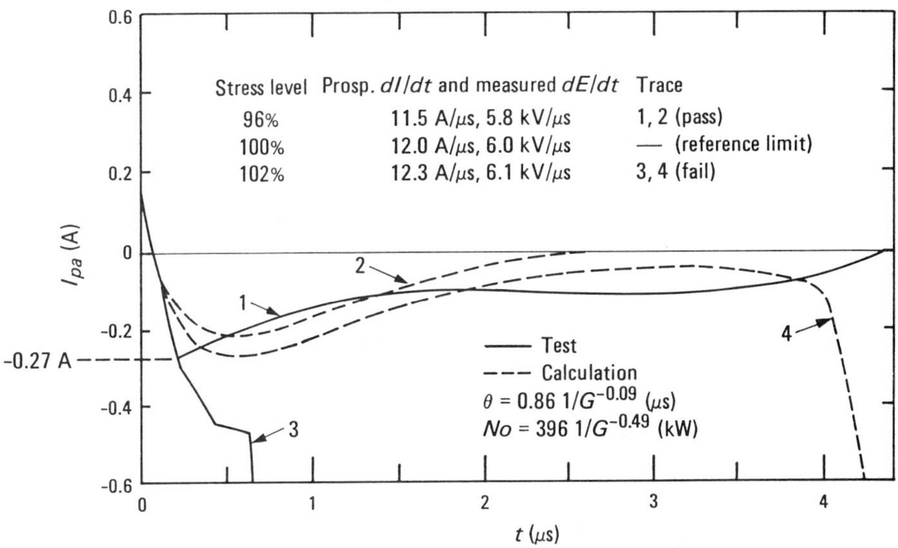

Fig. 4.45 Example of quantitative calculation of interrupting performance based on data obtained from residual current measurement. (From St-Jean et al., 1988. © IEEE.)

Residual current has also been measured to evaluate the interrupting performance of multibreak circuit breakers (Kobayashi et al., 1978). As a result, it has been clarified that thermal interrupting limits are not meaningfully affected by a small imbalance in static voltage distribution.

4.4.2 Transient Recovery Voltage

Transient recovery voltage (TRV) of circuit breakers depends on the type of interruption. Generally, the types of interruption employed are

1. Bus-terminal-fault interruption
2. Short-line-fault interruption
3. Out-of-phase interruption
4. Different-phase earthing interruption
5. Small-inductive/capacitive-current interruption

In measuring TRV, amplitude factor, frequency, rate of rise, crest value, time delay, and other factors are major measuring objects (Committee report, 1976).

Transient recovery voltages employed for short-circuit current interruption tests are specified by standards. In the case of breaker-terminal-fault interruption, TRV frequency must be several tens of kilohertz or below. As for TRV frequency on the line side for short-line-fault interruption, such frequencies are determined by distances up to the fault points, generally forming triangular waves of several hundred kilohertz.

For measuring the TRV on interruption tests at a high-power laboratory, *voltage dividers* are used. They must meet the following performance requirements:

1. Sufficient dielectric strength must be attained.
2. Thermal capacity must be large because circuit voltage is applied.
3. Proper response characteristics must be adopted to enable measurement up to high-frequency levels.

Voltage dividers can be classified into three types depending on their construction:

1. Resistance voltage dividers
2. Capacitance voltage dividers
3. Resistance-capacitance voltage dividers

In view of their thermal capacities, resistance voltage dividers fail to satisfy the performance requirements of voltage dividers for TRV measurement. Consequently, the other two types, capacitance voltage dividers and resistance-capacitance voltage dividers, remain as the types available. Capacitance voltage dividers possess limited capability for measurement of voltage involving dc components or low-frequency components. Therefore, it can be said that resistance-capacitance voltage dividers are generally the most suitable for measurement of TRV.

Figure 4.46 shows an example of a resistance-capacitance voltage divider that can be used at a high-power laboratory. Its construction is illustrated in Fig. 4.47. As shown in this figure, output voltage from the secondary circuit of the voltage divider is led into the measurement chamber with a coaxial cable, and further voltage dividing is accomplished inside this chamber. The total series of primary capacitors, primary damping resistors, and primary parallel resistors shown in Fig. 4.47 are 400 pF, 300 Ω, and 1500 MΩ, respectively. Dielectric strength is 750 kVac, 2550 $kV_{impulse}$; response time is 31 nsec (Haefely date unknown).

Figure 4.48 shows an example of the measurement of TRV with the above-described voltage divider. This is measurement on 90% SLF inter-

Fig. 4.46 Example of resistance-capacitance voltage divider at a high-power laboratory.

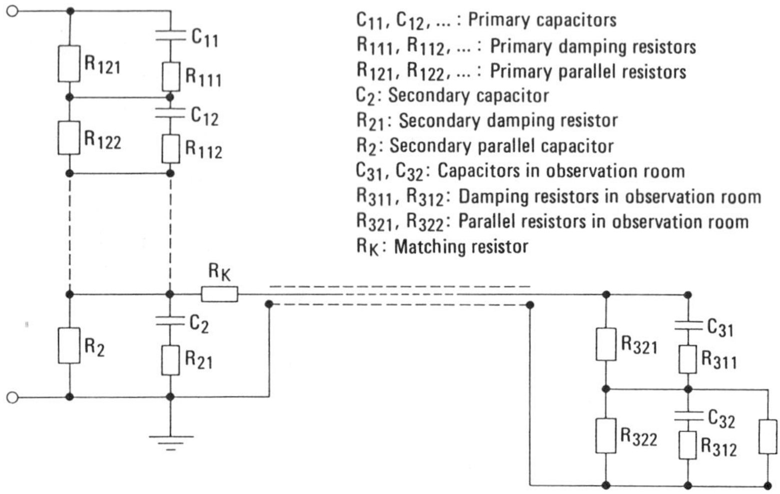

Fig. 4.47 Construction of resistance-capacitance voltage divider shown in Fig. 4.46. (From Haefely Manual.)

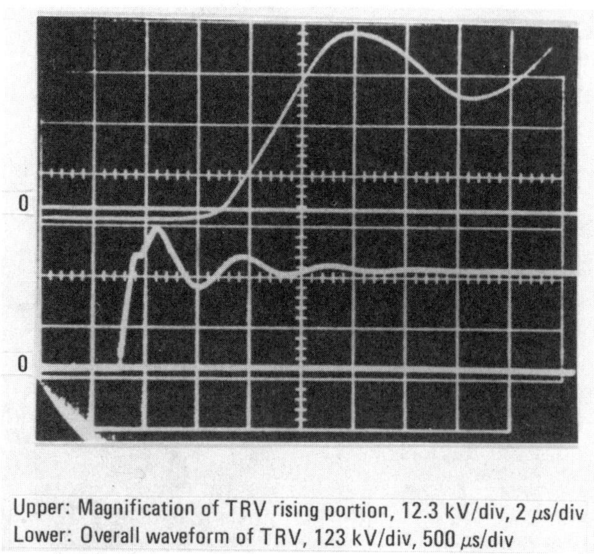

Fig. 4.48 Example of TRV waveform obtained by 90% SLF interruption test of 300-kV, 50-kA, 60-Hz SF_6 gas circuit breaker.

ruption—one of the cases of TRV involving the steepest rate of rise. The device subjected to the test was an SF_6 gas circuit breaker of 300 kV, 50 kA, 60 Hz. As is evident from the oscillogram shown at the top of this figure, steeply rising voltage is observable.

Waveforms on TRV obtained from interruption tests employing actual circuit breakers often differ from those of *inherent TRV* because of the effects of arcing voltage between circuit-breaker electrodes, residual current, or the capacitance of the circuit breakers themselves. Such inherent TRVs for interruption tests are specified by standards, and interruption tests must be conducted according to these standards. The following methods are available for measurement of inherent TRV (IEC, 1987).

1. Direct short-circuit breaking method
2. Power-frequency current-injection method
3. Capacitor current-injection method

With method 1, TRV is obtained by actually conducting an interruption test using a circuit breaker with nearly ideal performance—a unit that generates minimum arcing voltage and residual current while possessing its own limited capacitance. Often this type of test is conducted with reduced voltages by using vacuum circuit breakers.

As for method 2, instead of using circuit breakers, interruption is accomplished by using switching components that can be considered nearly ideal circuit breakers—for instance, diodes, thyratrons, or thyristors. By keeping a small ac current injected into the test circuits, the inherent TRV is clarified from the transient response of circuits during interruption.

In method 3, a capacitor is charged, and this charged capacitor is discharged to the test circuit to cause the current to flow in resonance with the frequency, which will depend on the values of the capacitance and inductance of the circuit. Then the inherent TRV is obtained by interrupting this resonant current similarly to method 2 through the use of a switching component that can be considered an ideal circuit breaker. Figue 4.49 shows the measuring method concerning the inherent TRV of artificial lines on SLF synthetic interruption test circuits by the capacitor current-injection method. Capacitor C (2 μF) shown in this figure is charged up to about several tens of volts, and the switch is closed. Then measurement is taken of resonant current I between C and inductance of the artificial lines, as well as of the voltage E of these lines, when resonant current I is interrupted with the high-speed switching diode dy.

In addition to TRV as mentioned above, which occurs during power-frequency current interruption, another matter of concern is *high-frequency surge voltage/current* that occurs during reignition or restriking of circuit breakers, especially when a small current is interrupted. Figures 4.50 and

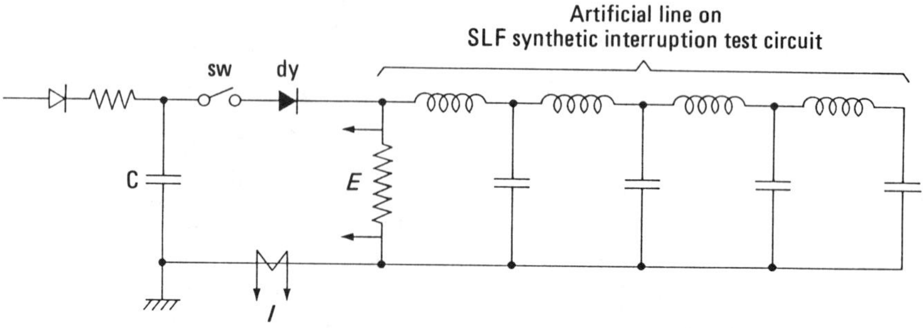

Fig. 4.49 Method of measuring inherent TRV on SLF test circuits—capacitor current-injection method.

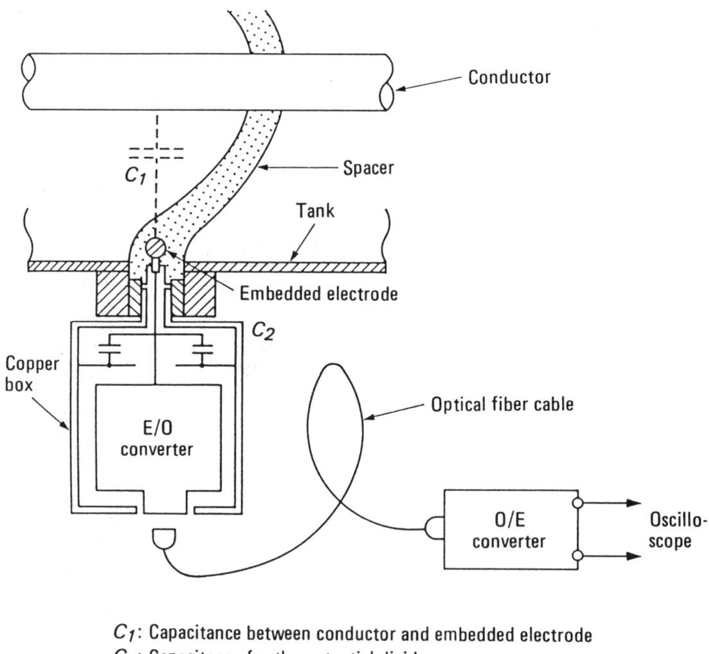

Fig. 4.50 Construction of optical-transmission-type high-frequency voltage divider using spacer. (From Kawada et al., 1987. © IEEE; and Nojima et al., 1987. © IEE.)

4.51 indicate methods of measuring such high-frequency surge voltage and current. Figure 4.50 shows the case of using a high-voltage-conductor-supporting *spacer* provided on the terminal of a tank-type SF_6 gas circuit breaker. This spacer actually acts as a voltage divider. A capacitance-type voltage divider is configured by using capacitance C_1 between a high-voltage conductor and a ring-shaped electrode embedded in the tank-side spacer interior as the primary-side capacitance of the voltage divider. Output voltage from capacitance C_2 of the voltage divider's secondary side is converted into light signals and transmitted to the measuring instrument. With this type of construction, measurements are possible up to a high-frequency voltage of 10 MHz.

Shown in Fig. 4.51 is a high-frequency-current-measuring device adopting the method of converting output from a *high-frequency current transformer* into light signals and transmitting these light signals with an *optical fiber cable.* This device permits direct measurement of current that flows through high-potential conductors.

The measuring devices were used for measuring the reignition high-frequency surge voltage and current during a shunt-reactor current-interruption test by using a 300-kV SF_6 gas circuit breaker. The test circuits used for this test are shown in Fig. 4.52; examples of the oscillograms obtained

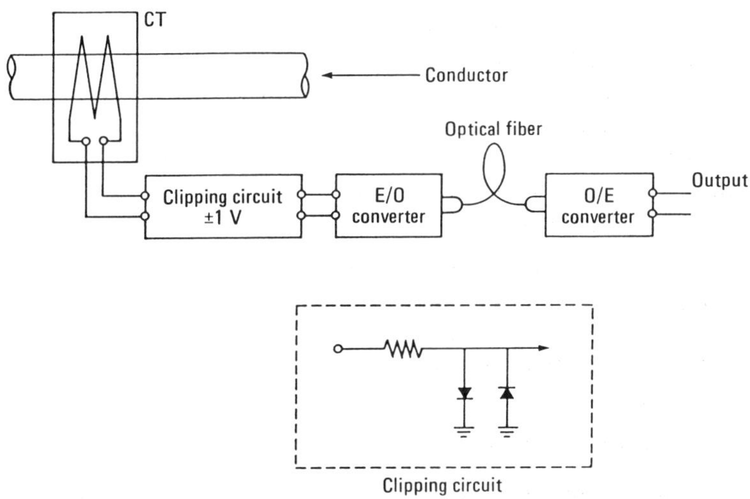

Fig. 4.51 Device for measuring high-frequency current for high-potential sections with output from current transformer converted to light. (From Kawada et al., 1987. © IEEE; and Nojima et al., 1987. © IEE.)

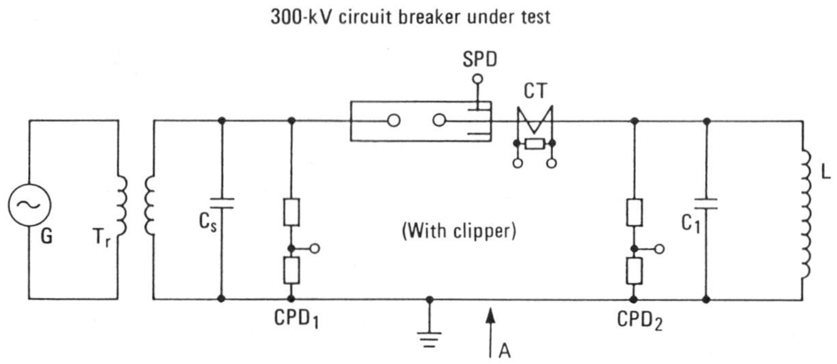

G: Short-circuit generator
CT: Current transformer
T_r: Transformer
SPD: Potential divider utilizing spacer
C_s: Supply-side capacitor
CPD: Capacitive potential divider
L: Load reactor
C_1: Load-side capacitor

Fig. 4.52 Reactor current-interruption test circuit and measuring points. (From Nojima et al., 1987. © IEE.)

are shown in Figs. 4.53 and 4.54 (Kawada et al., 1987; Nojima et al., 1987). When a circuit breaker is reignited, factors that disturb accurate voltage/current measurement are created. In other words, reignition of the circuit breaker causes the flow of high-frequency surge current, resulting in varying potential at the earthed point. Another adverse factor is that a voltage drop occurs on the lead wires because of the effects of stray inductance. The spacer-utilizing light-signal-transmitting voltage divider and the current-measuring device of the light signal transmission method, which were used in the experiment, offer the following features.

Optical voltage divider:

1. A spacer is provided on the circuit-breaker terminal, enabling voltage on this terminal to be measured.
2. Signals are transmitted in the form of light, so there is no interference in measurement by potential fluctuation at the earthed point.

Optical current transformer:

1. Measurements can be made by providing a current transformer on the circuit-breaker terminal of the high-potential section. This in turn enables measurement of high-frequency current flowing between the circuit-breaker electrodes.

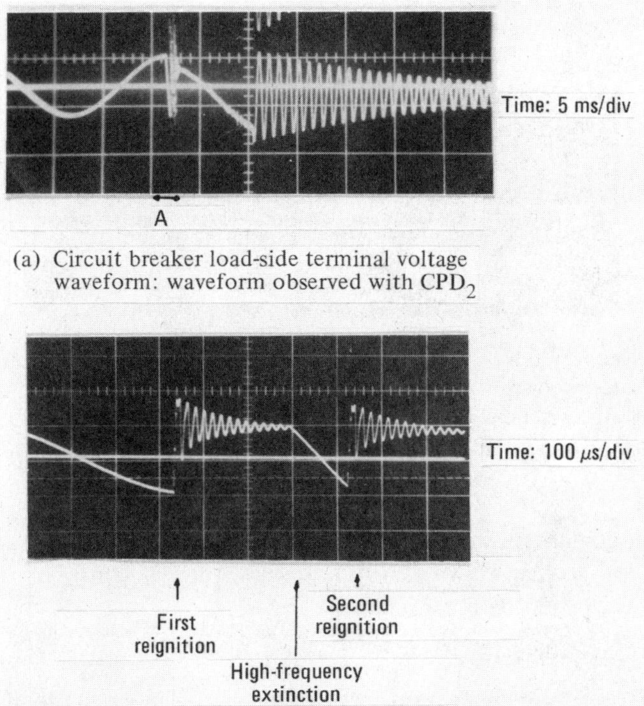

(a) Circuit breaker load-side terminal voltage waveform: waveform observed with CPD_2

(b) Voltage waveform of reignition and high-frequency extinction: waveform of part A in the above oscillogram observed with SPD

Fig. 4.53 Reignition high-frequency voltage waveform in reactor current interruption test shown in Fig. 4.52. Measurement by spacer-utilizing light-signal-transmitting voltage divider. (From Kawada et al., 1987. © IEEE; and Nojima et al., 1987. © IEE.)

2. Signals are transmitted in the form of light, so there is no interference in measurement by potential fluctuation at the measuring point.

In Fig. 4.53b, the high-frequency voltage obtained by the above-mentioned measurement during the reignition shown in part A of Fig. 4.53a is clearly recorded. In Fig. 4.54, we are able to note clear recording of high-frequency current during reignition. Figure 4.53 reveals the fact that reignition is caused repeatedly. Figure 4.54 indicates the occurrence of

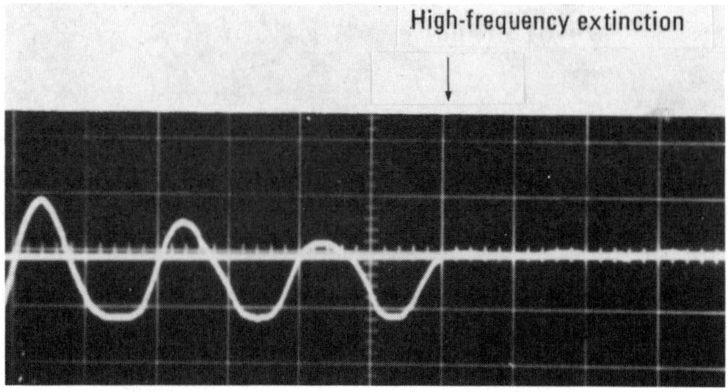

Fig. 4.54 Reignition high-frequency current waveform in reactor current interruption test shown in Fig. 4.52. Measurement by CT-output light-conversion transmitting current measuring device. (From Kawada et al., 1987. © IEEE; and Nojima et al., 1987. © IEE.)

high-frequency arc extinction, that is, the fact that high-frequency current is interrupted at the current-zero point.

4.4.3 Digital Processing System

Computers are being increasingly adopted for measurement of switching phenomena, and technical advances in this field are quite conspicuous recently. Industrial development has rapidly progressed to a level that enables us to purchase at moderate prices various types of modern measuring instruments incorporating minicomputers, personal computers, or microprocessors. By employing computers in this manner, the quality of measurements and the processing of measured data have been upgraded, and efficiency has also been remarkably improved, so that measurements that were formerly difficult are now easily possible.

As an example of such efficiency improvement and quality upgrading, we can cite a new measurement system employed at the high-power laboratory. Figure 4.55 is a block diagram of this measurement system (Yanabu and Yamashita, 1987). In the case of a low-speed phenomenon observable on power frequency, data are directly read into a measurement computer with an A/D converter through an amplifier. As for a high-speed phenomenon such as transient recovery voltage, signals are first converted into light when required for removal of noise and other purposes and are then transmitted to the observation room. In this observation room, light signals are converted back into electrical signals and are read into the measurement computer after being recorded into a high-speed *transient recorder.* The

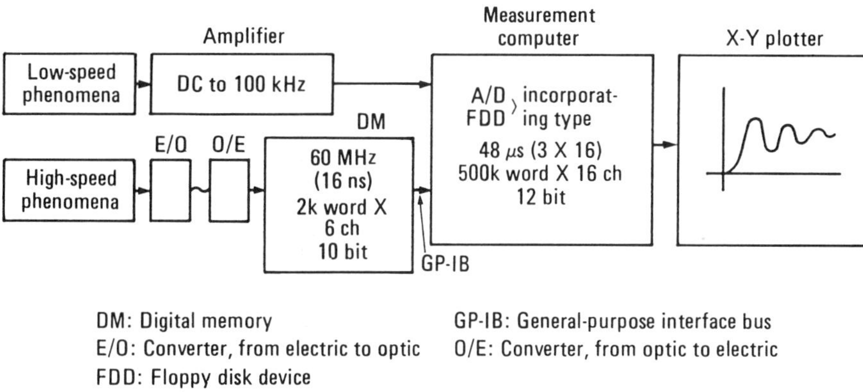

Fig. 4.55 Block diagram of measurement system at high-power laboratory. (From Yanabu et al., 1987. Copyright JIEE.)

measurement computer performs data processing, outputs the processed results to a plotter, and stores the data in the memory of a floppy disk device.

Actual switching phenomena at the high-power laboratory widely differ in variety. Waveforms obtained by measurement are all read on the display screen through manual movement of the cursor. Figure 4.56 shows an example of data processing for a three-phase short-circuit reclosing test and an example of system setting adopted for this test. Figure 4.56a indicates displays of symmetrical-portion and dc-component values read on low-frequency phenomena by clarifying the envelopes of ac voltage/current waveform crest values. Simultaneously displayed in this figure are opening time and closing time. As for the high-frequency phenomena shown in Fig. 4.56b, rise time, rate of rise, and delay time are clarified by reading crest values of transient recovery voltage and obtaining tangents of the wavefront section. The system setting used for measurement is shown in Fig. 4.56c.

Data read in this manner are collected and then summarized into tables and histograms by computers.

The next example we will cite is related to the case where measurement work, conventionally difficult to perform, has become quite easy through the utilization of digital processing. First, shown in Fig. 4.57 is a high-frequency current-interruption test circuit for an SF_6 gas circuit breaker (GCB). This test was conducted for the study of reignition high-frequency extinction ability (Zaima et al., 1990) during reactor current interruption. For measurement of high-frequency arcing voltage, a light-signal-transmit-

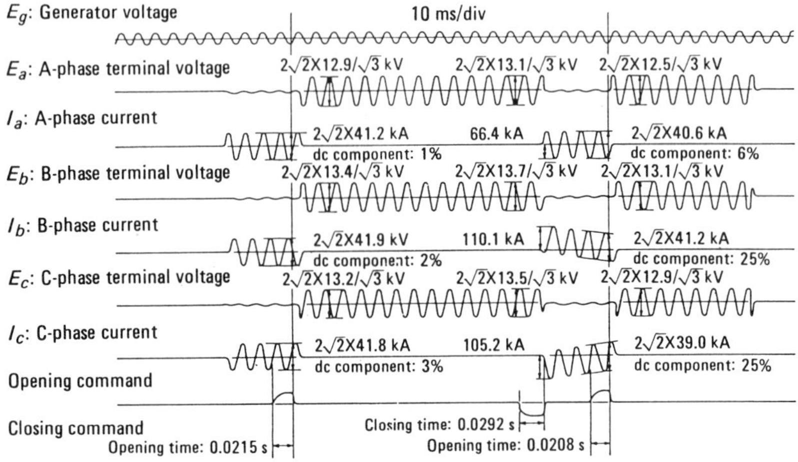

(a) Low-frequency phenomena

Rated voltage: 12 kV
Rated interrupting current: 40 kA

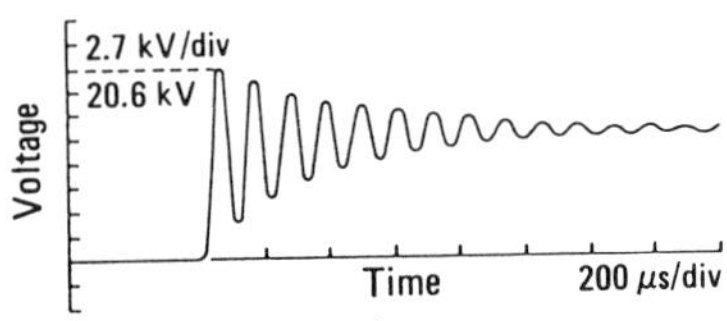

Specified values:
Crest value = 20.6 kV
Rise time = 51.0 μs
Rate of rise = 0.40 kV/μs
Delay time = 5.0 μs

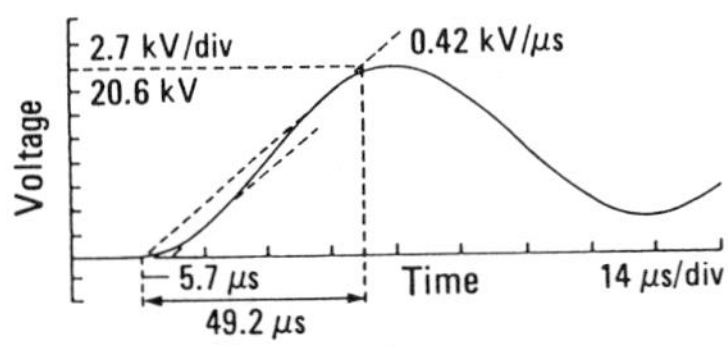

Measured values:
Crest value = 20.6 kV
Rise time = 49.2 μs
Rate of rise = 0.42 kV/μs
Delay time = 5.7 μs

(b) High-frequency phenomena
(Transient recovery voltage)

	Low-frequency phenomena	High-frequency phenomena
Sampling interval	50 μs	5 MHz (200 ns)
Measurement time	1 s	800 μs
No. of measured waveforms	9 ch	3 ch
Total No. of waveform data	180,000 points	12,000 points
Analysis time (including cursor operation)	1-2 min	0.5-1 min
Plotter processing time	2-5 min	Within 0.5 min
FDD storing time	Approx. 1 min	Approx. 10 s

(c) System setting

Fig. 4.56 Example of data processing for three-phase short-circuit reclosing test and system setting for this test. (From Yanabu et al., 1987. © JIEE.)

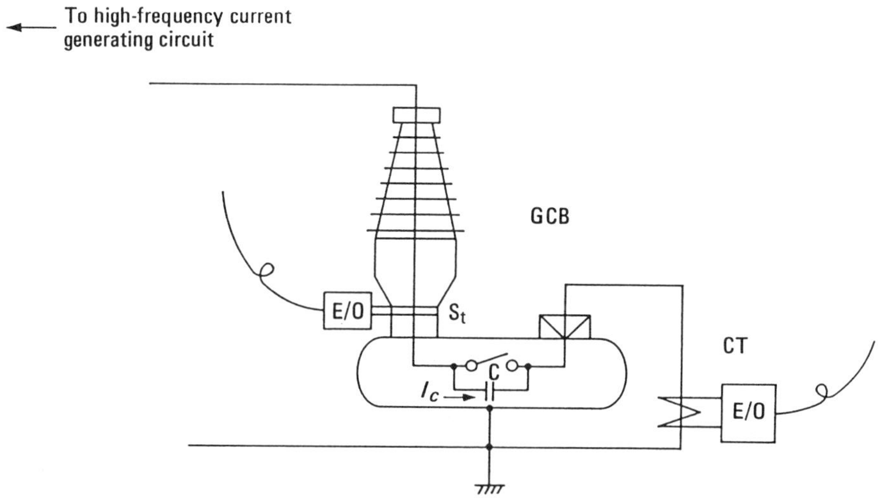

Fig. 4.57 High-frequency current-interruption test circuit and current/voltage measuring method.

ting voltage divider utilizing a spacer (Sp) is used, and for measurement of high-frequency arcing current, a light-signal-transmitting current transformer is used. Shown in Fig. 4.58a are original measurement waveforms. From this figure, we are able to observe that a high-frequency current of about 30 kHz is interrupted, and zero points of current and voltage waveforms shown in Fig. 4.58a do not reveal mutual congruity. Therefore, incidentally, the following errors are included in these measurement waveforms:

1. Voltage errors accrued from time constants possessed by the spacer-type voltage divider
2. Voltage errors resulting from the voltage drop of stray inductance of the circuit
3. Current errors created by the capacitance connected between the GCB electrodes

Figure 4.59 shows the equivalent circuit of the spacer-utilizing voltage divider of Fig. 4.50. By assigning E_1 to input voltage, E_4 to output voltage, and $\overline{E}_1$ and $\overline{E}_4$ to the Laplace transforms of these input/output voltages, the following equation can be established:

$$\overline{E}_1 = k\left(2\frac{C_1 + C_2}{C_1} + \frac{1}{sC_1R_2}\right)\overline{E}_4 \tag{4.2}$$

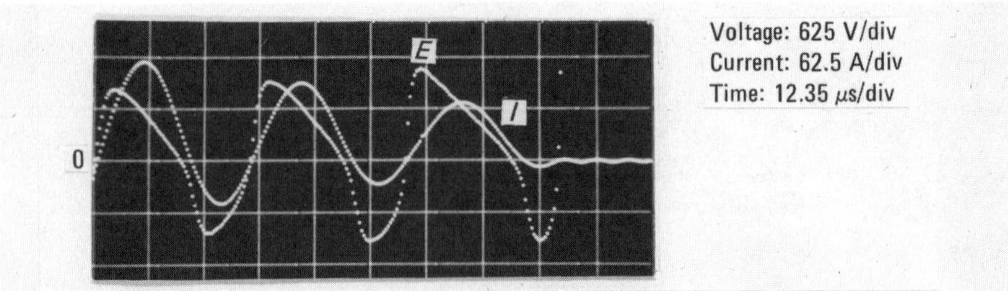

(a) Original high-frequency current/voltage waveforms obtained by measurement

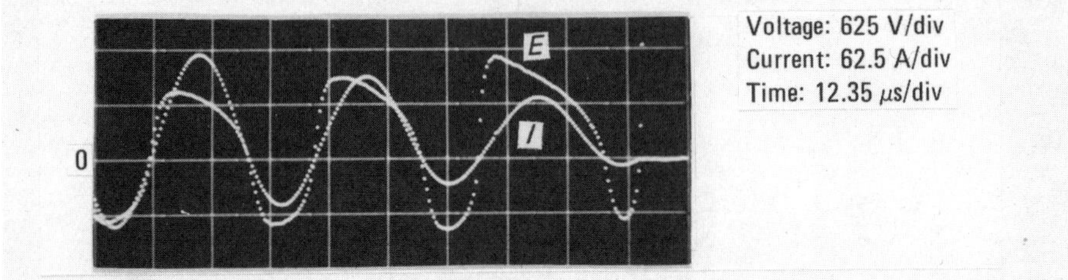

(b) Arcing voltage/current waveforms after compensation by digital processing

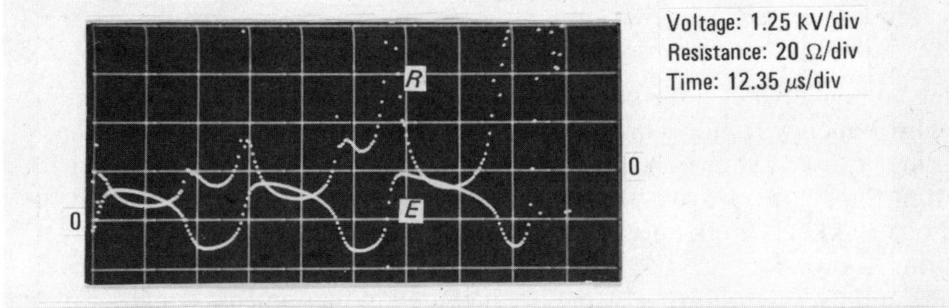

(c) Arcing resistance obtained through dividing voltage waveform by current waveform in Fig. 4.59 (b)

Fig. 4.58 Oscillograms obtained by measurement.

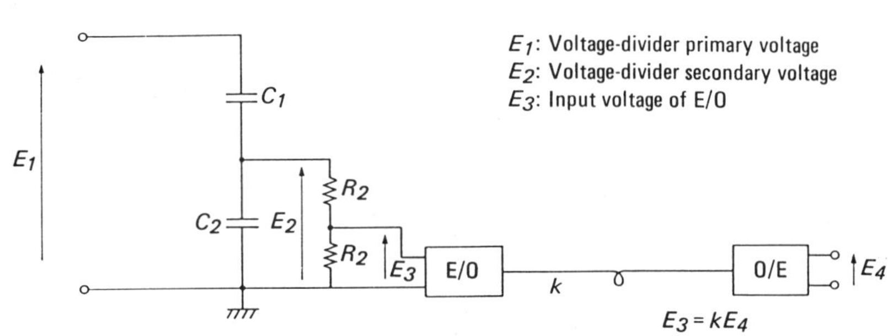

Fig. 4.59 Equivalent circuit of spacer-utilizing voltage divider.

where

S = Laplace transform operator

C_1 = capacitance of spacer

C_2, R_2 = capacitance and resistance of voltage-divider secondary-side circuits

k = signal transmission coefficient

In other words, the second term of Eq. (4.2) indicates that input voltage can be obtained by adding the integrating term for compensation of output voltage. This represents the first type of error listed above.

Output from the spacer-utilizing voltage divider includes, in addition to arcing voltage, voltage drop by stray inductance of the circuit as indicated by the equation

$$E_L = L\frac{dI}{dt}$$

Since I denotes high-frequency current, E_L becomes greater. While the magnitude of arcing voltage is small, around several kilovolts, this E_L cannot be ignored. Therefore, to clarify true arcing voltage values, E_L must be obtained by differential calculation of I, and the resultant value must be subtracted from the output voltage value of the voltage divider. This represents the second type of error listed above.

A capacitor is connected across electrodes of the test circuit breaker as shown in Fig. 4.57. Therefore, in addition to arcing current, the observation current includes capacitor current that can be calculated as

$$I_c = C\frac{dE}{dt}$$

Consequently, to obtain arcing current, the value of I_c must be clarified by subjecting the interelectrode voltage E to differential calculation. Again, the value obtained from the differential calculation must be subtracted from the observed current. This represents the third type of error.

Figure 4.58b shows the current/voltage waveforms after the three types of errors are compensated for by subjecting the observed waveforms to differentiation and integration followed by calculations through multiplication, addition, and subtraction. Unlike the original waveforms shown in Fig. 4.58a, the waveforms in Fig. 4.58b indicate the tallying of zero points between current and voltage. This tally proves that error compensation has been correctly achieved. By clarifying arcing current and arcing voltage in this manner, and through dividing this voltage by this current, arcing resistance as shown in Fig. 4.58c can be obtained. Also, it becomes possible to read the θ and N_0 values indicated in Eq. (4.1).

Unless digital processing is employed, these measurements will be extremely difficult to achieve.

4.5 DIAGNOSTICS (S. Nishiwaki)

4.5.1 Monitoring and Control Systems

Society has required an increasingly stabilized supply of power and conservation of labor in the operation and maintenance of power-supply facilities. To prevent operating accidents and accelerate the application of remedies after an accident, the introduction of equipment-monitoring and diagnostic systems is becoming increasingly common.

According to investigation results (CIGRE, 1981) of CIGRE WG13-06, the major causes of accidents are mechanical factors (70%), auxiliary and control circuits (19%), and electrical factors of main circuits (11%). Based on such percentages, primary accidents involving circuit breakers can be classified as (1) operational failures, including drives and control, or (2) insulation failures, including main contacts.

In this section certain examples of the monitoring and diagnosis of operating status regarding the above-mentioned operational and insulation failures are introduced by citing applications of hydraulically operated SF_6 gas circuit breakers for gas-insulated switchgear.

Major cases of operational failures and insulation failures are listed in Tables 4.11 and 4.12, respectively. For conducting diagnoses of these failures, monitoring on the following points is believed effective.

1. *Operational failures:* Circuit breaker operation time, hydraulic pump operation time, operating frequency, and oil-level/oil-pressure-drop alarm
2. *Insulation failures:* Partial discharge, gas pressure

Explanations will be given regarding case examples of monitoring on these points (Goto et al., 1989; Ohshita et al., 1989).

Figure 4.60 shows the structure of a GCB operation time monitoring system. This system is composed of the following devices:

1. A command-detecting sensor for detecting an operation command signal from a magnetic coupler installed in the control circuit of the circuit breaker
2. Two-fiber optical sensors for sending and receiving light, attached on a drive case and facing a light-reflecting plate mounted on the moving part of the circuit breaker to detect operating locations

Table 4.11 GCB Operational Failures

Failure	Cause of failure
(Delayed operation)	Delay in control mechanism operation
(No operation/no response)	Wire breakage in control circuit; terminal loosened
	Rise in minimum operating voltage
(Operating speed slowdown)	Decrease in drive valve operating speed
	Decrease in operating speed
(Stroke curve distortion)	Operation scatter of main control valve
	Bouncing at stroke end
(Bad cushioning)	Improper cushioning timing
	Extended switching time of main control valve
(Incomplete stroke)	Improper contact
	Coupling slipped out
(Oil pressure loss)	Pressure switch failure
	Oil leakage
(Oil level drop)	N_2 gas leakage from accumulator
	Oil leakage
(Prolonged operation time/ increased operation of hydraulic pump)	Leakage between high and low pressure in hydraulic parts directly after circuit breaker operation caused by defective valve seats
	Defective sliding seals or valve seats Hydraulic pump discharge loss

Source: Goto et al., 1989. © IEEE.

3. A time counter for measuring the time intervals between output pulses of the command and optical sensors

Figure 4.61 shows the structure of a GCB operation command detecting sensor. Shown in Fig. 4.62 is the installation status of the optical sensor. In Fig. 4.62, optical sensors are located at the On and Off ends of the circuit breaker, enabling output pulses of the optical sensor to be obtained for both stroke ends. For the timing of output pulses of the sensors, as shown in Fig. 4.60, a time counter measures the interval between the arrival of the operation command and the start of circuit-breaker operation, or the so-called starting time (T_1 and T_3 in Fig. 4.60), and the interval between the start and finish of operation of the circuit breaker, or the operation times for opening and closing (T_2 and T_4 in Fig. 4.60). Operation time and operating frequency of the hydraulic pump can be easily read from On/Off data of the pump-operating contactor. Also, oil levels can be monitored by the float switches, and oil-pressure levels can be monitored by the oil-pressure sensor.

Generally, partial discharge occurs when an abnormality is created on dielectric performance. It is important to monitor this partial discharge to

Table 4.12 GCB Insulation Failures

Failure	Cause of failure	
(Increase in internal partial discharge) → (SF_6 gas decomposition)	(Discharge due to foreign matter in tank)	Parts slipped out by circuit breaker operation
		Foreign matter infiltrated during assembly
	(Foreign matter deposited on insulator surface)	Metallic dust generated by circuit breaker operation
	(Partial discharge caused by foreign matter in insulator)	Tree-pattern discharge generation caused by metallic foreign matter embedded in insulator
	(Live-part shield floating)	Incomplete contact of shield fitting
		Shield fitting bolts loosened by circuit breaker operation
	(Floating caused by incomplete conduction contact)	Incomplete contact of conduction contacts, particularly looseness of bolted parts
	(Floating of electrode embedded in insulator)	Looseness of connecting bolts
	(Incomplete stroke of contacts)	Poor contact or floating contact
	(Partial discharge in gas)	Gas pressure loss
		Resistance loss on insulator surface caused by increased water content in gas
	(Insulator aging)	Void discharge and other failures caused by defects of insulator

Source: Goto et al., 1989. © IEEE.

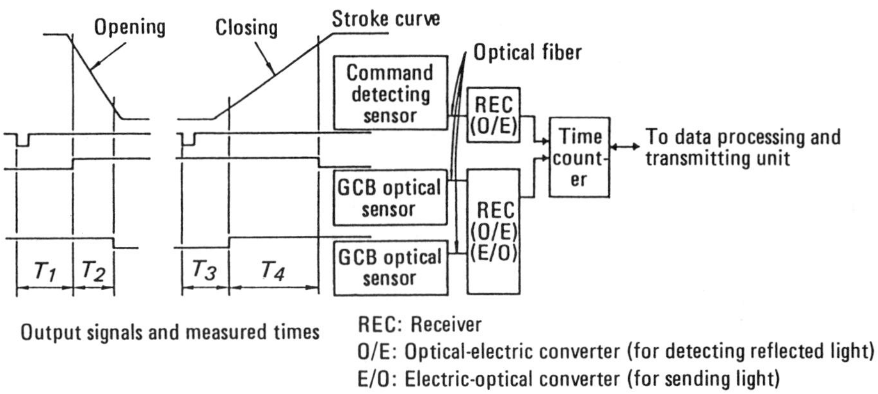

Fig. 4.60 Structure of GCB operation time monitoring system. (From Goto et al., 1989. © IEEE.)

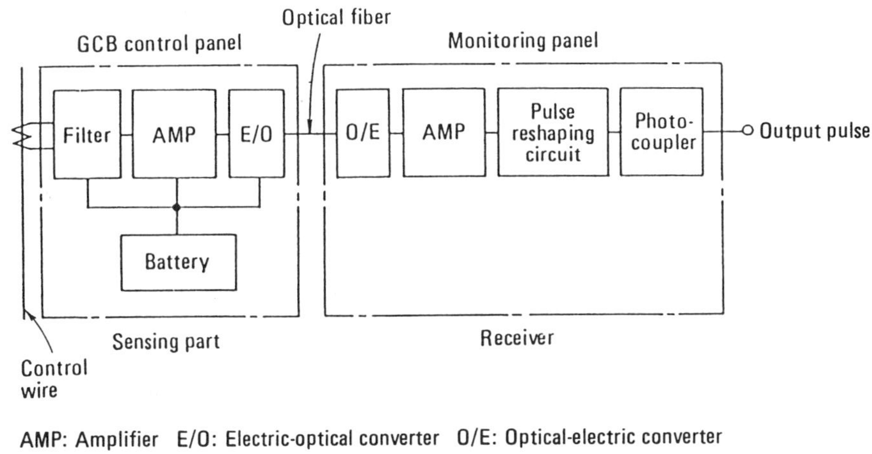

Fig. 4.61 Structure of GCB operation command-detecting sensor. (From Goto et al., 1989. © IEEE.)

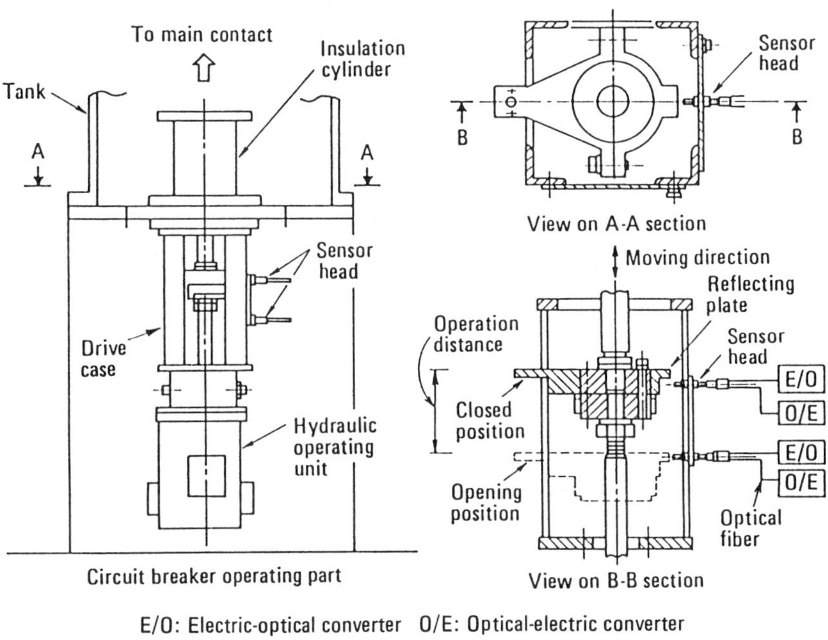

Fig. 4.62 Optical sensor installed. (From Goto et al., 1989. © IEEE.)

enable the prediction of insulation failures of circuit breakers. Partial discharge can be measured by

1. Detection of a radiating electromagnetic field caused by a partial discharge drifting out of the tank
2. Detection of a sound pulse traveling on the tank surface
3. Detection via floating electrodes in the tank

Among these measuring methods, the one that depends on the detection of sound pulses is susceptible to being influenced by rain or flying objects. On the other hand, the electric measuring method has the possibility of being adversely affected by noise from broadcast radio waves.

Figure 4.63 illustrates a measuring method utilizing a ring-shaped electrode embedded in a spacer that supports high-voltage conductors. This method enables measurement without the trouble of applying special devices to the circuit breaker itself. Since floating electrodes are provided in the tank interior, noise levels are also minimum. The lowest detection values are about 10 pC.

Gas pressure greatly affects dielectric characteristics or interrupting performance. In this method, by combining a piezoelectric pressure sensor with a temperature sensor, pressure values are subjected to temperature compensation and the values at 20°C are output.

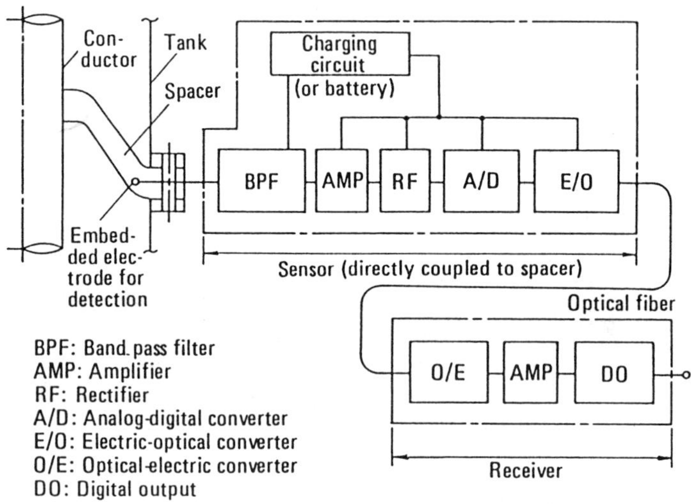

Fig. 4.63 Structure of partial discharge sensor. (From Goto et al., 1989. © IEEE.)

```
14:37, June 18, 1987
○#3 SHR line GCB (52)                                OFF
  Starting time                   Black phase : 21 ms
                                  Red phaase  : 21 ms
                                  White phase : 20 ms
  Operation time for opening      Black phase : 24 ms
                                  Red phase   : 24 ms
                                  White phase : 24 ms

                 At circuit breaker operation


5:00, June 18, 1987
○Gas pressure (values converted for 20℃)
  Gas system No. 5    Bus 1 side              : 3.0 kg/cm²·G
  Gas system No. 5    Bus 2 side              : 3.1 kg/cm²·G
○#3 SHR line  Partial discharge sensor output voltage
  Bus 1 side                      Black phase : 0.0 V
                                  Red phase   : 0.0 V
                                  White phase : 0.0 V
  Bus 2 side                      Black phase : 0.0 V
                                  Red phase   : 0.0 V
                                  White phase : 0.0 V

                      Periodic printout
```

Fig. 4.64 Example of printouts. (From Goto et al., 1989. © IEEE.)

Output from these sensors is displayed and recorded on the local monitor panel, which incorporates a microprocessor and a printer. When an abnormality is detected in the measured values, an alarm is issued. Figure 4.64 shows a typical printout.

REFERENCES

Section 4.1: Relevant Standards

ANSI/IEEE C37.04-1979, Rating Structure for AC High-Voltage Circuit Breakers Rated on a Symmetrical Current Basis.

ANSI/IEEE C37.06-1979. Preferred Ratings and Related Required Capabilities for AC High-Voltage Circuit Breakers Rated on a Symmetrical Current Basis.

ANSI/IEEE C37.09-1979. Test Procedure for AC High-Voltage Circuit Breakers Rated on a Symmetrical Current Basis.

ANSI/IEEE C37.010-1979. Application Guide for AC High-Voltage Circuit Breakers Rated on a Symmetrical Current Basis.

ANSI/IEEE C37.011-1979. Application Guide for Transient Recovery Voltage for AC High-Voltage Circuit Breakers Rated on a Symmetrical Current Basis.

ANSI/IEEE C37.012-1979. Application Guide for Capacitance Current Switching for AC High-Voltage Circuit Breakers on a Symmetrical Current Basis.

IEC56-1987. High-Voltage Alternating-Current Circuit-Breakers.

IEC427-1989. Synthetic Testing of High-Voltage Alternating Current Circuit-Breakers.

IEC Document 17A(Secretariat)291. Draft—High-Voltage Alternating-Current Switchgear and Controlgear—Inductive Load Current Switching Tests—High-Voltage Motor Switching Tests.

IEC Document 17A(Secretariat)297. Draft—High-Voltage Alternating-Current Switchgear and Controlgear—Inductive Load Current Switching Tests—Shunt Reactor Switching Tests.

Section 4.2

ANSI/IEEE (1981). C37.081-1981. IEEE Guide for Synthetic Fault Testing of AC High Voltage Circuit Breakers: Rated on a Symmetrical Current Basis.

Brasca, E., Dominioni, A. C., Rovelli, S., and Tiramani, A. R. (1983). Requirements and Performance of the New CESI High Power Facilities, *IEEE Trans. Power Appar. Syst., PAS-102:* 2224–2230.

Harada, T., and Yanabu, S. (1978). Generation and Measurement/Control of Extremely High Voltage and Current, *Trans. Inst. Measurement Automat. Control, 17*(1): 89–98.

Horiuchi, R., Iwamoto, S., Sunabe, K., and Fukagawa, H. (1989). Development of Expert-System for Assisting Short-Circuit Test of Circuit Breaker, *Trans. IEE Jpn., B-109*(5): 233–240.

Ishikawa, M., Ikeda, H., Yanabu, S., and Yamamoto, M. (1984). Numerical Study of Delayed-Zero-Current Interruption Phenomena Using Transient Analysis Model for an Arc in SF_6 Flow, *IEEE Trans. Power Appar. Syst., PAS-103*(12): 3561–3568.

Kawada, H., Goto, K., Hanamura, M., Yanabu, S., Nishiwaki, S., Ueda, T., and Murayama, Y. (1987). Switching Surge of Shunt Reactor Caused by SF_6 Circuit Breaker Operation, *IEEE Trans. Power Delivery,*

Section 4.3

CIGRE WG13.04 (1988). Requirements for Switching Tests of Metal Enclosed Switchgear, *ELECTRA, 110:* 15–19.

Fujii, T., Yamashita, S., and Nishikawa, A. (1969). Synthetic Short-Circuit Testing for Air Blast Breaker, *Tokyo Regional Conf. of IEE of Japan,* [6], *169:* 236. In Japanese.

Ibuki, K., Hosomi, M., Yoshizumi, H., and Yamamoto, S. (1988). Synthetic Test Method of Three Phase in One Tank Type Circuit Breaker, Associated Electrical Conference of Kansai Region in Japan, G3-19 G113.

IEC (1987). High-Voltage Alternating-Current Circuit Breakers, Publication 56.

JIEE (1982). Synthetic Testing Method for Circuit Breaker and Diagnostics of Interrupting Phenomena, Tech. Report II-131.

Kurosawa, Y., Kurasawa, N., Matsumoto, M., and Tsuda, H. (1987). Arc Extending Circuit for Vacuum Circuit Breaker, JIEE Conf. Switchgear and Protection Devices, SPD-87-13.

Kurosawa, Y., Takahashi, I., Saito, K., and Hirasawa, K. (1989). Four-Parameter Transient Recovery Voltage Circuit for Extra High Voltage Circuit Breakers, Report presented at the Colloquium of CIGRE SC 13.

Maury, E., and Vigreux, J. (1971). Design and Application of a Synthetic Testing Plant, IEEE Summer Symposium on High Power Testing, 6-5.

Schett, G. (1989). Modular Synthetic Test Circuit for Short Circuit Tests with High Power Breakers, CIGRE Colloquium SC 13.

Shimada, R., Tani, K., Kishimoto, H., Tamura, S., Ikeda, H., Tamagawa, T., Yanabu, S., and Matsushita, T. (1979). Synthetic Test Methods of High-Direct-Current Circuit Breakers, *Proc. IEE, 126*(10): 965–970.

St-Jean, G. (1974). A Method for Calculating Directly the Components of a Synthetic Circuit for the Testing of Ac Circuit Breakers, *IEEE Trans. Power Appar. Syst., PAS-93*(1): 429–443.

Yamamoto, M., Yamashita, S., Ikeda, H., and Yanabu, S. (1985). New Short-Circuit Testing Facilities to Cope with the Recent Development of GIS, *IEEE Trans. Power Appar. Syst., PAS-104:* 1.

Yamashita, S., Miyake, N., Suzuki, S., Ikeda, H., and Yanabu, S. (1989). Short Circuit Testing Method of Three-Phase-in-One-Tank-Type SF_6 Gas Circuit Breaker, *IEEE Trans. Power Delivery, 4*(1): 349–354.

Yamashita, S., Yamaguchi, T., Takahashi, K., and Miyake, N. (1976). Equivalent Making Test Method for Circuit Breaker, *JIEE Gen. Conf., 7:* 1349.

Yanabu, S., Tamagawa, T., Irokawa, S., Horiuchi, T., and Tomimuro, S. (1981). Development of HVDC Circuit Breaker and Its Interrupting Test, IEEE PES Summer Meeting, *81 SM:* 473–478.

Section 4.4

Blez, B., and Guilloux, C. (1989). Post-Arc Current in High Voltage SF_6 Circuit-Breakers When Breaking at Up to 63 kA, *IEEE Trans. Power Delivery, 4*(2): 1056–1062.

Haefly (date unknown). Combined Ohmic and Damped Capacitive Voltage Divider for Lightning Impulse Voltages Switching Impulse Voltage, Alternating Voltages and Direct Voltages (Manual).

IEC (1987). High-Voltage Alternating-Current Circuit-Breakers, Publication 56, Appendix GG.

Kawada, H., Goto, K., Hanamura, M., Yanabu, S., Nishiwaki, S., Ueda, T., and Murayama, Y. (1987). Switching Surge of Shunt Reactor Caused by SF_6 Circuit Breaker Operation, *IEEE Trans. Power Delivery, PWRD-2*(4): 1124–1130.

Kobayashi, A., Yanabu, S., Yamashita, S., Tomimuro, S., and Haginomori, E. (1978). Measurement of Voltage Distribution Phenomena in SF_6 Single Pressure Circuit Breaker, *IEEE Trans. Power Appar. Syst., PAS-97*(4): 1304–1312.

Murano, M., Nishikawa, H., Kobayashi, A., Okazaki, T., and Yamashita, S. (1975). Current Zero Measurement for Circuit Breaking Phenomena, *IEEE Trans. Power Appar. Syst., PAS-94*(5): 1890–1900.

Nojima, K., Nishiwaki, K., Okubo, H., and Yanabu, S. (1987). Measurement of Surge Current and Voltage Waveforms Using Optical-Transmission Techniques, *IEEE Proc., Pt. C, 134*(6): 415–422.

Standing Committee for Circuit Breakers, IEE of Japan (1976). Testing Methods of Circuit Breakers, *Technical Report from the Institute of Electrical Engineers of Japan,* Section II, No. 46. In Japanese.

St-Jean, G., Landry, M., Leclerc, M., and Chénir, A. (1988). A New Concept in Post-Arc Analysis Applied to Power Circuit-Breakers, *IEEE Trans. Power Delivery, 3*(3): 1036–1044.

Suzuki, K., Kobayashi, A., and Yanabu, S. (1980). Post-Arc Performance of Gas Blast Circuit Breaker for Air and SF_6, *Sixth Int. Conf. Gas Discharges and Their Applications,* September, pp. 5–8.

Yanabu, S., and Yamashita, S. (1987). LA/CAE Technical Progress at High-Voltage, High-Power Laboratory, Miniature Summary, *Rep. Inst. Electr. Eng. Jpn., 107*(8): 817–820. In Japanese.

Yanabu, S., Homma, M., Kaneko, E., and Tamagawa, T. (1985). Post Arc Current of Vacuum Interrupters, *IEEE Trans. Power Appar. Syst., PAS-104*(1): 166–172.

Zaima, E., Okabe, S., Nishiwaki, S., Ishikawa, M., Nakamoto, T., Murayama, Y., and Kawaguchi, Y. (1990). Reignition Surges at Reactor Current Interruption in Cable System GIS, *IEEE Trans. Power Delivery, 5*(2): 947–954.

Section 4.5

CIGRE (1981). *CIGRE Report* Sponsored by Study Committees, *ELECTRA, 79:* 21–71.

Goto, K., Sakakibara, T., Kamata, I., and Ikeda, S. (1989). On-Line Monitoring and Diagnostics of Gas Circuit Breakers, *IEEE Trans. Power Delivery, 4*(1): 375–381.

Ohshita, Y., Hashimoto, A., and Kurosawa, Y. (1989). A Diagnostic Technique to Detect Abnormal Conditions of Contacts Measuring Vibrations in Metal Enclosures of Gas Insulated Switchgear, *IEEE Trans. Power Delivery, 4*(4): 2090–2094.

APPENDIX
BASIC THERMODYNAMIC EQUATIONS

Koji Ibuki Mitsubishi Electric Corporation, Amagasaki, Japan

Pressure P, temperature T, volume of the gas per unit mass v, and the internal energy per unit mass u are properties that can describe the state of gas. One state (P_1, T_1, v_1, u_1) is changed into another state (P_2, T_2, v_2, u_2) by the input heat q_{12} and with the output work w_{12}.

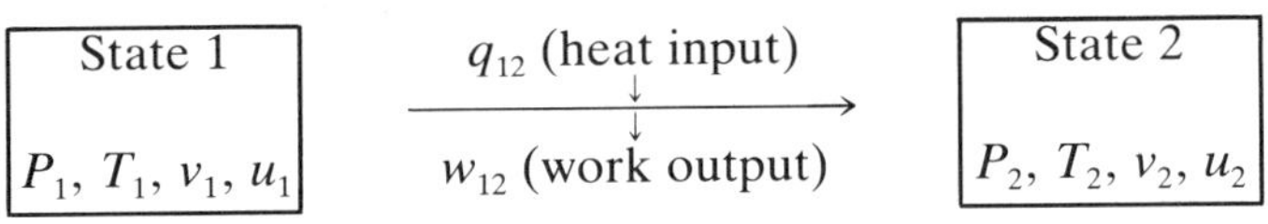

Energy conservation during this change requires

$$q_{12} = u_2 - u_1 + w_{12}$$

The work done by the gas to the outer system is $w_{12} = P\,dv$. This equation can be written in the derivative form

$$dq = du + P\,dv \tag{1}$$

Another property of the gas, *enthalpy* or *total thermal energy* per unit mass of gas, is defined as

$$h = u + Pv$$

This is expressed in derivative form as

$$dh = du + v\,dP + P\,dv \tag{2}$$

Therefore, Eq. (1) can be written as

$$dq\ (= du + P\,dv) = dh - v\,dP \tag{3}$$

Specific heat is defined as $c = dq/dT$, and the heat required to change the temperature is dependent on the state of the system. The specific heat at constant volume is represented by c_v, and specific heat at constant pressure is represented by c_p.

$$c_v = \left(\frac{\partial q}{\partial T}\right)_v = \left(\frac{\partial u}{\partial T}\right)_v$$

$$c_p = \left(\frac{\partial q}{\partial T}\right)_p = \left(\frac{\partial h}{\partial T}\right)_p$$

These properties of the gas are given for a unit mass of the material, but in the process of arc analysis the characteristic properties per unit volume are also important indices. ρh shows the total thermal energy sustained in a unit volume. $\rho(\partial h/\partial T)$, which is equal to ρc_p, is interpreted as the heat required to raise the temperature of a unit volume of gas.

The ratio of c_p to c_v (specific heat ratio γ) represents an important characteristic of gases. An adiabatic change of a gas satisfies the relation $V^{\gamma-1}T = \text{const}$, where V is the specific volume of the gas. Molecules formed with two atoms (e.g., nitrogen and oxygen) have a γ of approximately 1.4, while that of SF_6 is approximately 1.07 at room temperature.

INDEX